Automotive Power Systems

Automotive Power Systems

Dorin O. Neacşu

CRC Press
Taylor & Francis Group
Boca Raton London New York

CRC Press is an imprint of the
Taylor & Francis Group, an **informa** business

First edition published 2021
by CRC Press
6000 Broken Sound Parkway NW, Suite 300, Boca Raton, FL 33487-2742

and by CRC Press
2 Park Square, Milton Park, Abingdon, Oxon, OX14 4RN

Library of Congress Cataloging-in-Publication Data
Names: Neacşu, Dorin O., author.
Title: Automotive power systems / Dorin O. Neacşu.
Description: Boca Raton : CRC Press, 2020. | Includes bibliographical
references and index.
Identifiers: LCCN 2020020961 (print) | LCCN 2020020962 (ebook) | ISBN
9780367512965 (hbk) | ISBN 9781003053231 (ebk)
Subjects: LCSH: Automobiles--Electric equipment. |
Automobiles--Motors--Equipment and supplies.
Classification: LCC TL272 .N43 2020 (print) | LCC TL272 (ebook) | DDC
629.25/024--dc23
LC record available at https://lccn.loc.gov/2020020961
LC ebook record available at https://lccn.loc.gov/2020020962

ISBN: 978-0-367-51296-5 (hbk)
ISBN: 978-0-367-51297-2 (pbk)
ISBN: 978-1-003-05323-1 (ebk)

Contents

Preface

Vehicles are intrinsically linked to our lives. Since we have entered the new millennium, the expectations for comfort and safety in our vehicles are continuously rising. Meeting these expectations asks for more controllability of vehicle sub-systems, which can only be achieved with microcontroller software. In the software, the designer can include optimal features, treat case-based operation, or adapt to various conditions, all while employing the same electromechanical hardware. Electronics and software engineering applied to vehicle systems are the flashpoint for technological, economic, and social innovation, fueling imagination and technical debate, from elbow-deep laboratory experimentation to abbreviation-rich marketing meetings.

Starting in 1991, carmakers began to include more electronics to advance the performance of their vehicles. It began when California required all cars sold in the state to feature a self-diagnostic ability. By 1996, the US Federal government extended the requirement for all cars to include a standard port to connect to for system diagnosis. Technologies like Bluetooth, GPS, blind-spot detection, collision warning systems, and adaptive cruise control followed enthusiastically. The average vehicle today has numerous microcontrollers that help control a variety of functions for added comfort and safety operation. All this created a market for automotive semiconductors that is worth $36 billion.

The common assertion for electronics engineering used in vehicle calls for electric and hybrid vehicle technologies. Despite a huge media and regulatory presence, the electric vehicle has entered slowly into the mainstream market, with a tiny share of only about 1% of vehicle sales globally. Although manufacturing and selling electric vehicles help to comply with regulations, the money is made with engine technologies that have been around for over a century.

Meanwhile, without a comparable media or social presence, the process of vehicle electrification is more appealing to the young electronics engineer due to a sizeable employability market. Vehicle electrification aims to replace or improve numerous mechanical convenience systems with electrically actuated systems, while maintaining the engine technology for propulsion. Examples include applications for both chassis and body systems.

Collectively known as body electronics, comfort and convenience systems allow vehicle occupants to feel comfortable and safe inside the vehicle. Examples include power windows, soft-convertible systems, hard-convertible systems, power door locks, power seats, electric steps, electric mirrors, electric sunroof, auto-dimming rearview interior and side mirrors, remote keyless entry, power trunk or lift-gate systems. Each such gimmick requires one or more dc motors with either a simple on/off control or through communication with other features.

The chassis designates the complete car, less the body, and consists of the engine, power-transmission system, and suspension system. These systems are attached to a structurally independent frame. Electrification is herein achieved with power electronic converters used in brake systems, in steering systems, and in suspension systems.

The main challenge in vehicle electrification consists of replacing the engine-based mechanical, pneumatical, or hydraulic energy source with electrical energy through an electromagnetic device. Obviously, this raises other issues with the size and technology of the battery bank and its ability to accommodate the new load demands coming with electrification.

The two major processes of electrical propulsion and electrification are converging to create the all-electric vehicle. Unless man-made gas takes off and maintains a competitive combustion engine solution, built on long-established value chains, electric vehicles will become the norm in the future.

This book attempts to cover all the details of the electrification process, with a focus on the electronics and power electronics used within modern vehicles. It is aimed at young engineers in need of an introductory class in automotive power electronics. The book structure follows the architecture of the conventional engine-based vehicle, with the last chapter dedicated to an introduction into electric propulsion. Everything is described at an introductory level, yet covers all details from the components to system architecture.

The key component in both electrification and electric propulsion processes is provided by power electronics. Power electronics deals with the electronic processing of energy. This makes it a support technology for the automotive industry. All the automotive visions for the next decade (2020–2030) are built on top of power electronics. According to a recent report from the McKinsey Corporation (Ondrej Burkacky, Jan Paul Stein, Johannes Deichmann, "Automotive software and electronics 2030 - Mapping the sector's future landscape"), the power electronics industry associated to the automotive sector is expected to increase 15% CAGR (compound annual growth rate), which is the highest for a decade among all automotive technologies.

Since this expectation brings its own particulars, specific topologies, design practices, software control, or usage habits, it deserves to have a dedicated textbook for such a major application field.

The first part of the book describes the automotive applications at a system level, involving power electronics, with numerous examples. The second part explores further details of each component and can be seen as a brief course in basic power electronics, motor drives, or electrical material technologies. The author considers that the reader needs to first fully comprehend the core task, before tackling the power electronics solution.

Since young specialists in automotive power electronics systems are faced with more than building power electronics hardware, the book attempts a brief introduction to topics ranging from materials used in electrical circuits to control systems implemented in software controlling power converters.

Examples for the mathematical modeling of mechanical systems within a vehicle are presented in "Sections 4.9 Cruise control", "6.4. Automotive suspension", "6.3. Electronic Control of Power Steering" and "12.7. Ultrasonic motors".

Mathematical models are also presented for dc/dc power converters for either analog or digital implementation. Principles for the design of feedback control systems for dc/dc power converters are elaborated in a dedicated chapter based on mathematical models.

The novelty of the book – from other excellent works dedicated to automotive applications – consists of the details of the electronic and power electronic application, explained at an entry-level for the vehicle electrification topics. This is usually very difficult due to the complex, inter-disciplinary character of automotive applications. This book aims to challenge the reader to think about the problem and appreciate the power electronics solution rather than being provided with an inventory of results.

Since cars have been on the streets for over 100 years, each chapter uniquely starts with a list of historical milestones. Recognizing the engineering efforts spanning over more than a century ennobles the new R&D efforts of the new millennium. Focus on the history of electricity in vehicle applications is an attractive treat provided by the book.

The manuscript represents an improved and considerably evolved version of a lecture course offered within the Technical University of Iași, Romania, to MSc students.

Westford, Massachusetts and Iași, Romania

About the Author

The author has broad experience in the application of advanced power electronics concepts in industry from a 30-year career alternating between academic and industrial R&D positions.

Professor Dorin O. Neacşu received M.Sc. and Ph.D. degrees in electronics from the Technical University of Iasi, Iasi, Romania, in 1988 and 1994, respectively, and an M.Sc. degree in engineering management from the Gordon Institute for Leadership, Tufts University, Medford, MA, U.S.A., in 2005.

He was involved with TAGCM-SUT, Iasi, from 1988 to 1990, and with the faculty in the Department of Electronics, Technical University of Iasi, between 1990 and 1999. During this time, he held visiting positions at Université du Quebec a Trois Rivières, Canada, and General Motors/Delphi, Indianapolis, USA. In between 1999 and 2012, he was involved with in U.S. industry as an Electrical Engineer, Consultant, Product Manager, and Project Manager, and with U.S. academic activities at the University of New Orleans, Massachusetts Institute of Technology, and United Technologies Research Center. Since 2012, he has been an Associate Professor with the Technical University of Iasi, Romania, and was a multi-year Visiting Associate Professor with Northeastern University, Boston, MA, U.S.A. He has maintained a continuous stream of R&D publications since 1992 within various professional organizations around the world, has organized nine professional education seminars (tutorials) at IEEE conferences, and holds four US patents. He has published several books, most noticeably *Switching Power Converters – Medium and High Power* (Boca Raton, FL, USA: CRC Press/Taylor and Francis, editions in 2006, 2013, and 2017), and *Telecom Power Systems* (Boca Raton, FL, USA: CRC Press/Taylor and Francis, 2017). Other ISBN books or college textbooks have been published in the U.S.A., Canada, and Romania.

Dr. Dorin O. Neacşu received the 2015 "Constantin Budeanu Award" from the Romanian Academy of Sciences, the highest Romanian research recognition in electrical engineering, and the 2018 "Ad Astra Engineering Award" from the Romanian Researchers' Association. He is an Associate Editor of the *IEEE Transactions on Power Electronics* and *IEEE IES Magazine*, a Reviewer for the IEEE Transactions and Conferences, and a Member of various IEEE committees.

More about the Author at:

- ORCID = 0000-0003-0572-1838
- ThomsonReuters ResearcherID = 5276-2011
- https://www.linkedin.com/in/dorin-o-neacsu-82a842

1 Architecture of an Automotive Power System

HISTORICAL MILESTONES (ELECTRICITY IN CARS)

1769: Military tractor invented by Joseph Cugnot based on a steam engine with a parallel development of electric and engine propulsion
1830s: First electric cars—R. Anderson (Scotland); Stratingh & Becker (Holland)
1842: Davenport (US) uses the first non-rechargeable electric cells
1859: Lead-acid batteries invented by Gaston Plante
1899–1902: Highest point in electric cars—followed by decreased interest
1912: C. Kettering (GM) invented the electric starter—no need for hand crank
191x: Other use of electricity in engine cars: ignition and lighting
1912: An electric roadster sold for $1,750, while a gasoline car sold for $650.
1920s: Use of the battery charger
1950s: Change from 6 V to 12 V system in cars. Note that larger trucks, land-based military vehicles, or pleasure boating use 24 V.
1980s–2000s: Development of comfort systems requiring more electrical power.
Today, most vehicles have an installed power around 10 kW (without propulsion).

1.1 ARCHITECTURE OF THE AUTOMOTIVE POWER SYSTEM

The most-used architecture for the electrical distribution system is shown in Figure 1.1. The main source of electrical power is the *alternator*, which is an electrical generator actuated by the engine and able to produce voltage for the dc distribution bus. A *battery* is also connected to the same bus. Its main function occurs during the startup of the engine, when a large current is required to crank the engine with a *starter*. Therefore, the battery just feeds additional power into the system, and the alternator is seen as the main source of electrical energy.

Numerous small motors, solenoids, and actuators are connected to the electrical distribution bus. Their count is increasing year after year, with the adoption of more comfort, convenience, and safety systems. Based on their function, these electrical loads are grouped into body systems, chassis systems, powertrain systems, and other miscellaneous safety and convenience systems.

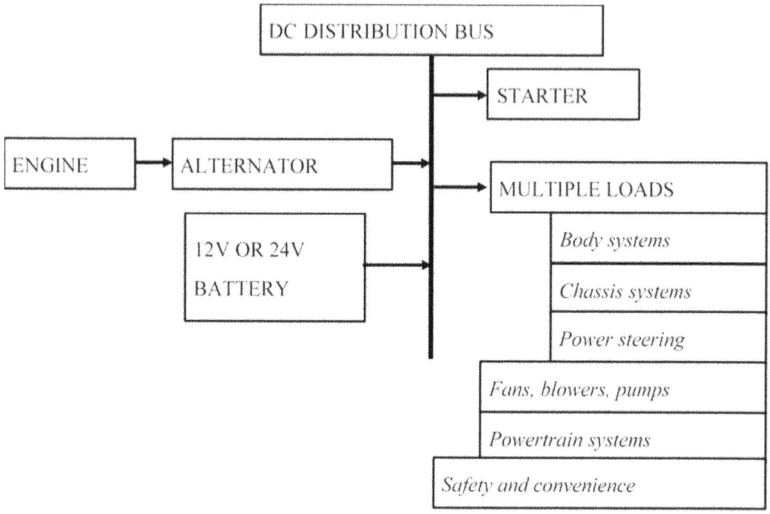

FIGURE 1.1 Typical architecture of an automotive electrical distribution system.

Electrical power produced by the alternator is not the only source of energy within a car. It is worthwhile to note that conventional cars use mechanical and hydraulic systems to transfer power and displacement to different components. Examples for the usage of these alternative sources of energy are presented throughout the book.

A modern concept applied to the architecture of electrical distribution system involves *drive-by-wire* (DbW) technology, which uses a central computer and various electrical actuators for the linkage of power. The main controller is similar to a video game controller in complexity and internal architecture. The driver can benefit from a user interface to fine-tune vehicle handling without changing anything in the car's mechanical components. Examples of sub-systems that use the *drive-by-wire* concept may include electronic throttle control, brake-by-wire, power steering, so on.

1.2 VOLTAGE USED FOR ELECTRICAL DISTRIBUTION SYSTEM

For the last 70 years, the electrical distribution system within auto vehicles has been built upon 14 V or 28 V batteries. The 14 V system is used for smaller cars, sedans, and sport-utility vehicles, while the 28 V system is used for large trucks and buses. This is mostly because the electrical motors used in large vehicles are also rated at higher power levels, which means the installed power is larger. More power means more current on the same voltage bus. Increasing the bus voltage allows for a reduction in current for the same installed power. A reduction in current allows for a reduction in calorimetric loss on the parasitic resistances of wires, connections, windings, and such. Nominal selection of 24 V as a multiple of 12 V allows for multiplication of the battery bays or components.

Historically, adopting the same voltage levels for multiple classes of vehicle allowed for the standardization and universal use of batteries and electrical components.

Characteristics of the 14 V power distribution system (also known as the 12 V system):

- Nominal voltage, 14.2 V
- Normal Operating Voltage in between 9 V (min) and 16 V (max)
- Jumper starts, 24 V
- Reverse polarity, maximum –12 V
- Lower voltage on cold cranking, 6.5 V

Characteristics of the 28 V power distribution system (also known as the 24 V system):

- Nominal voltage, 28.4 V
- Normal Operating Voltage in between 18 V (min) and 32 V (max)
- Jumper starts, 48 V
- Reverse polarity, –24 V
- Lower voltage on cold cranking, 13.3 V

The impressive development of comfort and safety systems intensified the load count and produced an increase in the installed power. In order to keep currents at lower levels, efforts have been targeted at an increase of the bus voltage. An industrial consortium led by *Massachusetts Institute of Technology* proposed a 42 V system in 1995, but it did not catch on and was abandoned for a while. However, their impact was seen in the development of certain components working on 42 V after a conversion from a central distribution bus of 12 V. Examples include the *General Motors*' BAS (belted alternator-starter) hybrids like *Buick LaCrosse* or *Saturn Aura* in the late 2000s.

The validity of an increased voltage concept led to another solution based on a hybrid 48 V dc system, more recently pushed by both *Delphi* and *Continental Corporations*. Both low-voltage 12 V dc and higher-voltage 48 V dc are proposed to co-exist in this architecture. Instead of replacing the electrical architecture entirely—like the auto industry contemplated doing with the 42 V systems—the up-and-coming 48 V dc designs complement it in what is essentially a mild hybrid arrangement. In consequence, a new electric motor and a 48 V battery are simply added onto the combustion engine, in addition to the conventional 12 V battery used for starting the vehicle.

1.3 THERMAL CHALLENGES FOR ELECTRICAL COMPONENTS

Automotive electrical and electronic equipment is rated and classified using several temperature ranges:

- Temperature range of equipment placed near the engine = –40 °C to 165 °C.
- Temperature range of equipment placed under the hood = –40 °C to 125 °C.
- Temperature range of equipment placed near passengers = 0 °C to 85 °C.

Moreover, in order to assess their performance, a combined thermal and electrical stress profile is used for testing. This is justified since failure can be produced by a combination of voltage and temperature conditions; for example, the engine cold cranking with a partially depleted battery. After acknowledging their importance, different regulatory agencies have established combined thermal and electrical stress profiles based on cyclic operation. The most important tests are:

- LV124: Electrical and electronic components in motor vehicles, usually up to 3.5 tons.
- LV148: Electrical and electronic components in newer motor vehicles, with 48 V electrical system.

These two tests (listed herein as examples) where put together by German Original Equipment Manufacturers (OEMs), but have nowadays several variations for almost all car manufacturers around the globe. An OEM is a company that produces parts and equipment that may be marketed by another manufacturer.

Other similar standards are VW 80000, BMW GS 95024-2, VDA 320, BMW GS 95026, VW 82148, MBN LV 124, so on, and even ISO standardized it as ISO 16750.

1.4 ABNORMAL VOLTAGES—SOURCES AND DEVICE RATINGS

1.4.1 Inductive Load

The automotive electrical system contains numerous solenoids or small electrical motors that are supplied through a power semiconductor switch, like a MOSFET. *A metal–oxide–semiconductor field-effect transistor* (MOSFET) is a field-effect transistor (FET) with an insulated gate. It is fabricated with the controlled oxidation of a semiconductor, typically silicon. The voltage applied to the gate determines the conductivity of the device. More details are provided in Chapter 10, which is dedicated to these devices.

The automotive loads are inherently inductive, and their study can be reduced to a MOSFET switching into an inductive load. The problem occurs when the switch turns-off since the inductance would tend to maintain the circulation of the current by creating a large self-induced voltage. Such voltage can augment the voltage applied to the drain-source circuit, and damage the semiconductor device.

In order to protect the MOSFET against the additional voltage induced by the inductive load, a diode is used to offer a path to the inductive current (Figure 1.2a). Since this diode conducts current in the absence of any active device operation, it is usually called the *freewheeling diode*. A disadvantage of the simple freewheeling diode approach is that it lengthens the decay of the current traveling through the inductor. Since the inductor can be the coil of a relay, a slow decay of the current can create problems such as "*sticking*" between the relay's contacts.

An alternative solution is to use a Zener diode, which gives a faster current ramp rather than an exponential decay seen with the freewheeling diode (Figure 1.2b).

Modern high-side switches frequently use a technique called *active clamping* that limits the drain-source voltage when switching inductive loads, in order to protect

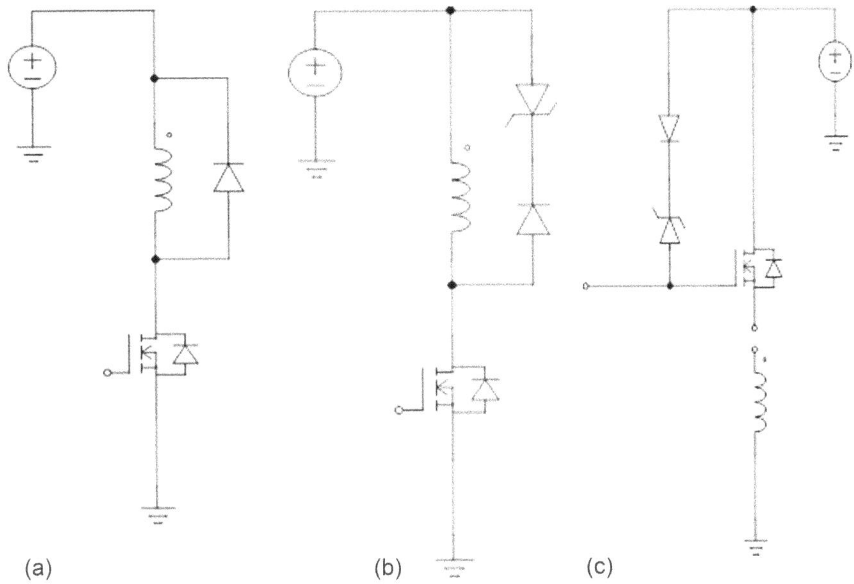

FIGURE 1.2 Protection solutions for inductive turn-off. (a) using a freewheeling diode; (b) using a Zener diode; (c) using an active clamping technique.

the MOSFET. This approach can also be found as a completely integrated solution (Figure 1.2c). During such clamping, the MOSFET dissipates more power than the load and needs to be designed for.

1.4.2 Transients on the DC Bus

Due to various transients and abnormal waveforms on the dc distribution bus, a front-end power stage is used to insulate the sensitive electrical and electronic loads from the wide variations of the dc distribution bus and to power the loads with a conditioned voltage rail. Since these converters (*front-end dc/dc converters*) need to work on various vehicles and loads, many original equipment manufacturers and organizations specify the immunity tests and the standardized test conditions for off-battery loads. Examples of worldwide accepted standards include ISO 16750-2 and ISO 7637-2.

The presence of diverse protection systems for the distribution bus has standardized their architecture, as shown in Figure 1.3. Various other protection circuits are considered within the actual design of the front-end converter.

Figure 1.4 illustrates possible abnormal voltage events on the distribution bus. Each such event provokes a different protection method and test procedure. These are shown in Table 1.1, along with the regulatory body.

Finally, the allowed voltage and duration limits used for the design are provided in Table 1.2. Knowledge of these values is used in the design of the front-end converters. The voltages seen with automotive dc/dc converters may be as shown in Figure 1.5. Note that the ISO 7630 standard requires bidirectional protection from –75 V dc to 87 V dc, for a 12 V dc distribution bus system.

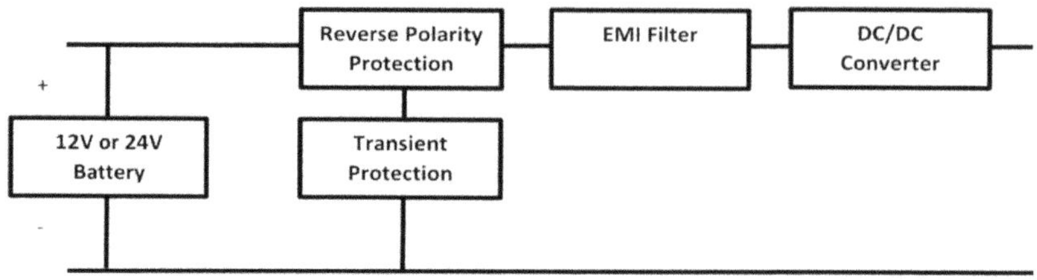

FIGURE 1.3 Protection of the distribution bus.

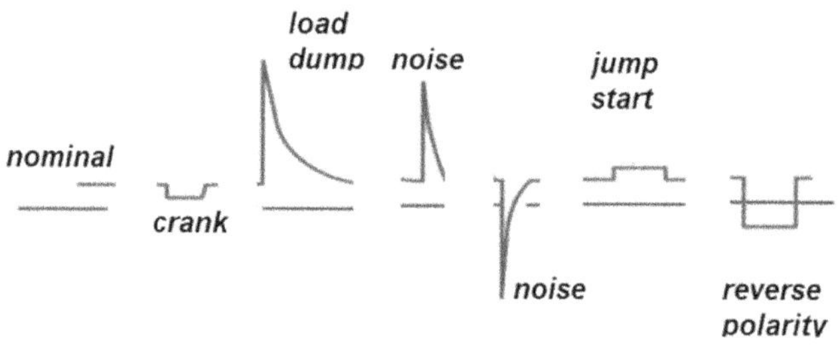

FIGURE 1.4 Possible sources of distortion on the dc bus.

TABLE 1.1
Details on the Definition of Distortion Sources on the Distribution Bus

Dump	Battery disconnection with alternator running, with the other load remaining on the alternator rails	ISO16750-2 (sec 4.6.4) FMC1278 CI 222
Starting profile	Simulates the disturbances during and after cranking	ISO16750-2 (sec 4.6.3) FMC1278 CI 230-231
Superimposed ac	Residual voltage ripple due to rectified sinusoid from a generator	ISO16750-2 (sec 4.4)
Superimposed ac	Superimposed pulses simulated sudden high current loads switching on the battery rail	FMC1278 CI 210 220 GMW3172 BMW3172
Reversed voltage	Reversed battery connection when using an auxiliary starting source	ISO16750-2 (sec 4.4)
Jump-start	DC voltage overstress due to a generator failure or jump-start using a 24 V battery	ISO 16750-2 (sec 4.3) FMC1278 CI 270

1.4.3 Reverse Voltage Protection

As mentioned in the previous section, the voltage distribution system can see negative voltages due to accidental connections. Various protection solutions can be used: A Schottky diode is the simplest solution, but it has a higher loss and is limited to low currents (Figure 1.6a). A P-type FET has the highest cost, and it needs a driving resistor and Zener clamp for a permanent conduction state when the voltage has the proper polarity (Figure 1.6b). Finally, a *smart diode* device includes a N-type FET, which provides lower R_{dson} and lower power loss (Figure 1.6c).

TABLE 1.2
Voltage and Duration Limits for Various Events on the Distribution Bus

Cause	Voltage spike	duration
Load dump—disconnect	<125 V	200–400 msec
Voltage regulator failure	18 V	Steady-state
Inductive load switching transient	80 … 300 V	<320 μsec
Alternator field decay	−100 … −40 V	200 msec
Ignition pulse, normal	<3 V	15 μsec
Ignition pulse, battery disconnect	<75 V	90 msec
Mutual coupling in harnessing	<200 V	1 msec
Accessory noise	<1.5 V	burst

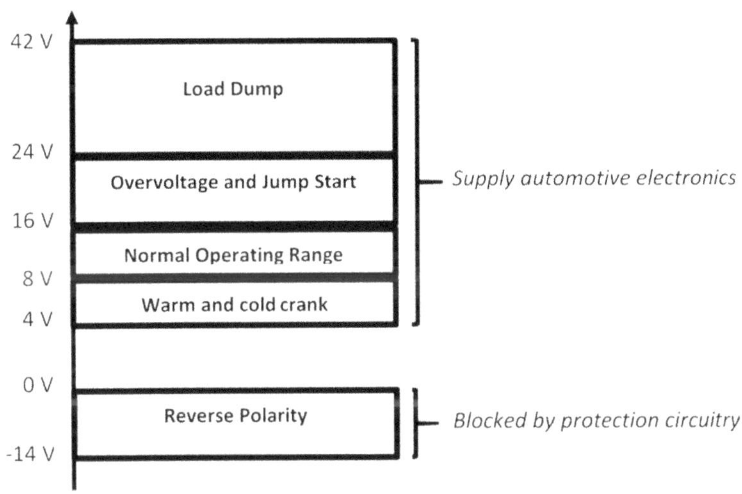

FIGURE 1.5 Actual voltage ranges within the automotive distribution system.

1.4.4 Mutual Coupling

Coupling between wires may indirectly produce transients. There are different types of coupling: magnetic, capacitive, and conductive (through wires). The possible mitigation includes the avoidance of long harnesses and the avoidance of ground loops.

1.5 REQUIREMENTS FOR THE ELECTRICAL ENERGY SYSTEM DESIGN

A series of requirements and best practices are included herein:

- The battery should have the minimum state of charge at the lowest temperature.
- Alternator output current depends on the engine speed and should be higher than the load current; otherwise, the battery will discharge.

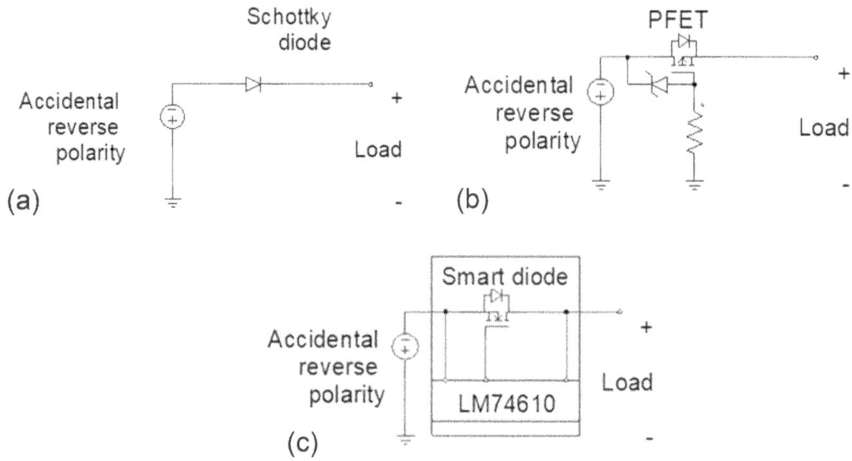

FIGURE 1.6 Protection circuits for the accidental reverse connection on the distribution bus. (a) A Schottky diode; (b) A P-type FET with a Zener clamp for a permanent conduction state when voltage has the proper polarity; (c) A smart diode built with a N-type FET.

- The time spent at idle operation should be reduced.
- The battery charging voltage should be higher in cold weather, and lower in warm weather to accommodate the electrical processes inside the battery.
- Charging balance calculation within the power management.
- Front-end converters should ramp up the operation at startup to avoid a sudden demand of current (inrush current) from the power converter and the dc distribution bus.
- A converter should be designed for a wide input voltage range so that it can also operate during a cold or warm crank, when the battery voltage and the dc distribution bus are low.
- Reduce the minimum dropout voltage at full load.
- Use the *Automotive Waterfall Test*, which is a standard automotive test where the input voltage of the dc/dc buck converter starts at the nominal level and is pulsed lower and lower. The Automotive Waterfall Test simulates the battery voltage to recover from a deep dropout of its nominal voltage.
- The designer has to be aware of battery life with quiescent current requirements:
  - A strict quiescent current is required to preserve battery life.
  - Always supplying components that stay active at engine turn-off, the off-battery dc/dc converter is required to have an excellent quiescent current and light load efficiency.
  - The quiescent current is required to be less than 100 μA at a module level and less than 30 μA at a front-end DC-DC converter level.
- The selection of the Switching Frequency and the EMI reduction are ensured on the automotive market by use of a synchronous buck converter, operated with a switching frequency either greater than 1.8 MHz or less than 530 kHz to avoid interference with FM/AM radio bands.

- The design should pass the CISPR 25 Class 5 conducted and radiated EMI standards.
- When switching below the AM band (<530 kHz), additional filtering is required to minimize emissions at the harmonics of the switching frequency.

1.6 DISTRIBUTION OF ELECTRICAL ENERGY

1.6.1 Fuses

The electrical energy produced by the generator and stored in the battery bank is distributed to various loads through several fuse boxes and through power cables rated according to the current being carried.

The electrical distribution system and the battery bank are protected against load malfunction with fuses. For convenience, fuses and relays are grouped in several *fuse boxes*. An example of a fuse box is shown in Figure 1.7, where the current rating is written on top of each fuse. A motor vehicle may have two or more fuse boxes. Fuses and relays will be presented in detail in Chapter 11.

1.6.2 Cables for Automotive Applications

All cables are almost always copper strands, insulated with PVC. Copper has a very low resistivity of about $1.7 * 10^{8}$ Ω·m, and it exhibits both ductility and malleability. Cables normally consist of multiple strands to provide greater flexibility. *Polyvinyl chloride* (PVC) is the world's third-most widely produced synthetic plastic polymer, after polyethylene and polypropylene. A polymer is a substance that has a molecular structure consisting of a large number of similar units bonded together, as many synthetic organic materials used as plastics and resins. PVC insulation has a very high resistivity of about $1.0 * 10^{15}$ Ω·m. PVC is resistant to temperature, oil spills, and water.

FIGURE 1.7 Example of a fuse box, located behind the glove compartment.

TABLE 1.3
Examples of Cable Ratings

Circuit	Typical Load	Cable Drop	Max. cable drop
Lighting circuit	<15 W	0.1 V	0.6 V
Lighting circuit	>15 W	0.3 V	0.6 V
Charging circuit	Nominal	0.5 V	0.5 V
Starter circuit	Nominal	0.5 V	0.5 V
Other circuit	Nominal	0.5 V	1.5 V

TABLE 1.4
Size Ratings for Cables

Cable strand/ diameter[mm]	Cross sectional area[mm2]	*AWG*	Continuous current rating [A]	Applications
9/0.30	0.6	*19*	5.75	Lights
14/0.25	0.7	*18–19*	6.00	Clock, radio, display
14/0.30	1.0	*17*	8.75	Ignition
28/0.30	2.0	*14*	17.50	Headlight
44/0.30	3.1	*12*	25.50	Main supply
84/0.30	5.9	*9*	45.00	Charging
37/0.90	23.5	*3*	350.00	Starter

Choice of cable size depends on the current drawn by the consumer. An undersized cable would heat up and eventually melt when conducting a large current. On the safe side, the larger the cable's diameter used, the smaller the voltage drop in the circuit, but the cable will be heavier. A copper cable with a larger diameter benefits from a smaller parasitic resistance (remember parallel resistors), hence a smaller voltage drop. This means a trade-off must always be sought between the allowable voltage drop and maximum cable size. Examples of voltage drops are shown in Table 1.3 and size ratings in Table 1.4.

In terms of standardization, cables are color-coded and carry terminal designation numbers for easier identification. While there are at least three systems used worldwide for motor vehicle applications, each manufacturer uses just a single system.

1.6.3 Harness Design

Cables are routed within a vehicle depending on the destination, and grouped for easier fabrication. This grouping of cables forms a *wiring harness*. The vehicle wiring harness has developed over the years from a loom containing just a few wires, to the looms used at present on top-end vehicles containing well over 1,000 separate wires. An example of a wiring harness is shown in Figure 1.8.

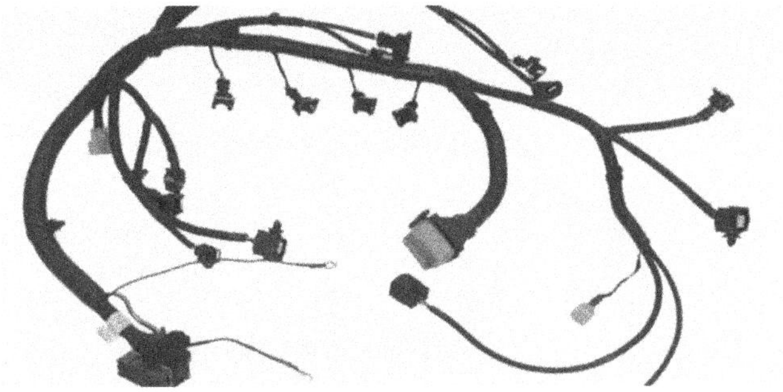

FIGURE 1.8 Wiring harness.

FIGURE 1.9 Making a wire harness

There are several ways to build a wiring harness. The most popular is still for the bundle of cables to be spirally wrapped in non-adhesive PVC tape, which is different from electrical adhesive tape (Figure 1.9).

When deciding on the layout of a wiring loom within the vehicle, many issues must be considered. Some of these are as follows.

- Cable runs must be as short as possible;
- The loom must be protected against physical damage;
- The number of connections should be kept to a minimum;
- The modular design must be appropriate;
- Accident damage areas need to be considered;
- Production line techniques should be considered;
- Access to main components and sub-assemblies must be available for repair purposes.

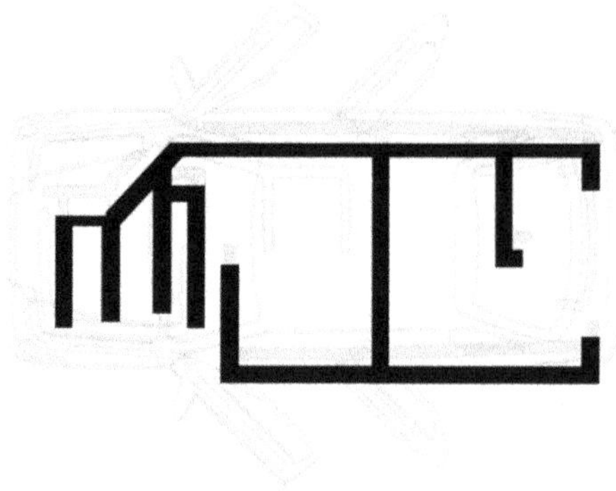

FIGURE 1.10 Routing wiring harness within the vehicle.

Typically, "H" and "E" wiring layouts are the most used (letters read as looking from the front of the motor vehicle). However, it is becoming the norm to have more than two main junction points as part of the vehicle wiring, with these points often being part of the fuse box (Figure 1.10).

1.7 REPRESENTATION OF THE ELECTRICAL CIRCUIT

Given the complexity of the electrical distribution system, standardized methods for the representation of electrical circuits in automotive applications have been developed. Depending on the use of information, several types of diagrams are accepted.

Conventional circuit diagrams show the electrical connections of a circuit but make no attempt to show the various parts in any particular order or position. By contrary, the *layout or wiring diagrams* attempt to show the main electrical components in a position similar to those on the actual vehicle. Due to the complex circuits and the high count of individual wires, some manufacturers use two diagrams—one to show electrical connections and the other to show the actual layout of the wiring harness and components.

For other goals, *terminal diagrams* only show the connections of devices and not the wiring. The terminal of each device is marked with a code. This code indicates the device terminal designation, the destination device code and its terminal designation, and, in some cases, the wire color code.

Finally, *current flow diagrams* have the page laid out to show current flow from top to bottom. These diagrams often have two supply lines at the top of the page marked 30 (main battery positive supply) and 15 (ignition-controlled supply). At the bottom of the page is a line marked 31 (earth or chassis connection). Note that the numerical notation is used by various manufacturers for the convenience of service workers.

1.8 CONCLUSION

Due to the complexity of automotive power systems, a structured architecture for the electrical distribution system is considered and presented herein. A series of abnormal voltage situations may occur during operation, and these have been outlined along with the most-used protection circuits.

Distribution of electrical energy is done through cables. The most-used types of cables and harnesses were discussed. Various ways to describe the electrical energy distribution circuitry were revealed.

Starting from the understanding of the electrical distribution system, components and sub-systems are described in the following chapters of the book.

BIBLIOGRAPHY

Choudhary, V. 2017. Designing the front-end DC/DC conversion stage to withstand automotive transients. *Analog Applications Journal AAJ.*

Denton, T. 2017. *Automobile Electrical and Electronic Systems.* 5th edition, Abingdon-on-Thames: Routledge.

Gerada, D., Zhang, H., Xu, Y., Calzo, L., Gerada, C. 2016. *Electrical Machine Type Selection for High Speed Supercharger Automotive Applications.* Paper Presented at 19th International Conference on Electrical Machines and Systems (ICEMS), Chiba, Japan, pp. 1–6.

Hamada, K. 2011. *Toyot's Activities on Power Electronics for Future Mobility.* Paper Presented in Proceedings of the 2011 14th European Conference on Power Electronics and Applications, Birmingham, pp. 1–1.

Hellenthal, B. 2012. *Power Electronics—Key to the Next Level of Automotive Electrification.* Paper Presented in Proceedings of the 2012 24th International Symposium on Power Semiconductor Devices and ICs, Bruges, Belgium, pp. 13–16.

Kanechika, M., Uesugi, T., Kachi, T. 2010. *Advanced SiC and GaN Power Electronics for Automotive Systems.* Paper Presented at International Electron Devices Meeting, San Francisco, CA, USA, pp. 13.5.1–13.5.4.

Neacşu, D. 2004. *Power Semiconductor and Control for Automotive Applications.* Tutorial Presented at IEEE APEC, Anaheim, CA, USA.

Nirmaier, T., Burger, A., Harrant, A., Viehl, A., Bringmann, A., Rosenstiel, W., Pelz, G. 2014. *Mission Profile Aware Robustness Assessment of Automotive Power Devices.* Paper Presented at EDAA Conference, Dresden, Germany.

Ostmann, A., Hofmann, T., Neeb, C., Boettcher, A., Manessis, D., Lang, K.D. 2012. *Embedded Power Electronics for Automotive Applications.* Paper Presented at 7th International Microsystems, Packaging, Assembly and Circuits Technology Conference (IMPACT), Taipei, pp. 163–166.

Sridhar, N. 2013. *Power Electronics in Automotive Applications.* Texas Instruments White Paper SLYY052.

Wondrak, W., Dehbi, A., Willikens, A. 2010. *Modular Concept for Power Electronics in Electric Cars*. Paper Presented at CIPS Conference, Nuremberg/Germany, paper 2.4.

APPENDIX #1 BRITISH STANDARD COLORS FOR CABLES

Color (alphabetical order)	Symbol	Usage
Black	B	All Earth/Ground connections
Blue	U	Headlight switch to dip switch
Blue/White	U/W	Headlight main beam
Blue/Red	U/R	Headlight dip beam
Blue/Yellow	U/Y	Fog lights
Brown	N	Main battery feed
Green	G	Ignition-controlled supply
Green/Brown	G/N	Reverse light
Green/Purple	G/P	Stoplights
Green/Red	G/R	Left-side indicator
Green/White	G/W	Right-side indicator
Light green	L	Instruments
Pink/White	K/W	Ballast resistor wire
Purple	P	Supply
Red	R	Sidelight main feed
Red/Black	R/B	Left-hand sidelight
Red/Orange	R/O	Right-hand sidelight
Slate	S	Electric windows
White	W	Ignition to ballast resistor
White/Black	W/B	Coil negative
Yellow	Y	Overdrive and fuel ignition

APPENDIX #2 EUROPEAN COLOR CODES FOR CABLES

Color (alphabetical order)	Symbol	Usage
Black	Sw	Reverse
Black/Yellow	Sw/Ge	Fuel injection
Black/Green	Sw/Gn	Ignition-controlled supply
Black/White/Green	Sw/Ws/Gn	Indicator switch
Black/White	Sw/Ws	Left-hand indicator
Black/Green	Sw/Gn	Right-hand indicator
Black/Red	Sw/Rt	Stoplights
Brown	Br	Earth/Ground
Brown/White	Br/W	Earth connections
Gray	Gr	Sidelight main feed
Gray/Black	Gr/Sw	Left-hand sidelights
Gray/Red	Gr/Rt	Right-hand sidelights
Green/Black	Gn/Sw	Fog light
Light green	LGn	Coil negative
Pink/White	KW	Ballast resistor wire
Red	Rt	Main battery feed

White	Ws	Headlight main beam
White/Black	Ws/Sw	Headlight switch to dip switch
Yellow	Ge	Headlight dip beam

APPENDIX #3 POPULAR TERMINAL DESIGNATION NUMBERS

Terminal number	Designation
1	Ignition coil negative
4	Ignition coil high voltage
15	Ignition switch positive
30	Input from positive battery
31	Earth/Ground connection
49	Input to flasher unit
49a	Output from flasher unit
50	Starter control from solenoid
53	Wiper motor
54	Stoplights
55	Fog lights
56	Headlights
56a	Main beam
56b	Dip beam
58L	Left-hand sidelights
58R	Right-hand sidelights
61	Warning light
85	Relay winding out
86	Relay winding in
87	Relay contact input
87a	Relay contact/break
87b	Relay contact/make
L	Left-hand light indicator
R	Right-hand light indicator
C	Warning indicator

2 Batteries

HISTORICAL MILESTONES

1911: Charles Kettering's DELCO team invented and built the first starter for Cadillac's CEO, Henry Leland.
1918: The first standardized battery produced by Battery Council International.
1920s: Car batteries became widely used after cars became equipped with electric starters.
1950s: Migration from 6 V to 12 V.
1971: The sealed battery, which did not require refilling, was invented.

2.1 FUNCTIONS OF THE STORAGE BATTERY

This chapter introduces storage batteries for conventional engine-based vehicles, rated at either 14 Vdc or 28 Vdc. The electric and hybrid vehicles, as described in Chapter 14, require a larger size battery bank to support the vehicle's propulsion, usually configured for a higher dc voltage, in the range of 250–400 Vdc.

The storage battery is the heart of the electrical circuit. The battery must operate the starting motor, ignition system, electronic fuel injection system, and other electrical devices for the engine during engine cranking and starting, even at temperatures as low as –18 °C. This can be understood from Figure 2.1: the available power from the battery must be above the power required for the starting of the engine. This condition becomes a problem under –18 °C, where the battery availability is reduced at around 50% of the power available at an ambient temperature of 27 °C.

The battery must also supply all the electrical power for the vehicle when the engine is not running (Figure 2.2). It helps the charging system provide electricity when current demands are above the output limit of the charging system during certain transients in vehicle operation. It is worth noting that it is the charging system that feeds the loads, and not the battery directly. The charging system (usually, an *alternator*) must act as a voltage stabilizer able to smooth current flow through the electrical system. The battery must store electrical energy for extended periods, independent of weather, and temperature swings.

The type of storage battery most used in automotive applications is a *lead–acid cell-type battery*.

2.2 CONSTRUCTION OF A LEAD–ACID CELL-TYPE BATTERY

Figure 2.2 illustrates the internal structure of the conventional lead–acid cell-type battery. The 14 Vdc automotive battery has six elements called cells. Each battery element (cell) is made up of negative plates, positive plates, separators, and straps. Both positive and negative plates are a stiff mesh framework coated with porous lead. Lead-oxide (also named lead-peroxide, PbO_2) is used on the positive plates,

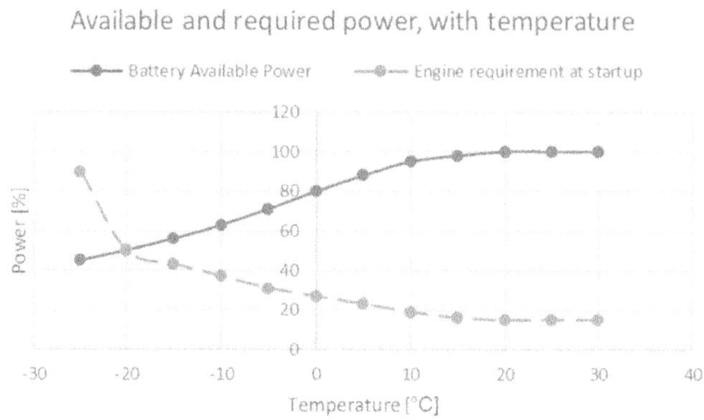

FIGURE 2.1 Power considerations for batteries at an engine's startup.

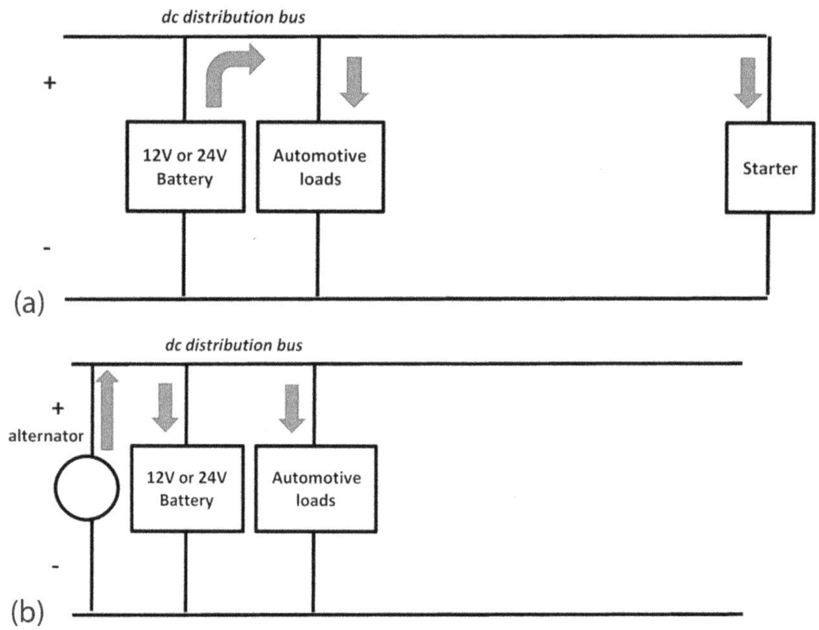

FIGURE 2.2 Roles of the storage battery: (a) engine not running, (b) engine running.

while a spongy, highly porous lead (Pb) on the negative plates. This is also shown in Figure 2.3.

The plates are insulated from each other by suitable separators, which are microporous rubber, fibrous glass, or plastic-impregnated material. The separators are thin and porous so that the electrolyte will flow easily between the plates. The side of the separator that is placed against the positive plate is grooved so that the gas that forms during charging will rise to the surface more readily.

The plates are submerged in a sulfuric acid solution, called *electrolyte*. The electrolyte is a diluted sulfuric acid (H_2SO_4), usually in a ratio of ~60% water and ~40% sulfuric acid.

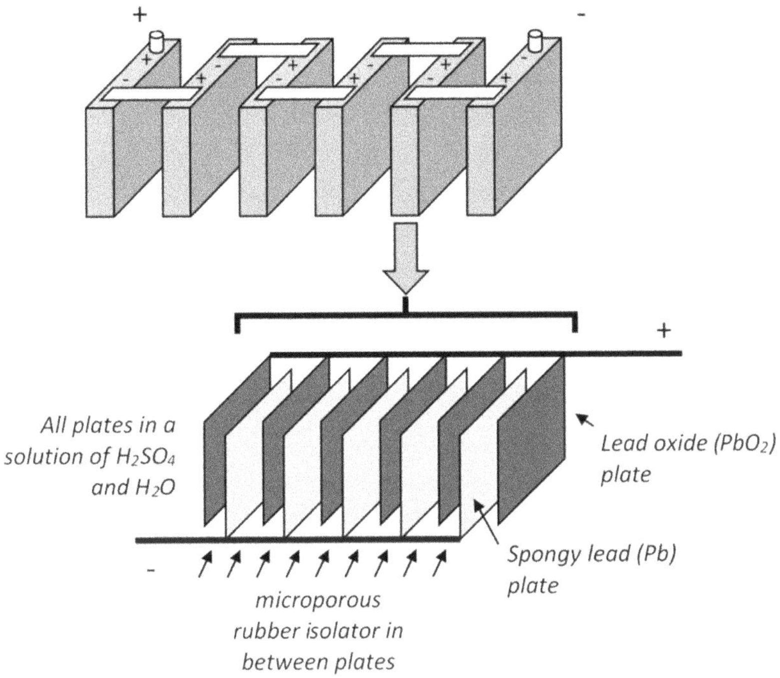

FIGURE 2.3 Construction of a battery.

The battery case is made of hard rubber or a high-quality plastic. The case must withstand extreme vibration, temperature change, and the corrosive action of the electrolyte. Stiff ridges or ribs are molded in the bottom of the case to form a sediment recess for the flakes of active material that drop off the plates. The battery caps are called *vent plugs*, and they allow gas to escape, prevent the electrolyte from splashing outside the battery, and serve as *spark arresters*. A spark arrester is a device which prevents the emission of flammable debris from combustion sources.

The positive and negative plates assume electrical voltages with respect to the electrolyte. The cell voltage – at rest – is approximately 2 V dc, increasing during charging and decreasing, when the cell is subjected to a load. When a large current is demanded, the electrolyte loses ions. During this discharge process, the PbO_2 (positive) and Pb (negative) plates combine with H_2SO_4 from the electrolyte to form $PbSO_4$ (lead sulfate) and water. Lead sulfate is a white solid, which appears white in microcrystalline form, and it is often seen in the plates/electrodes of the batteries, as it is formed when the battery is discharged. Lead sulfate is poorly soluble in water. It can be seen that the sulfuric acid depletes and the electrolyte is gradually converted into water when the storage battery discharges.

The specific gravity of the electrolyte can be used to sense the charging state of the battery as it decreases from 1.28 (1.28 times as heavy as water) toward 1.00 (just water) when the battery is discharged. The accuracy of this relationship depends upon battery design as well as electrolyte stratification and battery wear.

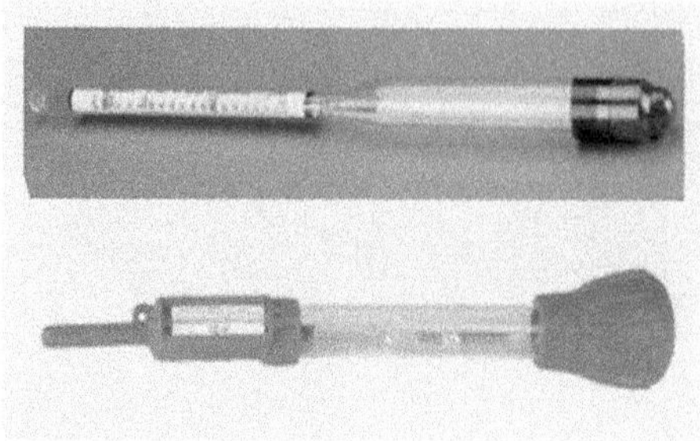

FIGURE 2.4 Hydrometer products.

2.3 HYDROMETER READINGS

The specific density of the electrolyte is measured with a *hydrometer*. This device makes use of Archimedes' principle: a solid suspended in a fluid is buoyed by a force equal to the weight of the fluid displaced. Figure 2.4 exemplifies different hydrometer products.

A *hydrometer* consists of a sealed hollow glass tube with a wider bottom portion for buoyancy, a ballast such as lead or mercury for stability, a narrow stem for measuring. The liquid to test is poured into a tall container, often a graduated cylinder and the hydrometer is gently lowered into the liquid until it floats freely. The point at which the surface of the liquid touches the stem of the hydrometer correlates to its specific density.

When investigating the specific density of the battery's electrolyte, one should also consider its dependency on temperature by following the guidance below.

- For every ten-degree change in temperature below 80° F, 0.004 should be subtracted from the specific gravity reading;
- For every ten-degree change in temperature above 80° F, 0.004 should be added to the specific gravity reading.

Beyond using the hydrometer, other test procedures have also been elaborated for battery testing.

2.4 VOLTAGE LEVEL TEST

The *battery voltage test* is used on maintenance-free (sealed) batteries because these batteries do not have caps that can be removed for testing with a hydrometer. Section 2.8 will provide more detailed information about the sealed maintenance-free batteries.

The procedure follows the next steps.

- Connect the voltmeter or battery tester across the battery terminals.
- Turn on the vehicle headlights or heater blower to provide a light yet consistent load.

- Read the voltmeter or tester.
- A well-charged battery should have over 12 V.
- If the voltmeter reads approximately 11.5 V, the battery is not charged adequately, or it may be defective.

Analogously, a single-cell voltage test can be performed for conventional batteries where access to each cell is possible. The procedure follows the next steps.

- Use a low voltage reading voltmeter with special cadmium (acid-resistant metal) tips.
- Insert the tips into each cell, starting at one end of the battery and work your way to the other.
- Test each cell and read the voltage measurement.
- If the cells are low, but equal, recharging will usually restore the battery.
- If cell voltage readings vary more than 0.2 V, the battery is bad.

2.5 CAPACITY

The capacity of a battery is measured in *Ampere–hours* and its rating is required by law in Europe, under the EU Battery Directive 2006/66/EC. The *Ampere–hour capacity* is equal to the product of the current in Amperes and the time in hours during which the battery is supplying current. For instance, a 10 Ah battery can be used for 10 hours at 1 Ampere before being discharged.

Despite this textbook definition with an *Ampere–hour* multiplication, the actual meaning refers to a 20-hour figure, that is considered for a possible discharge at load current. For instance, a 100 Ah battery would discharge for 20 hours at 5 A before becoming depleted.

The capacity of a cell depends upon many factors:

- The area of the plates in contact with the electrolyte;
- The quantity and specific gravity of the electrolyte;
- The type of separators;
- The general condition of the battery, expressed with factors like degree of sulfating, plates buckled, separators warped, sediment in the bottom of cells, and so on;
- The final limiting voltage.

A *battery load test* is also named a *battery capacity test*. The battery load test measures the current output and performance of the battery under the full current load. Calculation of the current drawn during the test is performed before load-testing a battery. The designer has to calculate how much current draw should be applied to the battery for the test. If the Ampere–hour rating of the battery is given, the battery is loaded to three (3x) times the Ampere–hour rating. For example, if rated at 60 Ah, test the battery at 180 A (because $60 \times 3 = 180$).

However, the majority of batteries nowadays are rated in Society of Automotive Engineers (SAE) *cold-cranking A* (*denoted with cold-cranking-amp* (CCA)), instead of Ah. To determine the load test current, the cold-cranking current rating should be divided by two. For example, a battery with 400 cold-cranking-amps rating should be loaded to 200 A (400 ÷ 2 = 200).

A step-by-step test procedure is herein included.

- Connect the battery load tester.
- Apply calculated current as load.
- Hold this load current for 15 seconds.
- Read the voltmeter while the load is applied.
- If the voltmeter reads 9.5 V or more at room temperature, the battery is good.
- If the battery reads below 9.5 V at room temperature, battery performance is poor. This condition indicates that the battery is not producing enough current to run the starting motor properly.

A series of numerical ratings have been developed by the *Society of Automotive Engineers* and *Battery Council International* (BCI). First, the *cold-cranking rating* determines how much current, in Amperes, the battery can deliver for 30 seconds at 0 °F (–18 °C) while maintaining terminal voltage of 7.2 V dc or above (that is 1.2 V per cell). This rating indicates the ability of the battery to crank a specific engine based on starter current draw and at a specified temperature.

For example, one manufacturer recommends using a battery with 305 cold-cranking-amps for a small four-cylinder engine but a different 450 cold-cranking-amp battery for a larger V-8 engine. A more powerful battery is needed to handle the heavier starter current draw of the larger engine.

A second numerical rating proposed by *Society of Automotive Engineers* and *Battery Council International*, refers to the *reserve capacity rating* (in minutes), and it represents the time needed to lower battery terminal voltage below 10.2 V dc (that is 1.7 V per cell) at a discharge rate of 25 Amperes, from the battery fully charged, at 80 °F.

This reserve capacity rating will appear written on the battery as a time interval in minutes. For instance, if a battery is rated at 90 minutes and the charging system fails, the operator has approximately 90 minutes (1.5 hours) of driving time under minimum electrical load before the battery goes completely dead.

For a practical example of using these ratings, Figure 2.5 shows a battery product along with its specification.

Another parameter shown in Figure 2.5 and defined by *Battery Council International* refers to the *group size*. The complete definition is explained in Table 2.1. For instance, the battery in Figure 2.5 belongs to the group size 34.

2.6 BATTERY CHARGERS

Obviously, one needs to push current into the battery for charging it when it is not connected to a working alternator. The question is: *How much current and for how long?*

First, electronic chargers were based on a Si diode able to rectify a voltage following a grid-connected variable-voltage transformer. The variable-transformer was

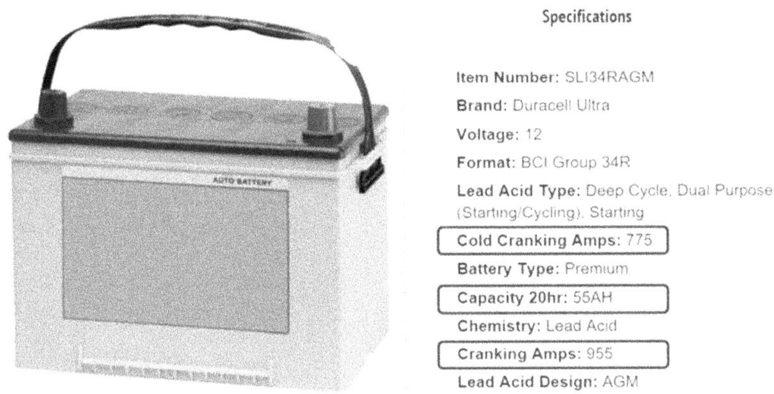

FIGURE 2.5 Example of a battery product.

used to set the variable current through the battery. An old recommendation ("*rule of thumb*") was that the battery should be charged at a tenth of its Ampere–hour (Ah) capacity for about 10 hours or less. Another recommendation was to set a rate at 1/16 of the reserve capacity, again for up to 10 hours. Finally, another suggestion was to set a charge rate at 1/40 of the cold start performance figure, also for up to 10 hours. All three concepts can be found within the following example for the theoretical charging characteristics in Figure 2.6.

Using a simple Si diode rectifier-based equipment within a shop can be quickly done with one of the following methods.

- *Constant current* – apply a constant current until the battery is charged;
- *Constant voltage* – apply a higher voltage and expect the current to yield depending on the current state of battery charging.

The charging procedure is optimized with the following steps.

- Clean and inspect the battery thoroughly before placing it on charge;
  - Use a solution of baking soda and water for cleaning;
  - Inspect for cracks or breaks in the container;
- Connect the battery to the charger;
  - Connect positive post to positive (+) terminal and negative post to negative (–) terminal. Positive terminals of both battery and charger are marked; those unmarked are negative. Also, the positive post of the battery is, in most cases, slightly larger than the negative post;
  - Ensure that all connections are tight.
- Check if vent holes are clear and open. Do not remove battery caps during charging.
- Check the electrolyte level before charging begins and during charging.
  - Add distilled water if the level of electrolyte is below the top of the plate.

TABLE 2.1
BCI Classification by Size

BCI Group Size	L (mm)	W (mm)	H (mm)	L (inches)	W (inches)	H (inches)
21	208	173	222	8 3/16	6 13/16	8 3/4
21R	208	173	222	8 3/16	6 13/16	8 3/4
22F	241	175	211	9 1/2	6 7/8	8 5/16
22HF	241	175	229	9 1/2	6 7/8	9
22NF	240	140	227	9 7/16	5 1/2	8 15/16
22R	229	175	211	9	6 7/8	8 5/16
Group 24 Batteries	260	173	225	10 1/4	6 13/16	8 7/8
24F	273	173	229	10 3/4	6 13/16	9
24H	260	173	238	10 1/4	6 13/16	9 3/8
24R	260	173	229	10 1/4	6 13/16	9
24T	260	173	248	10 1/4	6 13/16	9 3/4
25	230	175	225	9 1/16	6 7/8	8 7/8
26	208	173	197	8 3/16	6 13/16	7 3/4
26R	208	173	197	8 3/16	6 13/16	7 3/4
Group 27 Battery	306	173	225	12 1/16	6 13/16	8 7/8
27F	318	173	227	12 1/2	6 13/16	8 15/16
27H	298	173	235	11 3/4	6 13/16	9 1/4
29NF	330	140	227	13	5 1/2	8 15/16
33	338	173	238	13 5/16	6 13/16	9 3/8
Group 34 Battery	260	173	200	10 1/4	6 13/16	7 7/8
Group 34R Battery	260	173	200	10 1/4	6 13/16	7 7/8
35	230	175	225	9 1/16	6 7/8	8 7/8
36R	263	183	206	10 3/8	7 1/4	8 1/8
40R	277	175	175	10 15/16	6 7/8	6 7/8
41	293	175	175	11 3/16	6 7/8	6 7/8
42	243	173	173	9 5/16	6 13/16	6 13/16
43	334	175	205	13 1/8	6 7/8	8 1/16
45	240	140	227	9 7/16	5 1/2	8 15/16
46	273	173	229	10 3/4	6 13/16	9
47	246	175	190	9 11/16	6 7/8	7 1/2
48	306	175	192	12 1/16	6 7/8	7 9/16
49	381	175	192	15	6 7/8	7 3/16
50	343	127	254	13 1/2	5	10
51	238	129	223	9 3/8	5 1/16	8 13/16
51R	238	129	223	9 3/8	5 1/16	8 13/16
52	186	147	210	7 5/16	5 13/16	8 1/4
53	330	119	210	13	4 11/16	8 1/4
54	186	154	212	7 5/16	6 1/16	8 3/8
55	218	154	212	8 5/8	6 1/16	8 3/8
56	254	154	212	10	6 1/16	8 3/8
57	205	183	177	8 1/16	7 3/16	6 15/16

(Continued)

TABLE 2.1 (CONTINUED)
BCI Classification by Size

BCI Group Size	L (mm)	W (mm)	H (mm)	L (inches)	W (inches)	H (inches)
58	255	183	177	10 1/16	7 3/16	6 15/16
58R	255	183	177	10 1/16	7 3/16	6 15/16
59	255	193	196	10 1/16	7 5/8	7 3/4
60	332	160	225	13 1/16	6 5/16	8 7/8
61	192	162	225	7 9/16	6 3/8	8 7/8
62	225	162	225	8 7/8	6 3/8	8 7/8
63	258	162	225	10 3/16	6 3/8	8 7/8
64	296	162	225	11 11/16	6 3/8	8 7/8
Group 65 Batteries	306	190	192	12 1/16	7 1/2	7 9/16
70	208	179	196	8 3/16	7 1/16	7 11/16
71	208	179	216	8 3/16	7 1/16	8 1/2
72	230	179	210	9 1/16	7 1/16	8 1/4
73	230	179	216	9 1/16	7 1/16	8 1/2
74	260	184	222	10 1/4	7 1/4	8 3/4
75	230	179	196	9 1/16	7 1/16	7 11/16
76	334	179	216	13 1/8	7 1/16	8 1/2
78	260	179	196	10 1/4	7 1/16	7 11/16
85	230	173	203	9 1/16	6 13/16	8
86	230	173	203	9 1/16	6 13/16	8
90	246	175	175	9 11/16	6 7/8	6 7/8
91	280	175	175	11	6 7/8	6 7/8
92	317	175	175	12 1/2	6 7/8	6 7/8
93	354	175	175	15	6 7/8	6 7/8
95R	394	175	190	15 9/16	6 7/8	7 1/2
96R	242	173	175	9 9/16	6 13/16	6 7/8
97R	252	175	190	9 15/16	6 7/8	7 1/2
98R	283	175	190	11 3/16	6 7/8	7 1/2

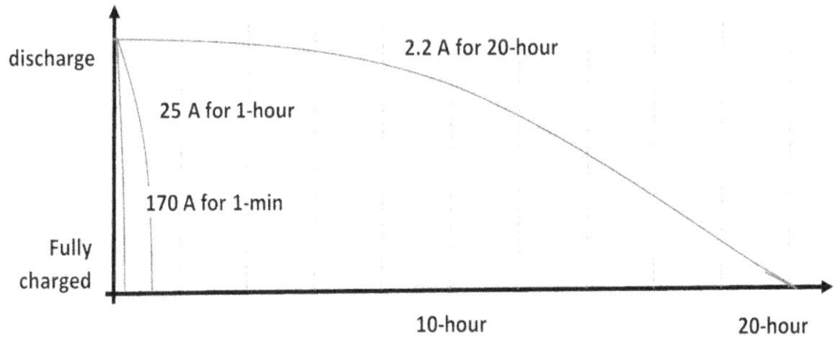

FIGURE 2.6 Theoretical charging characteristics.

- Keep the charging room well ventilated since batteries on charge release hydrogen gas. A small spark may cause an explosion.
- Take frequent hydrometer readings of each cell and record them. You can expect the specific gravity to rise during the charge. If it does not rise, remove the battery and dispose of it as per local hazardous material disposal instruction.
- Reduce the charging current if excessive gassing occurs. Some gassing is normal and aids in remixing the electrolyte.
- Do not remove a battery until it has been completely charged.

Modern electronic chargers are able to precisely control a constant charging current through power management and electronic power converters, connected from the grid to the battery. Such simple power stage circuits are built and controlled with dedicated integrated circuits (ICs) or microcontroller programs. Using an electronic charger allows a more efficient charging, with an optimized alternance of constant current and voltage charging methods.

The charging procedure using a power electronic converter follows the steps illustrated in Figure 2.7. If the initial voltage is very low, a low trickle current is applied to bring up the battery voltage to an acceptable level (step in Figure 2.7a). A larger current is next applied for a bulk charge (step in Figure 2.7b). When the voltage reaches its maximum level, a float charge current is pass through to maintain the voltage and counterbalance leakage (step in Figure 2.7c). Obviously, this complex procedure is based on sensors and microcontroller software and it is possible with modern power converters only.

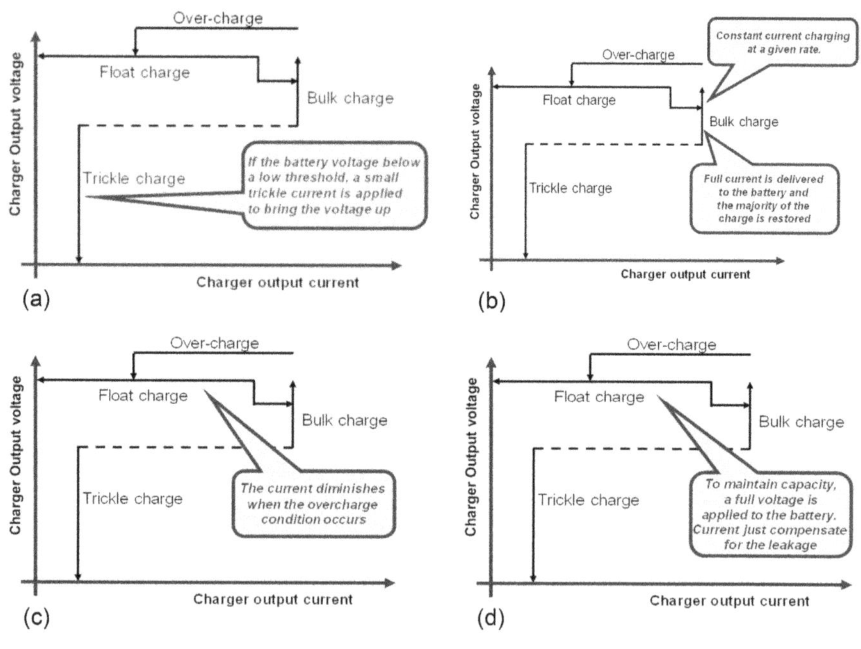

FIGURE 2.7 Step-by-step procedure for battery charging.

2.7 ELECTRICAL CHARACTERISTICS OF LEAD–ACID BATTERIES

For an electronic control engineer, it is important to understand the *electrical characteristics (model) of a battery* before designing the electronic charger.

2.7.1 Internal Resistance

The battery's internal resistance is defined as for any source of electrical energy and it can be determined experimentally. A numerical example is herein included along with a step-by-step procedure.

- Connect a voltmeter across the battery and note the open-circuit voltage (example 12.7 V).
- Connect an external load to the battery, and measure the current, say 50 A.
- Note again the on-load terminal voltage of the battery, under full load, for example, 12.2 V.
- A calculation determines the internal resistance: R_i = (U–V)/I, where U = open-circuit voltage, V = on-load voltage, I = current, R_i = internal resistance.

For this example, the result of the calculation is 0.01 Ω.

2.7.2 Efficiency

The efficiency of a battery can be defined in several ways.

The *Ampere–hour efficiency* is defined as:

$$Ah_{\text{efficiency}} = \frac{Ah_{\text{discharging}}}{Ah_{\text{charging}}}\% \tag{2.1}$$

This is used with actual time intervals for charging and discharging during actual usage. A possible reference is that, at the 20-hour, this Ah-efficiency can be as much as 90%. It is more often used in a reversed form, as a charge factor, usually of a value around 1.1.

The *energy efficiency* is another possible definition which is expressed as

$$EE = \frac{P_d \cdot t_d}{P_c \cdot t_c} \tag{2.2}$$

where P_d = discharge power into the loads, t_d = discharge time, P_c = charging power from the source/charger, t_c = charging time.

A typical result of this calculation is about 75%. This is lower than the *Ah-efficiency* because it takes into account the higher voltage required to force the charge into the battery.

The state of charge of the battery also determines charge efficiency. With the battery at half charge or less, the charge efficiency may be over 90%, dropping to near 60% when the battery is above 80% charged. Unfortunately for the automotive battery, when an electronic load is drawing a high current from the battery, the voltage drops and may trigger the circuit's undervoltage protection. This means the battery needs to be somewhere over 50% charged to avoid the electronics cutting out due to low voltage.

Furthermore, if a battery is only partially charged instead of going toward the full-charge status, the efficiency is reduced with each charge. If the battery never reaches full charge, repetitively, the lifetime of the battery is reduced and failure due to aging is accelerated.

2.8 NEW TECHNOLOGIES FOR SEALED AND MAINTENANCE-FREE BATTERIES

Due to the high-volume market for automotive batteries, new technologies have been explored over the years. In the mid-1970s, researchers developed a maintenance-free lead–acid battery that can operate in any position. The liquid electrolyte is gelled into moistened separators and the enclosure is sealed. Safety valves allow venting during charge, discharge, and atmospheric pressure changes.

Gel batteries are attractively advantageous since they do not leak and are more resistant to poor handling. Because a gel is used as electrolyte, the speed of the chemical reaction is hence reduced and the *cold-cranking capacity* (CCC) of this type of battery is lower than the equivalent-sized conventional battery. They can be used as specialty batteries, mostly as energy storage units.

Another new technology targets *Long-service batteries.* These use lead–antimony (PbSb) for the positive plate grids and lead–calcium (PbCa) for the negative plate grids, which produces a significant reduction in water loss and an increase in service life. Due to production of a lower quantity of water, these batteries can therefore be sealed and operated without service.

Despite the developments in sealed lead–acid batteries, these still have a *drawback during a long non-operating interval.* As an improvement, *alkaline cells* withstand better heavy discharge and overcharging, alternating with long non-operating periods. The alkaline batteries used for vehicle applications are generally the nickel–cadmium (Ni–Cd) type. A fully charged Ni–Cd cell contains a nickel–oxide–hydroxide positive electrode plate, a cadmium negative electrode plate, a separator, and an alkaline electrolyte (potassium hydroxide). Nickel–oxide–hydroxide is the inorganic compound with the chemical formula NiO(OH). It is a black solid that is insoluble in all solvents but attacked by base and acid. These batteries use, as the electrolyte, a mix of potassium hydroxide (KOH) and water (H_2O). At charging, the oxygen is moving from the negative plate to the positive plate, and the reverse is happening when discharging. When fully charged, the negative plate becomes pure cadmium and the positive plate becomes nickel hydrate. Another difference from conventional batteries demonstrates that a relative density reading will not indicate the state of charge since the electrolyte does not change during the reaction. The cell voltage of a fully charged *alkaline cell* is 1.4 V dc and falls rapidly to 1.3 V dc as soon as discharge starts. The cell is discharged at a cell voltage of 1.1 V dc.

The nickel–cadmium battery rapidly lost rechargeable market share in the 1990s, to Ni–MH and Li–ion batteries. Neither technology has proven its merits for automotive applications. The Ni–MH (nickel–metal-hydride) battery technology shows some promise for electric vehicle use (high voltage storage). Next section explores other energy storage solutions for electric and hybrid vehicles, therefore no-cranking applications.

2.9 OTHER POSSIBLE STORAGE OF ELECTRICAL ENERGY

2.9.1 SUPERCAPACITORS

The *supercapacitors* (often called *ultracapacitors*) feature a novel structure of capacitor, that is different from ceramic or electrolytic capacitors. This structure is based on two layers. Hence also the name of *double-layer capacitor.* The supercapacitor uses electrodes and electrolytes, not unlike a battery, and it is able to deliver a capacitance thousands of times larger than an electrolytic capacitor. Comparative performance of an ultracapacitor is shown in Table 2.2.

While ultracapacitors did not receive acceptance as drop-in replacements for conventional lead–acid batteries, a mainstream application imposes its use along the conventional lead–acid batteries for an engine start–stop electronic system. An engine running at idle when there is no vehicle displacement spends gas and damages the overall efficiency. Gas can be saved with a repetitive start–stop of the combustion engine, avoiding the idle running. At each start of the engine, a high peak current is demanded from the battery. While the conventional battery is ideal for supporting this peak current demand, it also has a limited lifetime cycle available for start–stop cycles. Figure 2.2 has previously shown that the storage battery needs to both support engine start–stop and the other loads when the alternator does not work. Hence, a combination of the conventional battery and an ultracapacitor was proposed to successfully satisfy both demands.

First, *Continental Automotive* designed-in the ultracapacitor technology for the voltage stabilization system supporting start–stop in *Lamborghini* and *Peugeot-Citroën* models. The same voltage stabilization system (VSS) conceived by *Continental Automotive* with ultracapacitors from *Maxwell Technologies* (now part

TABLE 2.2
Performance of Various Solutions for Energy Storage

Function	Supercapacitor	Lithium–ion battery
Charging	1 ... 10 seconds	10 .. 60 minutes
Lifetime (cycles)	1 million or 30,000 hours	>500
Cell voltage	2.3 … 2.75 V	3.6 … 3.7 V
Specific energy [Wh/ kg]	5	100 ... 200
Specific power [W/ kg]	<10,000	1,000 … 3,000
Cost per Wh	$2	@0.5 … $1.0

of *Tesla Automotive*) was adopted by *General Motors* as a standard feature on 2016 Cadillac ATS and CTS sedans and ATS coupes.

Engine start–stop is one function that helps manufacturers achieve fuel economy improvements in either conventional or hybrid-electric vehicles. The control system turns off the engine when the car comes to any temporary full stop and restarts the engine when the foot is taken off the brake. Engine start–stop prevents idling, saves fuel, and reduces emissions, satisfying the requirements for both improved fuel economy and emission control. In this respect, ultracapacitors are integrated into a battery-based engine start–stop system and deliver burst power to restart the engine, relieving the vehicle's battery of high current, repetitive cycling, that often shortens the battery lifetime. Within this architecture, the ultracapacitor allows the starter to be separated from the 12-V battery and other electrical loads, thereby eliminating voltage sags during start–stop function. This qualifies as a voltage stabilization for the dc distribution bus. Meanwhile, the conventional lead–acid battery supplies the other vehicle's electrical loads. This solution is described in Figure 2.8, where the ultracapacitor is also charged from the alternator through a dc/dc converter. The ultracapacitor offers high peak power while the conventional battery offers long-term energy. The ultracapacitor does not have a lifetime limitation for deep repetitive cycles.

Other applications of ultracapacitors in association with the conventional lead–acid battery concern chassis systems like active suspension, braking, and power steering. Chapter 6 presents details of the electrification of these systems. Electrification is driving new requirements for energy storage to support all the electrical systems added as loads to the dc distribution system. Such requirements need to assure both a high-rate discharge and a regeneration capability. Due to high peak demands in these chassis systems, the ultracapacitor technology serves as a dedicated energy source for the application.

2.9.2 Fuel Cell

The *fuel-cell* system represents an electrochemical device which combines liquid combustible (fuel) with oxygen to produce electricity, heat, and water. Fuel-cell systems are similar to a battery when concerning the chemical reaction. The hydrogen

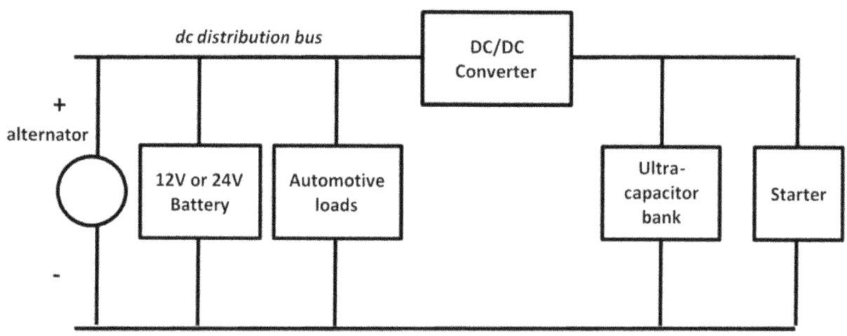

FIGURE 2.8 Using ultracapacitors for start–stop of modern vehicles.

combustible (fuel) is stored in a pressurized container, while the oxygen is taken from free air. Since there is no burning process, the only waste is water. The fuel-cell systems are reliable and can be 30% cheaper than batteries. They need a long time to start, usually around 5 seconds. Thus, another battery or ultracapacitor is needed within the same system, for this short interval.

There is a version of fuel cell which works with methanol. A *reformator* is converting methanol into hydrogen, then the fuel cell.

The fuel-cell systems did not really succeed in automotive applications as a replacement for conventional lead–acid batteries. Even if product solutions have been designed to replace a 12 V lead–acid battery, its use did not prove to have substantial merit. The fuel-cell systems are sometimes used in larger propulsion systems working at higher voltages.

Some success has been achieved with their possible application in public transportation. Fuel cells for "*urban transport*" vehicles typically use 20 × 10 kW stacks (200 kW) operated on a 650 V bus. The so-called streetcars (or tramway, trolley-bus), are using supply voltages in between 600 V dc and 825 V dc. The 600 V dc is the typical European Western model, while traditionally, Eastern European solutions were attempted with 825 V dc or 750 V dc. In order to use the same equipment with local electrical energy sources, fuel-cell systems have been adapted to such dc voltage bus. The same powertrain can be used with a large battery or fuel cell, obviously with a lack of cold-cranking requirement, which simplifies their design and adoption.

2.10 CONCLUSION

The sole storage source of electrical energy within an auto vehicle consists of a battery. Given its importance within the electrical system of a motor vehicle, numerous standards for ratings, charging, or testing have been elaborated over the years. The conventional lead–acid battery remains the top choice for the storage of electrical energy, especially when used for starting the combustion engine in any Ambiental conditions.

Other solutions used either as stand-alones or in association with the conventional lead–acid battery were introduced in this chapter.

REFERENCES

Anon. 2009. *MAX1640/MAX1641 – Adjustable-Output, Switch-Mode Current Sources with Synchronous Rectifier.* Maxim Integrated Corporation, Datasheet.

Anon. 2019. *Battery Group Size.* Internet reading, https://www.batteryequivalents.com/bci-battery-group-size-chart.html, Last reading December 9, 2019.

Anon. 2020. *Maxwell Ultracapacitor Automotive Solutions*, Internet reading, https://www.mouser.com/pdfDocs/Maxwell_Auto_Brochure.pdf. Last reading March 11, 2020.

Bakran, M. 2009. *A Power Electronics View on Rail Transportation Applications.* Paper Presented at 13th European Conference on Power Electronics and Applications, Barcelona, Spain, pp. 1–7.

Baumhöfer, T., Waag, W., Sauer, D.U. 2010. *Specialized Battery Emulator for Automotive Electrical Systems.* Paper Presented at IEEE Vehicle Power and Propulsion Conference, Lille, France, pp. 1–4.

Denton, T. 2017. *Automobile Electrical and Electronic Systems.* 5th edition, Abingdon-on-Thames: Routledge.

Gueguen, P. 2015. *How Power Electronics Will Reshape to Meet the 21st Century Challenges?* Paper Presented at IEEE 27th International Symposium on Power Semiconductor Devices & IC's (ISPSD), Hong Kong, pp. 17–20.

Hellenthal, B. 2012. *Power Electronics – Key to the Next Level of Automotive Electrification.* Paper Presented at 24th International Symposium on Power Semiconductor Devices and ICs, Bruges, Belgium, pp. 13–16.

Li, J., Mazzola, M., Gafford, J., Younan, N. 2012. *A New Parameter Estimation Algorithm for an Electrical Analogue Battery Model.* Paper Presented at 27th Annual IEEE Applied Power Electronics Conference and Exposition (APEC), Orlando, FL, USA, pp. 427–433.

Neacşu, D. 2004. *Power Semiconductor and Control for Automotive Applications.* Tutorial Presented at IEEE APEC, Anaheim, CA, USA.

Niemiec, H., Szelest, M. 2011. *Combining Various Modeling Techniques for Power Electronic Systems in Automotive Applications.* Paper Presented in Proceedings of the 18th International Conference Mixed Design of Integrated Circuits and Systems – MIXDES, Gliwice, Poland, pp. 442–445.

3 Starter—Alternator

HISTORICAL MILESTONES

1832: French inventor, Hippolyte Pixii, was the first to construct an electric dynamo. His early version was actually an alternator, but he did not know how to handle the alternating current, so he created a commutator to produce dc current.

1911: Charles Kettering's DELCO team invented and built the first starter for Cadillac's CEO, Henry Leland.

1914–1920: Vincent Bendix was awarded a series of US patents for an engagement mechanism allowing the pinion gear of the starter motor to engage or disengage the flywheel of the combustion engine automatically when the starter is actuated or when the engine fires, respectively.

1919: Ford upgraded the Model T to include an electric starter.

1920s: Batteries and electrical starters go hand-in-hand!

1940s: Vehicle alternators were perfected and first used by the military during WWII, to power radio equipment on specialist vehicles.

1950s: Dodge started selling Leece-Neville alternators as options in their taxi and police cars, and later had them as an option in a power wagon.

1954: The Siba Electric company was formed in Germany to supply a combined starter generator unit for use on motorcycles, scooters, and mopeds (Dynastart).

1957: Ford also put optional alternators in police cars and ambulances.

1950s: Generalized usage of alternators.

1960: Chrysler Corporation introduced alternators within the standard series-production Valiant model (first production series usage).

1960: Until the 1960s, automobiles used dc dynamo generators with commutators. Alternators were used with silicon diode rectifiers after 1960.

1970s: Electronic voltage regulator was standard equipment on all passenger cars

2000s: Most typical passenger vehicle and light truck alternators use Lundell or "clawpole" field construction.

3.1 ALTERNATOR'S ROLE

An *alternator* is an electrical generator that converts mechanical energy to electrical energy in the form of alternating current. It is called an alternator to point out its difference from a dc generator. It must deliver enough current to the electrical distribution system to avoid battery discharge.

The power installed within an automotive distribution system varies depending on how many convenience and safety systems are installed and the type of vehicle. Several examples of loads seen within the electrical system of a modern car are shown in Table 3.1. Due to the continuous process of the setting into practice of new convenience systems, as well as the process of vehicle electrification, the installed power is increasing year after year, model after model. *Vehicle electrification* refers to the replacement of mechanical and pneumatical actuators with electrical motor drives on account of efficiency, size, and, especially, controllability.

This increase in installed power has historically evolved as follows.

- 100 W in the 1920s
- 500 W in 1950s
- 1.8 kW in 1985
- 4.0 kW in 2000
- 10.0 kW (without propulsion), 25 kW (with propulsion) in current vehicles

Figure 3.1 illustrates the role of the alternator within the electrical system. One must understand that the alternator is used during engine operation for electrical energy supply, with a side task of recharging the battery. The battery's role is to occasionally supplement the alternator, not to substitute.

3.2 CONSTRUCTION OF AN ALTERNATOR

The alternator concept assumes that an ac generator produces ac current that is rectified into the dc distribution system. The operating efficiency is defined as the *power to weight ratio*, and it increases as a function of the rotational speed. The electrical system designer should consider a conversion ratio as high as possible between the engine crankshaft and the alternator because the engine speed during nominal operation is lower than that of the electrical machine used as an alternator. A high-speed electrical generator is also smaller in size and has a better electrical efficiency. Typical conversion factors are 1:2 to 1:3 for automobiles, and 1:5 for large trucks. The limiting effect against a higher speed operation on bearings, collector rings, and carbon brushes should also be considered along with additional design factors like engine vibration and high-temperature operation.

Alternators are designed to supply charge voltages of 14 V dc (or 28 V dc, for larger vehicles), that is the same voltage levels with batteries within the same electrical distribution system. The ac generator produces a voltage that is a converter from ac to dc through a three-phase diode rectifier, which also prevents battery discharge toward the alternator due to the unidirectional conduction of diodes. Usually, alternators are self-excited 12- or 16-pole synchronous ac machines. The ac windings are wound in the stator's slots while the dc excitation winding is housed within the rotor and powered through conductor rings and carbon brushes. This setup is sketched in Figure 3.2.

The most used in production is the *conventional Lundell alternator*. The conventional *Lundell alternator* has a series of well-known drawbacks, and this choice limits maximum output power at several kW. This means it is unable to handle larger loads within modern vehicles, especially at idle speeds. The conventional alternator has another problem with the internal heating of the coil windings. *Lundell alternators* are today built in a power range up to 5 kW and at a speed range of up to 18,000 rpm.

TABLE 3.1
Examples for Possible Electrical Loads and Their Ratings

Continuous load	**Power [W]**	**Current with 14 V battery**	**Current with 28 V battery**
Fuel pump	70.00	5.00	2.50
Fuel injection	70.00	5.00	2.50
Ignition	30.00	2.00	1.00
Instruments	10.00	1.00	0.50
TOTAL	180.00	13.00	6.50
Prolonged loads	**Power [W]**	**Current with 14 V battery**	**Current with 28 V battery**
Headlights main beam	200.00	15.00	7.50
Headlights dip beam	160.00	12.00	6.00
Side and tail lights	30.00	2.00	1.00
Dashboard lights	25.00	2.00	1.00
Radio–cassette–CD, audio	15.00	1.00	0.50
Rear plate lights	10.00	1.00	0.50
TOTAL	440.00	33.00	16.50
Intermittent loads	**Power [W]**	**Current with 14 V battery**	**Current with 28 V battery**
Seat heater	200.00	14.00	7.00
Electric windows	150.00	11.00	5.50
Radiator cooling fan	150.00	11.00	5.50
Seat movement	150.00	11.00	5.50
Sunroof	150.00	11.00	5.50
Heated rear window	120.00	9.00	4.50
Auxiliary lamps	110.00	8.00	4.00
Cigarette lighter	100.00	7.00	3.50
Front wipers	80.00	6.00	3.00
Blower motor	80.00	6.00	3.00
Heater	50.00	3.50	2.00
Indicators	50.00	3.50	2.00
Rear wiper	50.00	3.50	2.00
Brake lights	40.00	3.00	1.50
Horns	40.00	3.00	1.50
Rear fog lights	40.00	3.00	1.50
Reverse white lights	40.00	3.00	1.50
Interior lights	10.00	1.00	0.50
Electric mirrors	10.00	1.00	0.50
TOTAL	1,620.00	118.50	60.00

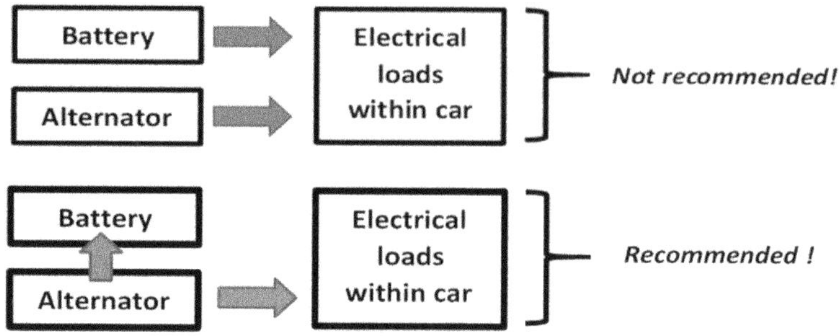

FIGURE 3.1 Alternator within the electrical distribution system.

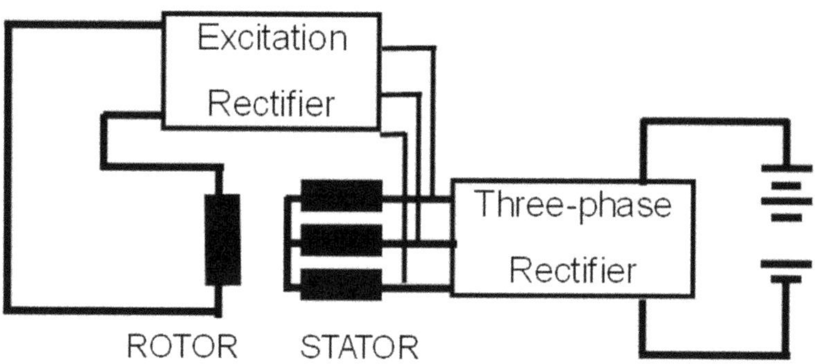

FIGURE 3.2 Electrical circuit associated with an alternator.

The mechanical-to-electrical conversion efficiency for the 14 V Lundell alternator is in the vicinity of 50%, which means that every watt of output electrical power is accompanied by a watt of heat-generating loss, requiring 2 W of mechanical input power from the engine shaft. It also shows abruptly decreasing efficiency beyond the 2 kW barrier. Making the alternator physically bigger adds substantial weight, making installation harder.

The construction of the conventional Lundell alternator is shown in Figure 3.3.

The rotor is based on a *field coil* that is wound around a large bobbin and surrounded by two interlocking "*clawfoot*" iron shells. Whereas a design with six claws is suggested herein, other designs may consider a different number of claws on each shell. The number of claws would define the rotational speed for the magnetic field.

The field coil herein represents the rotor, which is defined as the moving component of an electromagnetic system in the electric motor. The field coil is supplied with dc voltage and the field current is much smaller than the output current of the alternator; for example, a 70 A alternator may need only 7 A of dc field current. This is because the power applied to the field coil plus the mechanical power equals loss power plus produced power. This field coil is electrically connected to two copper slip rings, which the regulator applies power to through two copper and carbon brushes. The coil is used to control the charging of the alternator.

When the rotor is energized with a dc voltage, the top shell becomes a magnetic "*North*" and the bottom shell becomes a magnetic "*South*" (remember, "right-hand-rule" for magnetic fields?).

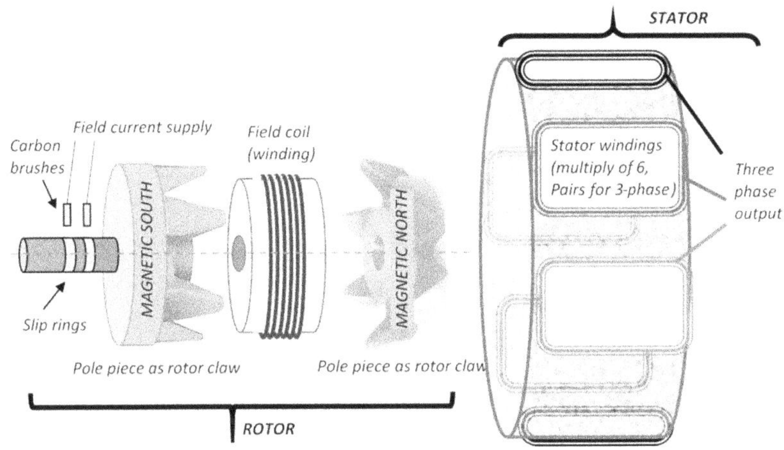

FIGURE 3.3 Structure of a Lundell alternator and location of windings.

A stationary set of three-phase windings is wrapped around a steel core called a *stator*. Figure 3.3 illustrates a distributed winding for clarity. The six coils in Figure 3.3 are connected in three pairs; each pair of coils are placed at 180° on the circumference and connected in series. In actuality, more than six coils may be used, with some overlapping, in order to provide for a smoother induced voltage.

Automotive alternators are usually belt-driven at 5–10 times the crankshaft speed. As the rotor spins with the belt, the clawfoot design of the rotor produces the alternating between North and South poles in the vicinity of each wire part of the stator winding system. This induces in the stator windings a set of alternating currents of variable frequency. The geometrical position of the stator windings secures the appropriate phase shift for the definition of a three-phase system of voltages.

One cycle of alternating current is produced each time a pair of field poles passes over a point on the stator. The relation between speed and frequency is

$$N = \frac{120 \cdot f}{P} \tag{3.1}$$

where f is the frequency of the generated current in Hz (cycles per second), P is the number of poles (2, 4, 6, …), and N is the rotational speed of the rotor in revolutions per minute (RPM). For the design in Figure 3.3, $P = 12$ (six North and six South claws).

A typical electrical characteristic is shown in Figure 3.4.

3.3 ELECTRONIC CONTROLS FOR ALTERNATOR

3.3.1 General Requirements

Modern solutions for alternator construction include electronic voltage regulators to deal with wide fluctuations in the alternator speed and load. For instance, the voltage is higher in the winter since the battery is more difficult to charge, and lower during the summer. A four-quadrant single-phase PWM inverter can be used

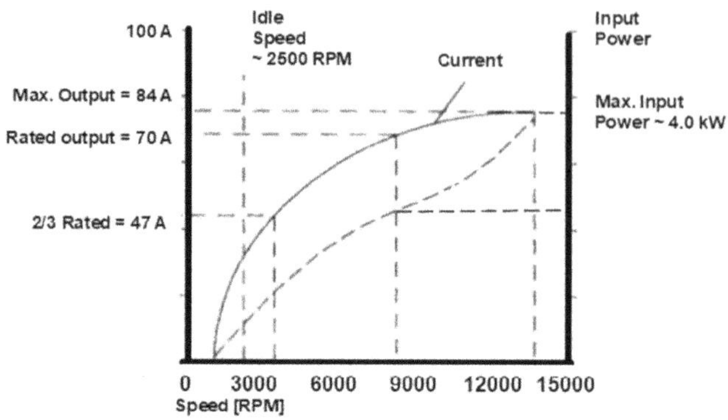

FIGURE 3.4 Electrical characteristics of an alternator, for a 24 V dc bus and a 50% efficiency.

bi-directionally, for both starter and alternator functions. It can start the engine from the battery and charge the battery from the engine during operation. The inverter is rated using the highest current that is achieved at engine start with a *starter function*.

Performance can be further improved using a higher speed generator. Using an operation at a higher speed allows for a 20% higher electrical efficiency and significantly quieter action. New designs generate between 90% and 100% of their rated amperage at idle speeds.

The problem of Lundell alternators relates to generating high output power at lower engine speeds. A solution to this constraint is a variable drive ratio, but this is unfortunately fraught with mechanical problems. The currently used solution is tending toward alternators capable of much higher maximum speeds, which allows a greater drive ratio and hence greater speed at lower engine speed, given as revolutions/minute.

3.3.2 Closed-Loop Regulation of Voltage

To prevent both the overcharging of the vehicle's battery and the production of overvoltages in the electrical distribution system, the regulated voltage should be kept below the gassing voltage of the lead-acid battery. A figure of 14.2 ± 0.2 V dc was traditionally used as the maximum accepted voltage for batteries of nominal 12 V ratings.

Accurate voltage control is vital with the ever-increasing use of electronic systems as loads. Newer electronic loads require better control of the dc distribution bus voltage and prevention of overvoltage that could damage electronic components. Using closed-loop electronic controls has also enabled the widespread use of sealed batteries, since the possibility of overcharging is minimal.

The most-used closed-loop control solution is illustrated in Figure 3.5 and it is based on electronic control of the rotor current for the alternator. A switch controlled with pulse width modulation applies a train of pulses to the rotor inductance. This inductance filters the current into a quasi-dc quantity, adjustable by the control of the pulse width. If the bus voltage drops, the excitation current is increased with a

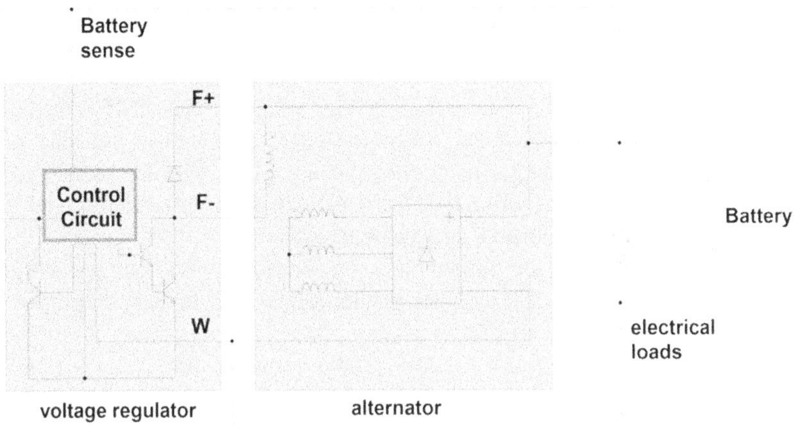

FIGURE 3.5 Electronic control of the rotor current of an alternator.

wider pulse width in control of the switch. An integrated circuit typically sets up this controller.

The stator will be subjected to the mechanical rotation and the magnetic field produced by the rotor. In this solution, the set of three-phase voltages produced in the stator windings is also diode-rectified and helps to charge the battery with a quasi-dc current. A special circuit disables the charger function during the startup sequence when the battery supports the starter.

3.3.3 Alternator Requirements for 48 V Systems

Translation of the electrical distribution system from 14 V dc to 48 V dc automotive systems has been analyzed from technical and economical perspectives. It is noteworthy the implication of this conversion on the dc/dc power supply industry where most existing technologies for CMOS or DMOS integrated controllers are working up to 40–45 V dc.

The first systems to benefit from the conversion to 48 V are the auxiliary motor-based systems. Examples include herein the chassis systems like electric power steering systems, the electric brakes, and the engine cooling fans.

The 48 V dc distribution bus used in dual (or hybrid) 14/48 V dc systems would primarily be used for the start–stop functions or braking with energy recovery. The same MOSFET-based inverter configuration would be able to work both as a starter and as a generator.

3.3.4 Using a Switched-Mode Rectifier to Increase Output Power

The available output power is limited by the output voltage and rotational speed when a diode rectifier is used for ac–dc conversion, as previously shown in Figure 3.2. The ac generator produces a back-emf voltage defined as follows.

$$V_{rms} = 4.44 \cdot f \cdot \Phi \cdot N \left[\text{V}\right] \tag{3.2}$$

where f represents the frequency, Φ represents the flux per pole, and N represents the number of turns per phase. This form is similar to the voltage in a transformer, and can be herein considered as a no-load (open-circuit) voltage.

The phase circuitry of the stator is completed with this back-EMF voltage, a series inductance, a series loss resistance, followed by the diode rectifier and load. A simple Kirchhoff equation can be considered for this ac circuit, with the rectifier-side voltage bound by the dc distribution bus. Obviously, this is a simplified model, neglecting saliency effects, iron losses, and magnetic saturation.

Figure 3.6 shows the dc output power generation for various speeds and output voltages, and Figure 3.7 arranges results differently to show dc output voltage

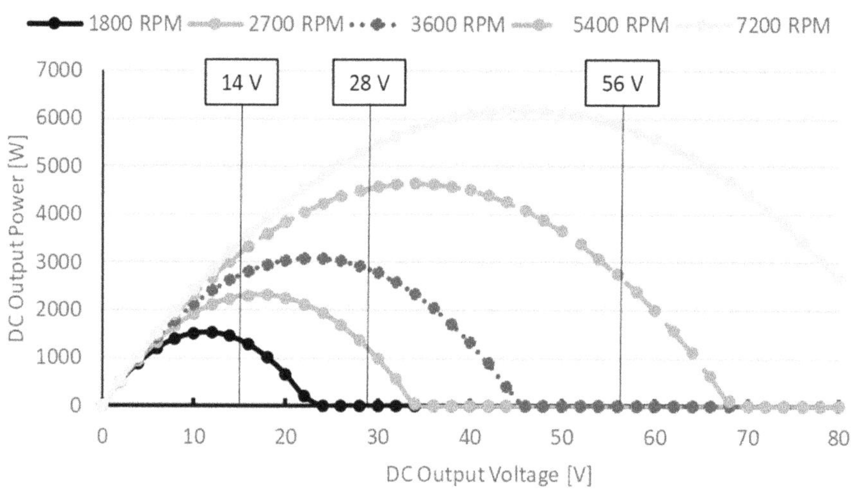

FIGURE 3.6 dc output power dependency on dc output voltage, shown at different rotational speeds.

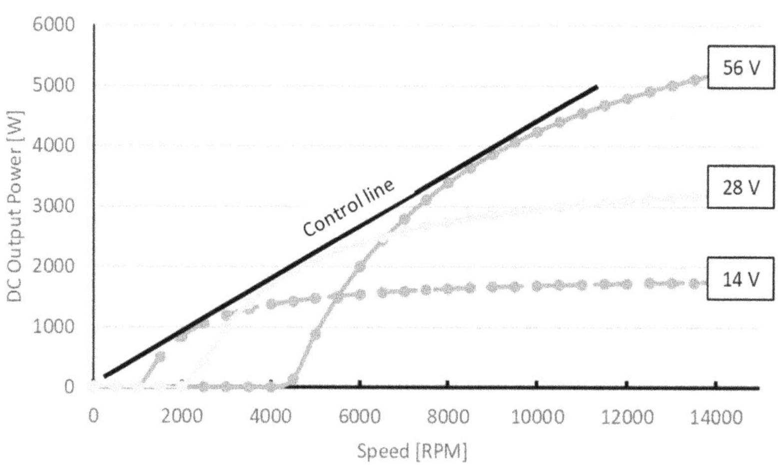

FIGURE 3.7 dc output voltage dependency on alternator speed.

dependency on alternator speed. The field current has been maintained at its maximum throughout the analysis. Numerical values are just an example, peculiar to a certain design. However, the shapes are always similar.

It can be seen that a low rotational speed of 1,800 RPM can allow for the use of this alternator for the 14 V battery bus only, since the maximum dc voltage is in the 20–29 V. In this case, the back-EMF voltage yields smaller than the ac voltage reflected at the rectifier's input by the dc distribution bus. There is insufficient back-EMF voltage to generate a current into the rectifier. Higher rotational speed can allow for the production of power at higher dc (output) voltage levels and accommodate higher dc bus voltages.

This increase in available power delivered by the rectifier is not due to a higher dc output voltage alone. For instance, the same stator can be rewound to raise the output voltage, but this would lower the output current, thus not ultimately improving the dc output power availability.

Higher output power can be obtained with a switched-mode rectifier (SMR) used instead of the diode rectifier. The versions of SMR in Figures 3.8(a) or (b) use power semiconductor switches to form a boost converter able to adjust the output voltage.

While the dc distribution bus voltage is kept constant, the voltage produced by the alternator can be varied in order to maximize output power at any speed. The SMR provides a continuous transition between different characteristic voltage curves in

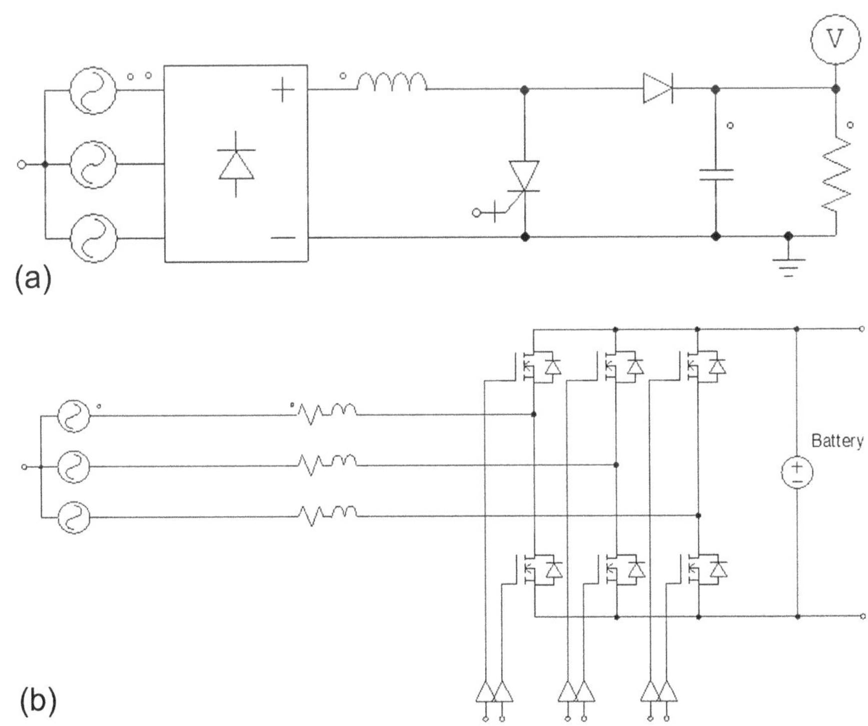

FIGURE 3.8 SMR topologies: (a) simple use of a boost converter after the diode rectifier, (b) full-bridge switched-mode rectifier.

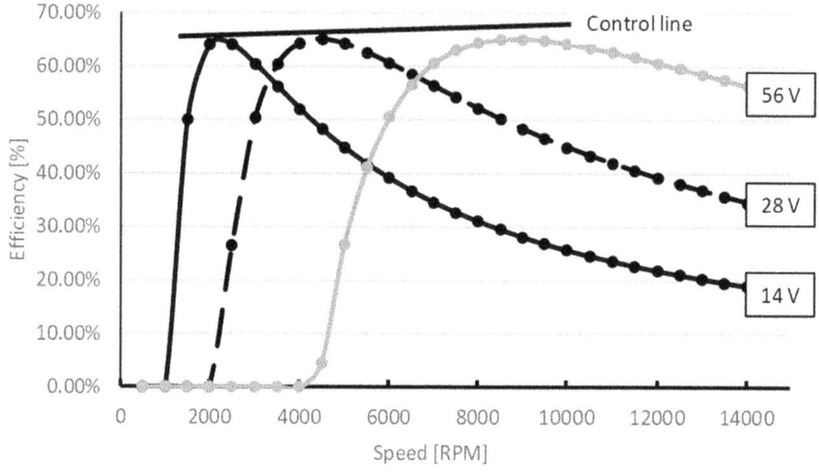

FIGURE 3.9 Efficiency of the Lundell alternator operated with a switched-mode rectifier.

Figure 3.7, allowing maximum power to be generated even at low speeds. A result of this technology assures the operation with maximum efficiency at any speed, Figure 3.9.

3.4 OTHER ELECTRICAL MACHINE INSTEAD "ALTERNATORS"

The most-used solution, the conventional Lundell alternator, is a synchronous machine.

Other solutions explored in the past include a high-voltage high-power generating system based on a permanent magnet alternator. Such a system would not require excitation given the presence of the permanent magnets. Unfortunately, the solution is expensive since the inverter rating should include the wider voltage range produced by a system with magnets.

Another alternative, the *synchronous reluctance machines* are more sensitive to noise and vibrations and not a well-defined technology.

Much cheaper alternative would be an *induction machine.* If only the alternator (generator) function is considered from an induction machine, an alternative implementation is based on a separate diode rectifier for battery charging and a lower power inverter used for its own dc bus control and reactive power delivery. A special construction of the induction machine should be herein considered to separate the active and reactive power transfer with a dual system of windings. Such a solution should expect to achieve 20% better efficiency than the Lundell alternator.

3.5 STARTER SYSTEMS

The starter has the role of producing a minimum rotating speed of about 100 *revolutions/minute* (RPM) to be able to facilitate the engine start. This minimum speed depends on the rated voltage of the starting system, the lowest possible temperature at which it must still be possible to start the engine, and the engine cranking resistance. The latter is defined as the torque required to crank the engine at its starting limit temperature, herein including the initial stalled torque. The minimum speed is also influenced by the voltage drop between the battery and the starter, and the starter-to-ring gear ratio.

On the other hand, the required minimum speed depends on the engine and its ignition system. Table 3.2 reviews possible setups used in motor vehicles. The electrical circuit used for connection of the starter at the electrical distribution system is shown in Figure 3.10 and the characteristics are illustrated in Figure 3.11.

The higher the torque produced from the same dc bus, the better the starter. A current as large as possible is therefore required to produce such a high torque. For a given battery, the current is limited just by the internal resistance of the battery and the line resistance of the motor (starter), Figure 3.10. The rated power of the motor corresponds to the power drawn from the battery less copper loss due to the resistance of the circuit, iron loss due to eddy currents being induced in the iron parts of the motor, and friction loss. The required power level governs the selection of the starter. Table 3.3 and Table 3.4 schematically shows several situations that occur when selecting the starter.

This type of graphical representation of the characteristics can be used for design. For example, considering a four-cylinder 2-liter engine requiring 480 N·m to overcome static friction and 160 Nm to maintain the minimum cranking speed of 100

TABLE 3.2
Engine and Ignition System Configurations

Engine	Minimum cranking speed [revolutions/min]
Reciprocating engine (piston engine) with spark ignition	60–90
Rotary engine, with spark ignition	150–180
Diesel with glow plugs	60–140
Diesel without glow plugs	100–200

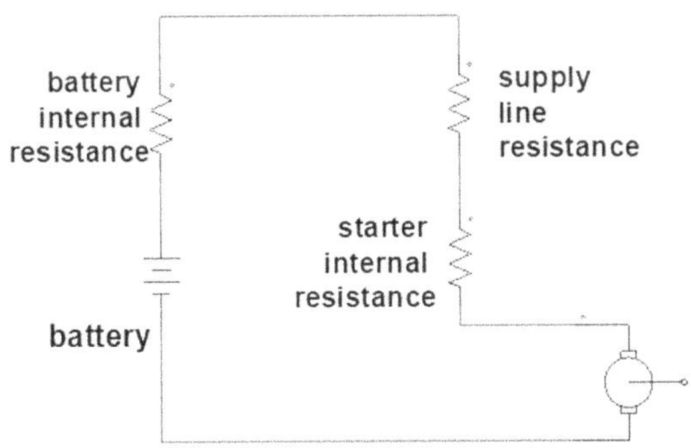

FIGURE 3.10 Electrical circuit for a starter.

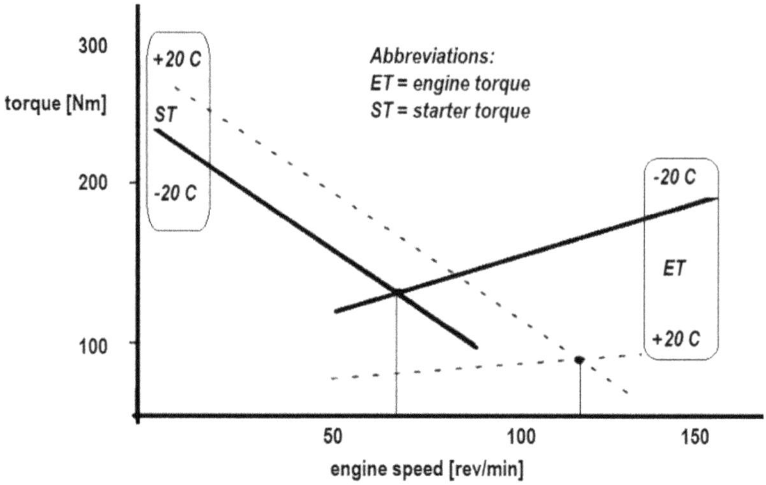

FIGURE 3.11 Torque-speed characteristics of a starter, with dependency on temperature.

TABLE 3.3
Selection of the Starter Based on Engine Requirements: Power versus Engine Capacity

	Output power [W]	
Engine capacity [liter]	**min. battery**	**max. battery**
0.8 .. 1.5	400	600
1.2 .. 1.8	500	800
1.6 .. 2.4	750	1,000
2.2 .. 3.2	900	15,00
3.0 .. 5.6	1,400	17,00
5.4 .. 9.0	1,800	25,00

TABLE 3.4
Selection of the Starter Based on Engine Requirements: Required Torque per Liter with Number of Engine Cylinders

Torque per liter [Nm]	Cylinders
12.50	2
8.00	4
6.50	6
6.00	8
5.50	12

revolutions/minute. The engine uses a starter pinion-to-ring gear ratio of 10:1. The electrical motor used for the starter must therefore be able to produce a maximum stalled torque of 48 N·m (design data) and a driving/starting torque of 8 Nm × 2 liters = 16 N·m.

In order to have 100 revolutions/minute (RPM) on the engine side, the electrical machine needs to rotate at 1,000 revolutions/minute due to the pinion-to-ring gear ratio of 10:1. For transformation of units, 1,000 (10x100) revolutions/minute equals to 1,000·2·pi/60 = 104.66 rad/sec.

The power developed by the starter through the pinion, at this speed, with a torque of 16 Nm, yields as 16 × 104.66 = 1,674.66 W. The actual electrical power may need to be double that, 3.3 kW, due to poor efficiency of the starter motor.

Observing Tables 3.3 and 3.4 allows for a selection of the starter on the fifth row, and a battery with 460:490 CCA start performance, which means the first available battery above 3.3 kW/7.2 V = 458 CCA, for operation in a worst-case start. Figure 3.12 illustrates the choice for this design example.

3.6 STARTER CONSTRUCTION

The starter's construction has improved over the years in order to reduce its weight. In the 1970s, the starter system weighted 12 kg, while modern systems, after the 2000s, are at 5 kg.

The actual electrical motor can be a battery-driven dc series-wound motor with large initial torque, or motors with permanent magnet excitation. Most starter designs use a four-pole four-brush design and switchgear. The electrical motor is operated at high-speed and a reduction allows for a lower-speed startup, with increased torque.

FIGURE 3.12 Battery selected for the design example.

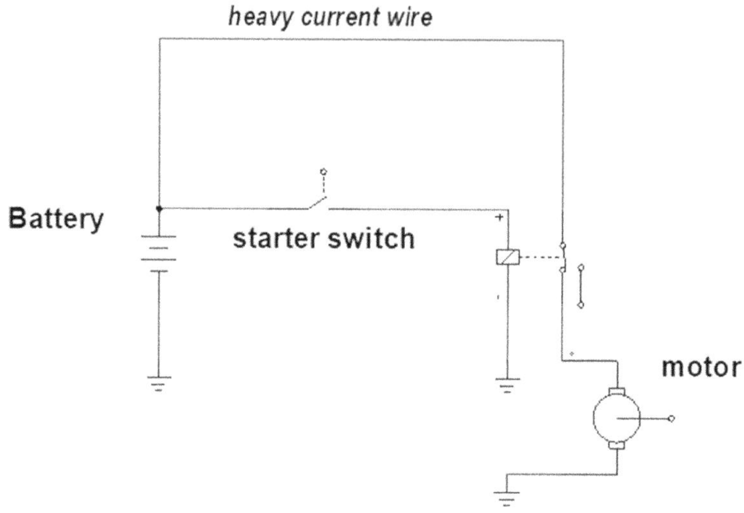

FIGURE 3.13 Electrical connection of the starter.

This is achieved with a pinion-to-ring gear. Considering a conservation of power allows the speed to be lower and the torque to be higher.

Another requirement considers the engine's starting speed being way lower than the speed during continuous operation. If the connection between starter and engine remains permanent, the excessive speed at which the starter would be driven by the engine during continuous operation would destroy the electrical motor almost immediately.

The system also includes a control unit for the starter and associated wirings.

Therefore, the starter system is made up of an electric motor, a pinion, and a clutch. The pinion drive needs to initially engage the engine's crankshaft to transfer movement from the electrical motor to the engine. The clutch needs to separate the pinion from the engine as soon as the engine runs faster than the starter to avoid damage to the electrical motor.

The usual electrical connection of the starter is shown in Figure 3.13. A spring-loaded key switch operates the starter and controls ignition and accessories. A relay conveys the starting action to a power switch, which controls the heavy current of the motor. The design is constrained by the maximum allowed circuit resistance, due to wires and terminals. In this respect, it is generally accepted that a maximum volt drop of 0.5 V should be allowed between the battery and the starter when operating, that is a resistance of around 2 mΩ.

3.7 INERTIA STARTER

The inertia starter was popular in the past. The principle of the inertia starter is illustrated in Figure 3.14, where the brushed dc motor is not shown in detail. The operation principle can be understood by playing with a screw and a nut while applying a torque successively to screw or nut.

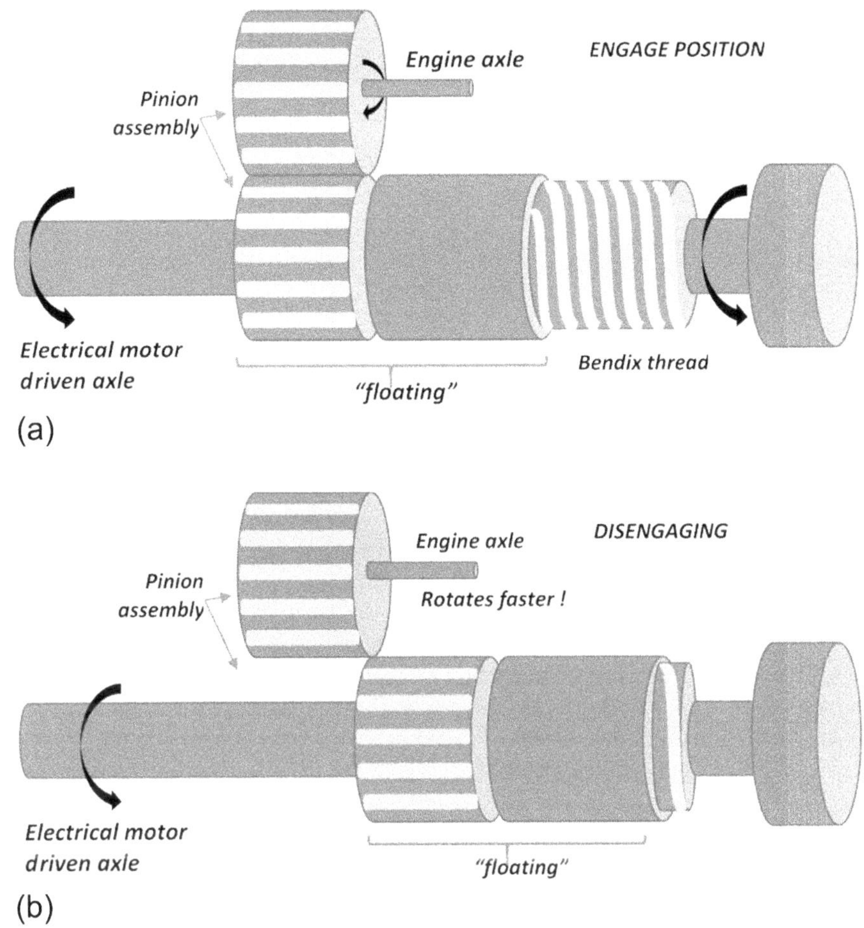

FIGURE 3.14 Inertia starter: (a) engaged position, (b) disengaged position.

In Figure 3.14, the starter engages with the flywheel ring gear by means of a small pinion. When the starter is operated, the armature will cause the sleeve splined on to the armature shaft to rotate inside the pinion. The pinion remains still due to its inertia and, because of the screwed sleeve rotating inside it, the pinion is moved/unscrew to mesh with the ring gear and forces the engine to spin slowly. When the engine fires and runs under its own power, the pinion is driven faster than the armature shaft. This causes the pinion to be screwed back along the sleeve and out of engagement with the flywheel.

3.8 PRE-ENGAGED STARTERS

The violent engagement and disengagement of an inertia starter can cause heavy wear on the gear teeth. To overcome such a problem, the *pre-engaged starter* was introduced, which has a solenoid mounted on the motor. It is the most used today

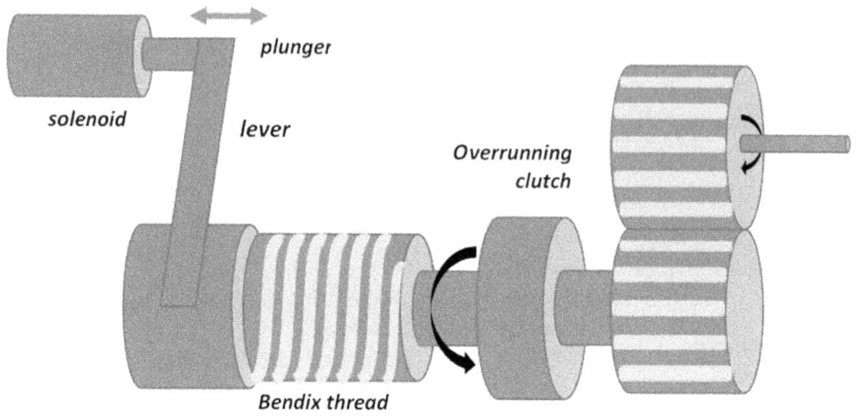

FIGURE 3.15 Pre-engaged starter.

and it is easily recognizable due to a double-cylinder shape from the solenoid sitting atop the brushed dc motor. The principle of the pre-engaged starter is shown in Figure 3.15.

The starter motor is fitted with a relay valve (solenoid), which controls the movement of the pinion. When the starter motor is activated, the solenoid moves the plunger, and the lever causes the pinion to move along the shaft until it comes into contact with the toothed ring gear. The pinion is then interlocked with the ring gear before rotating to crank the engine. As the engine starts operation and the speed is increased, the pre-engaged starter motor is protected from overspeed by a freewheel clutch. This is similar to the freewheel gear of a bicycle where the wheels can turn without the movement from pedal after some speed has been reached. At this stage, the solenoid is deactivated and the return spring retracts the pinion to its disengaged position through the action of the plunger and lever.

3.9 PERMANENT MAGNET STARTERS

The permanent magnet starters are a newer solution, and they are similar to pre-engaged starters. The main difference being the replacement of field windings and pole shoes with high-quality permanent magnets. In other words, the brushed dc machine is replaced with a permanent magnet motor. The reduction in weight is about 15% while the cost is increased.

3.10 TYPICAL TORQUE CHARACTERISTICS

Figure 3.16 illustrates the typical characteristics of a dc machine used as a starter. This example is based on a 900 W dc motor used at 12 V dc, with a loss resistance of the distribution system of 10 mΩ and a torque constant of 0.025 [Nm/A]. Note that the maximum power (defined as a torque-speed product) of this motor is developed at mid-range speed, but maximum torque is at zero speed.

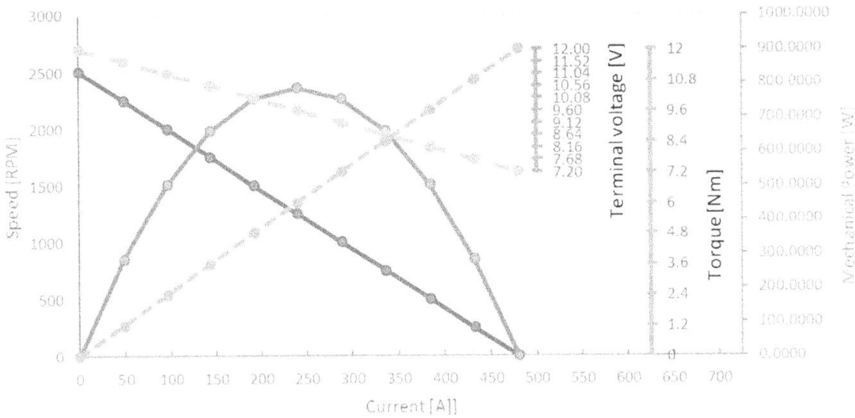

FIGURE 3.16 Typical characteristics of a dc machine used as a starter.

3.11 INTEGRATED STARTER ALTERNATOR

Since any vehicle needs both a starter and an alternator, which do not work simultaneously and have reverse conversion functions, the idea of merging the two devices has always existed in the automotive world. A device called a "*dynastart*" was used on a number of vehicles from the 1930s through to the 1960s. This device was a combination of the starter and a dynamo. The device, directly mounted on the crankshaft, was a compromise and hence not very efficient.

An alternative method nowadays is known as an *integrated starter alternator damper* (ISAD). It consists of an electric motor, which functions as a control element between the engine and the transmission, and can also be used to start the engine and deliver electrical power.

The electric motor can also replace the mass of the flywheel and it is also able to act as a damper/vibration absorber unit. The damping effect is achieved by a rotation capacitor. A change in relative speed between the rotor and the engine due to the vibration, causes one pole of the capacitor to be charged. The effect is to take the energy from the vibration.

This arrangement offers several advantages: cranking speeds up to 700 revolutions/minute. A stop–start function is possible as an economy and emissions improvement since the engine will fire up in about 0.1–0.5 seconds due to the high-speed cranking. The motor can also be used to aid with the acceleration of the vehicle. This allows a smaller engine to be used for the same vehicle, or to enhance the performance of a designed-in standard engine.

The electronic starter incorporates a static relay on a circuit board integrated into the solenoid switch. This will prevent cranking when the engine is running.

Other "smart" features can be added to improve comfort, safety, and service life.

- Starter torque can be evaluated in real-time to tell the precise instant of engine start.

- The starter can be simultaneously shut off to reduce wear and noise generated by the freewheel phase.
- Thermal protection of the starter components allows optimization of the components to save weight and to give short circuit protection.
- Electrical protection also reduces damage from misuse or system failure.
- Modulating the solenoid current allows redesign of the mechanical parts allowing a softer operation and weight reduction.

3.12 CONCLUSION

Conversion of mechanical energy into electrical energy is achieved with an *alternator*. This is an ac generator and needs to be followed by a rectifier in order to provide dc voltage to the dc distribution bus. While most practical solutions use a simple diode rectifier, a more advanced solution introduces control of the produced voltage with a boost converter.

The conventional alternator solution is based on a Lundell alternator, which is a synchronous generator. Several other machine options are investigated in this chapter.

The second major electromechanical conversion within a motor vehicle is achieved with a *starter*. The battery's electrical energy is used to supply an electrical motor able to spin the engine in any ambient conditions for a startup. Both *inertia* and *pre-engaged* starter solutions are explored.

Design examples and technical limitations complete this chapter dedicated to the major electromechanical energy conversion within a motor vehicle. Such conversion can be seen as a linkage between the electrical system and the engine's operation.

REFERENCES

Chen, S., Lequesne, B., Henry, R., Xue, Y., Ronning, J. 2002. Design and testing of a belt-driven induction starter-generator. *IEEE Transactions on Industry Applications*, 38(6):1525–1533.

Comnac, V., Cernat, M., Mailat, A. 2008. *Optimal Use of the 14 V Alternator in 42 V Automotive Supply Systems*. Paper Presented at 13th International Power Electronics and Motion Control Conference, Poznan, Poland, pp. 1748–1754.

Denton, T. 2017. *Automobile Electrical and Electronic Systems*. 5th edition, Abingdon-on-Thames: Routledge.

Lorilla, L.M., Keim, T.A., Lang, J.H., Perreault, D.J. 2005. Topologies for future automotive generators. Part I. Modeling and analytics. Paper Presented at IEEE Vehicle Power and Propulsion Conference, Chicago, IL, USA, pp. 74–85.

Lorilla, L.M., Keim, T.A., Lang, J.H., Perreault, D.J. 2005. *Topologies for Future Automotive Generators – Part II: Optimization*. Paper Presented at IEEE Vehicle Power and Propulsion Conference, Chicago, IL, USA, pp. 831–837.

Murray, A., Wood, P., Keskar, N., Chen, J., Guerra, A. 2001. A 42V *Inverter/Rectifier for ISA Using Discrete Semiconductor Components*. Paper Presented at Future Transportation Technology Conference, Costa Mesa, CA, USA.

Naidu, M., Walters, J. 2003. A 4-kW 42V induction machine based automotive power generation system with a diode bridge rectifier and a PWM inverter. *IEEE Transactions on Industry Applications*, 39(5):1287–1293.

Neacşu, D. 2004. *Power Semiconductor and Control for Automotive Applications.* Tutorial Presented at IEEE APEC, Anaheim, CA, USA.

Stoia, D., Cernat, M. 2009. *Design of a 42 V Automotive Alternator with Integrated Switched-Mode Rectifier.* Paper Presented at 8th International Symposium on Advanced Electromechanical Motion Systems & Electric Drives Joint Symposium, Lille, France, pp. 1–8.

Whaley, D.M., Soong, W.L., Ertugrul, N. 2004. *Extracting More Power Din Lundell Alternator.* Paper Presented at Australasian Universities Power Engineering Conference, Brisbane, Australia.

4 Body Systems

HISTORICAL MILESTONES

1922: The first retractable hard-top system invented in 1922.

1934: The first power-operated retractable hard-top was produced by Peugeot, followed in 1939 by Plymouth (mechanically operated convertible roof).

1939: Packard became the first automobile manufacturer to offer an air conditioning unit in production series cars (the Bishop and Babcock Weather Conditioner).

1941: Chrysler manufactured a version of the hard-top convertible car with the Chrysler Thunderbolt.

1940: The first power windows were introduced in the 1940 Packard 180 series automobiles and used a hydro-electric mechanism.

1941: Ford followed and installed power windows on the Lincoln Custom limousines and sedans.

1948: Modern cruise control (also known as a speedostat or tempomat) was developed by the inventor and mechanical engineer Ralph Teetor.

1958: First cruise control systems were introduced in production models of the Chrysler Imperial, New Yorker, and Windsor.

1960: Cruise control was a standard feature on all Cadillacs.

1959–1971: The power MOSFET and microprocessor.

The MOSFET (MOS field-effect transistor, or MOS transistor), invented by Mohamed M. Atalla and Dawon Kahng at Bell Labs in 1959, led to the development of the power MOSFET by Hitachi in 1969, and the single-chip microprocessor by Federico Faggin, Marcian Hoff, Masatoshi Shima, and Stanley Mazor at Intel in 1971.

1960: Brushless dc motors were made possible by the development of solid-state electronics.

1990–2000: Advent of embedded units.

Multiple microprocessor-based systems are used to control the operation of small motors, under the generic name of electronic control unit (ECU). Possible ECU are door control unit, engine control unit, electric power steering control unit, human-machine interface, powertrain control module, transmission control unit, or powertrain control module, seat control unit, speed control unit, telematic control unit, transmission control unit, brake control module (ABS or ESC), and battery management system.

4.1 INTRODUCTION TO BODY SYSTEMS

Power electronics applied to body systems are identified with the application of mechatronics systems to motor vehicles. Electrical movement of a variety of systems within the car is studied in this chapter. These systems are well known; they are reviewed herein for understanding the mechatronics implications. These systems are very similar to each other, and these systems are very similar and are used in the applications of mechatronics in the automotive industry. They use one/several permanent magnet motors, together with a bidirectional power supply, and one/several sensors. All these small mechatronics systems are tied to the *Body Control Module* that is a microcontroller managing the interactions between functions.

Mechatronics is a multidisciplinary field of engineering that includes a combination of mechanical engineering, electronics, computer engineering, telecommunications engineering, systems engineering, and control engineering. The word "mechatronics" was invented or introduced by the Japanese engineer Tetsuro Mori in 1969. The word represents a combination of the terms "mechanical" ("mecha" for mechanisms, i.e., machines that "move") and "electronics." The term is identified with the idea of adding some electronic control to body motions.

In automotive electronics, *Body Control Module* or *body computer* is a generic term for an electronic control unit responsible for monitoring and controlling various electronic accessories in a vehicle's body. This may include comfort and safety systems, as well as active stabilizer systems.

The *active stabilizer system* responds in real-time to driving conditions based on actual or anticipated body motions. Such a system contains one or two controlled active stabilizer bar modules, a hydraulic pump, a hydraulic manifold with integral valves and a pressure sensor, a controller, vehicle sensors (steer angle, lateral acceleration, vehicle speed, and others). The power electronic setup contains the hydraulic pump that can be electronically controlled.

Fully integrated circuits (ICs) with integrated power MOSFETs and protection circuitry can be used up to 1–3 A. For larger power levels, ICs including motor control and gate drivers with external MOSFETs are used. The major design problem for control systems with integrated circuits relates to operating in harsh battery conditions such as start–stop and cold-crank.

Collectively known as *body electronics*, the comfort and convenience systems allow vehicle occupants to feel comfortable and safe inside the vehicle. Several examples of systems are reviewed in this chapter. They include power windows,

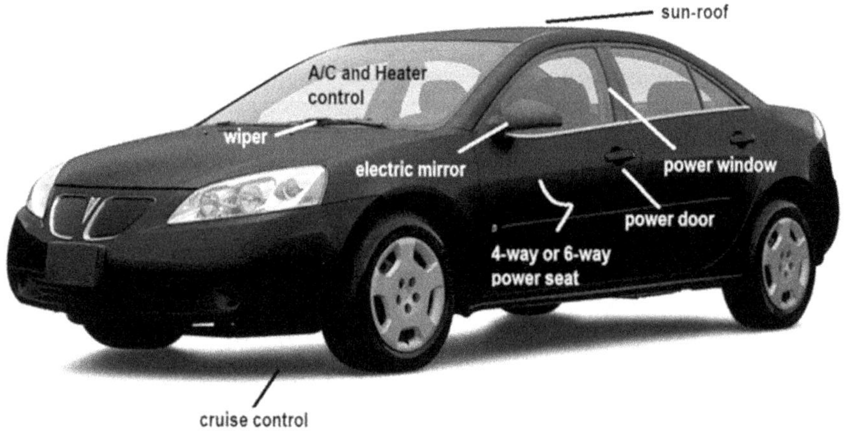

FIGURE 4.1 Possible body systems within a vehicle.

soft-convertible systems, hard-convertible systems, power door locks, power seats, electric steps, electric mirrors, electric sunroof, auto-dimming rearview interior and side mirrors, remote keyless entry, power trunk or lift-gate systems, so on (Figure 4.1). Each such gimmick requires one or more dc motors with a simple on/off control or through communication with other features.

The integrated circuit manufacturers (like TI, Allegro, and so on) came up with a set of power electronics solutions which work as building blocks, suitable for different applications to the body systems. Depending on integration level, such solutions are discussed toward the end of this chapter.

4.2 POWER WINDOW (ELECTRICAL WINDOWS)

This is the most common application in body systems. Figure 4.2 shows the basic setup. A small electric motor is attached to a worm gear and several other spur gears to create a large gear reduction able to provide enough torque for lifting the window while keeping it level. The motor can be actuated with relays/switches. More advanced systems imply a power electronic motor drive and an advanced control circuit. The low-power motor drive requires a power stage with a current limit of around 1 A that can easily be achieved within a single integrated circuit.

Besides the on/off operation of the motor, the digital control of the motor is preferred because it may include special features like one shot up or down, inch up or down, a lazy lock which allows the car to be fully secured by one operation of a remote infrared key, back-off when the ECU reverses the motor until the window is fully open if the rate of change of speed in the motor is detected as being below a certain threshold when closing, and so on. Some of these special features require correlation with other functions, and need to be controlled with the microcontroller-based *Electronic body control unit*.

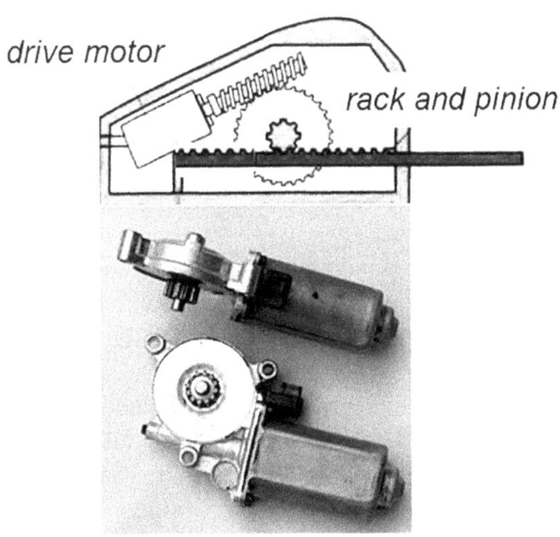

FIGURE 4.2 Power window gear and motor.

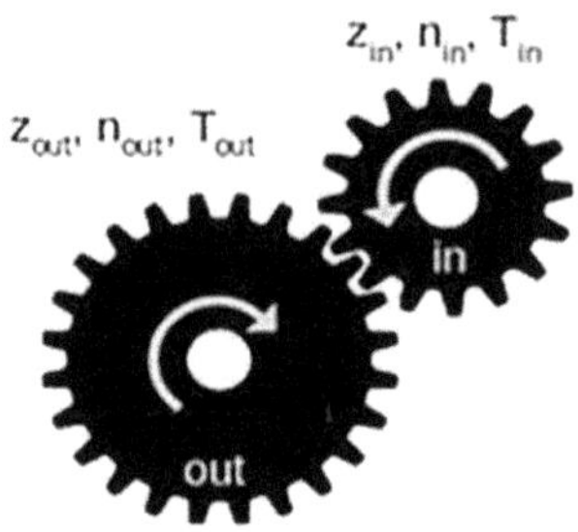

$$i = \frac{z_{out}}{z_{in}} \quad \text{(gear ratio)}$$

$$n_{out} = \frac{n_{in}}{i} \quad \text{(output speed)}$$

$$T_{out} = T_{in} \cdot i \quad \text{(output torque)}$$

FIGURE 4.3 Movement transfer through gears, where "*i*" herein represents the gear ratio.

While electrical motors rotate at thousands of rotations per minute (RPM), the actual movement within the body system application requires a considerably slower rotational speed. This means that all systems require a speed conversion from the electrical motor to the application, which is achieved with gears. Since gears are often used within mechatronic systems between the electrical motor and the mechanical system, a review of their physics laws is included herein, with the help of Figure 4.3.

A simple gear consists of two toothed wheels (called gears) meshed together. The *gear ratio* is the transfer ratio through the gears. Gear ratio is the ratio between the number of teeth of the output gear (z_{out}) and the number of teeth of the input gear (z_{in}), not unlike the ratio of the number of turns within an electrical transformer. Since movement is transmitted through gears, the relationship between the speed of the actuator and the speed of the mechanical load is discussed next. The speed (often given in RPM) is decreased when moving to a gear with more teeth since more movement on the circumference is required before a complete turn. A decrease in speed produces an increase in torque since their product is conserved when considering the same power before and after the gears, also seen as input and output of the system. This principle is often used in mechatronics systems as the electrical motors are easier to make at higher rotational speed than it is actually required by the mechanical application. In this case, the window needs to move relatively slow, whereas the electrical motor rotates at a higher speed. Conversely, the torque applied to the window yields larger than the one produced at the motor axle.

Besides the actuation of the window, it is important to devise an automatic stop circuit. While mechanical solutions exist along with solutions able to identify a sharp increase in the motor current when reaching the end of the trip, an elegant solution is encouraged with a non-intrusive Hall sensor.

Therefore, dedicated control integrated circuits include a Hall sensor device able to turn the output off when a South magnetic pole of sufficient strength is sensed. Figure 4.4 illustrates this principle. This device also includes a voltage regulator able to accept input battery voltage.

The Hall effect integrated circuit processes the sensor information with a linear amplifier, and a CMOS Class A amplifier (single transistor) output structure.

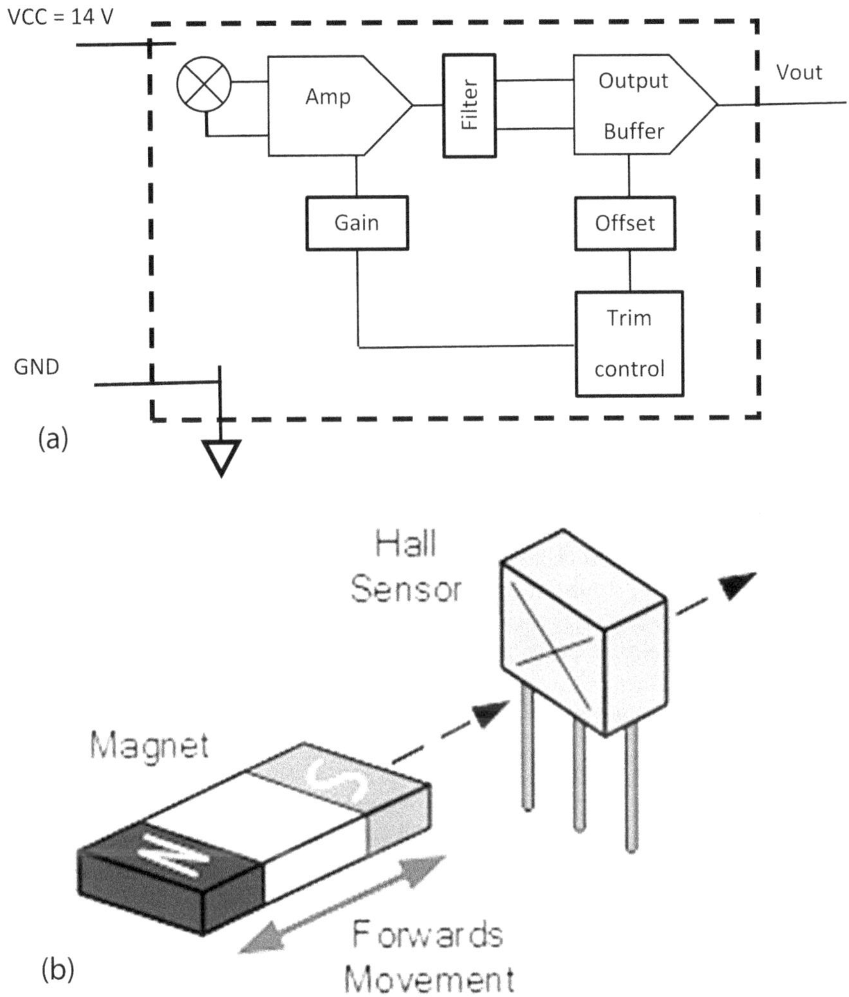

FIGURE 4.4 Integration of the Hall sensor inside the integrated circuit: (a) integrated circuit schematic, (b) principle of the Hall sensor usage in window drive.

Integrating the Hall circuit and the amplifier on a single chip minimizes many of the problems normally associated with low voltage level analog signals.

4.3 POWER LOCK DOORS

The principle of electric power lock doors is illustrated in Figure 4.5.

A small electric motor turns a series of *spur gears* for a gear reduction. Spur gears or *straight-cut gears* are the simplest type of gear and consist of a cylinder or disk with teeth projecting radially. Though the teeth are usually not straight-sided, the edge of each tooth is straight and aligned parallel to the axis of rotation. The spur gears are used for speed reduction, which in turn increases the torque on the axle.

FIGURE 4.5 Principle of power lock doors.

After speed reduction, a rack-and-pinion gear-set is connected to the actuator rod. The rack is able to convert the rotational motion of the motor into the linear motion needed to move the lock. A centrifugal clutch ensures that a manual turn of the lock is not turning the electrical motor.

The control of the motor revolves around the same Hall effect sensor for detecting the end of the trip. Another solution includes a combination of a low cost, low-power motor driver, and a separate Hall effect direction sensing device. In either case, the motor drive and power control IC are rated around 1 A.

Several systems—including power window systems, power lock systems, power mirrors, courtesy light, or speakers—are located inside the door, and lots of wires would need to pass through or near the hinge system and into the dashboard, toward a central computing unit. This is shown in Figure 4.6, where a rubber boot is used as

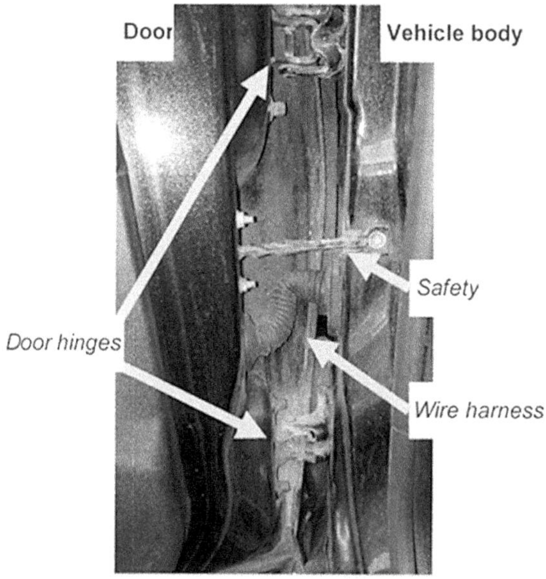

FIGURE 4.6 Wiring the door systems.

a conduit. To avoid bending the wires, certain manufacturers are routing wires up or down so that they twist (instead of bend) when the door is opened or closed.

Alternatively, a local microcontroller inside the door controls all these local motors and communicates with a central computer and the other doors through a serial communication interface. The preferred communication interface is a *Controller area network* (CAN) bus. This is a robust vehicle bus standard designed to allow microcontrollers and devices to communicate with each other, directly, without a host computer.

4.4 SOFT-TOP CONVERTIBLE

A soft-top convertible car uses a soft material, like canvas or vinyl, for the top of the vehicle. Soft tops are preferred by many drivers because these also give the "convertible look." This is because soft-convertible tops usually differ from a car's exterior color, making it more visible as a convertible. Another advantage when compared to hard-top convertibles consists of the minimal cargo space lost with storing the top. On the downside, the material used for soft-top usually lasts 5 to 7 years.

Since the principle is very similar to the electrical window, this chapter will not offer too much detail. An electrical motor turns a set of gears on each side of the car in order to raise and lower the roof. The gear engages a bracket that has gear teeth cut into it (like the gear in the power window system) and it is connected into the main structure of the roof. As the gear turns, it moves the roof. A scissors-type mechanism is formed by a set of arms and brackets that are linked together by pins. It folds down into itself when the roof is open, and expands to form the structure of the roof when the roof is closed.

The same type of Hall sensors for the end of the drive are used for the soft-top convertible drive as an alternative to a purely mechanic switch.

The power level of the motor drive used to move the soft-top is larger than for a power window system, usually in the range of 5–10 A. This is achievable with MOSFET single or three-phase converters.

4.5 HARD-TOP CONVERTIBLE

A hard-top roof is a safer choice, with better insulation from elements or external noise, yet more expensive. Other drawbacks relate to cargo space lost for storage, the operation required at lower speeds, or more expensive repairs. For instance, the fastest hardtop retraction reported so far is achieved by 2021 Aston Martin Vantage Roadster in 7 seconds.

A hard-top retractable roof is achieved with a complex system of motors and sensors able to produce the main sliding movement and to open the trunk for storage.

Each motor drive includes the same automatic control from the Body Control Unit computer; the electronic driver powered from a power supply up to 24 V dc (usually a battery) and a system of sensors able to determine the end of the drive range.

The actual design and combination set of motors differ from one manufacturer to another, and it is difficult to distinguish a generic solution. However, the power electronics contribution resides in motor drives with MOSFET inverters.

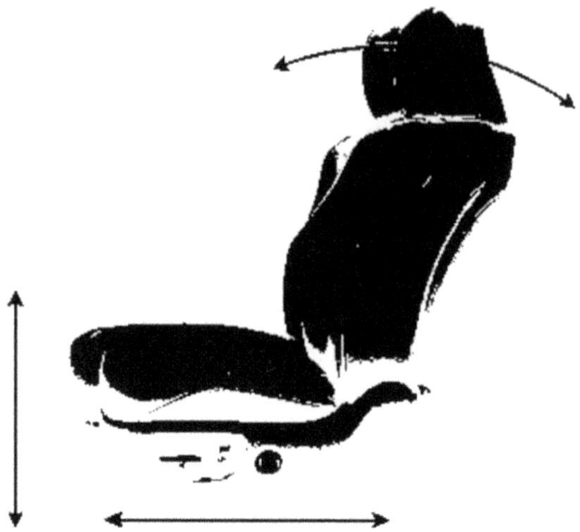

FIGURE 4.7 Three directions of movement for an automotive seat.

4.6 POWER SEATS

Power seats including both automatic movement and heating are the most complex systems in contemporary automotive body systems. For a complete power seat, movement is expected in multiple directions: front to rear, cushion height rear, cushion height front, backrest tilt, headrest height, or lumber support. The set of functions is illustrated in Figure 4.7.

Each such function requires at least one motor drive. Modern power seat systems may have up to seven motors per seat. Due to the complexity of the complete solution, various systems are actually used in practice. A possible classification would distinguish between:

- A *four-way (two-axis) power seat* with a motor for the fore-and-aft motion of the seat, and a motor for the up-and-down motion of the seat;
- A *six-way power seat* with a motor, a transmission, three clutches (one for the fore-and-aft motion of the seat, one for up-and-down motion of the front of the seat, and one for the up-and-down motion of the back of the seat), and multiple sets of cables that engage various mechanisms. For any motion, the motor and one or more of the clutches engage. For tilt (or rotate), there is some combo motion, with the seat front going up and the seatback going down.

Because advanced motor drive solutions require digital control and protection, additional functions have been introduced, taking advantage of the novel software capabilities. These include the opportunity to set position in a memory for automatic re-positioning if the seat has been moved. Obviously, this is possible with a digital control only. A variable resistor is often used as a position sensor and it has the resistance value proportional to the actual position on each direction and provides feedback to an electronic control unit.

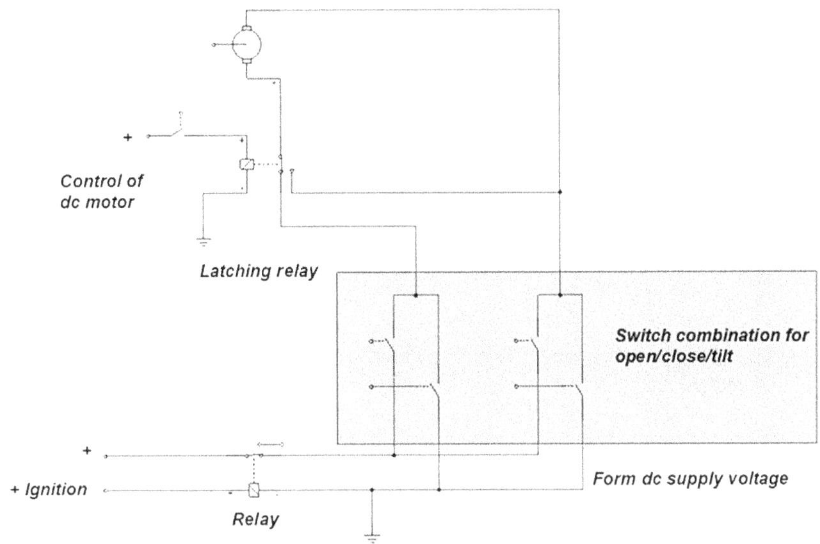

FIGURE 4.8 Electrical circuit for a motor drive used to control a sunroof.

4.7 ELECTRIC SUNROOF

The roof needs to slide, tilt, and stop in the closed position. To achieve this, several components are required, including a motor, a micro switch, and a latching relay. The latter operates as a normal relay, but it locks into position when energized.

The electrical circuit is shown in Figure 4.8.

The operation is similar to any motor with a reverse circuit. The microswitch is mechanically positioned such as to operate when the roof is in a closed position. A microswitch is a miniature snap-action switch, that is actuated by very little physical force, through the use of a tipping-point or "over-center" mechanism. A rocker switch allows the driver to adjust the roof. A rocker switch is an on/off switch that rocks when pressed, which means one side of the switch is raised while the other side is depressed, much like a rocking horse rocks back and forth. It is mostly known from light activation. The switch provides the supply to the motor to run it in the chosen direction. The roof will be caused to open or tilt. When the switch is operated to close the roof, the motor is run.

4.8 ELECTRIC MIRRORS

This is a fairly recent application. The electric adjustment of mirrors uses a system much the same as the seat adjustment system. Two small motors are used to move the mirror vertically or horizontally. Many mirrors also contain a small heating element on the rear of the glass. The heater is operated for a few minutes when the ignition is first switched on and can also be linked to the heated rear window circuit.

An example of the electrical circuit when controlled with a microcontroller is shown in Figure 4.9.

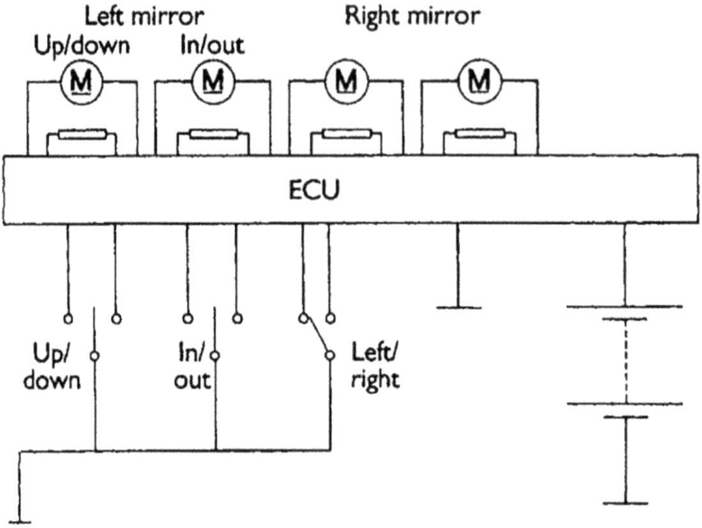

FIGURE 4.9 Circuit for electrical mirrors.

In Addition to the three-dimensional positioning of the external mirror, new features have been added recently. These transform the body of the external mirror into an assembly similar to a robotic arm. For instance, after the vehicle has turned off its engine, the external mirror automatically aligns with the vehicle's body in order to minimize the volume external to the car and to avoid damage to the mirror assembly. In another accomplishment, the mirror automatically flips downwards and changes its view angle toward the asphalt when the vehicle drives backward. This way the driver can see near the rear wheels rather than along the body for a better parking decision. When the car is again in forward drive mode, the mirror comes back to the setting predefined by the driver as the most optimal for highway driving.

4.9 CRUISE CONTROL

All previous applications have an on/off action, while the cruise control is an example of a closed-loop control (Figure 4.10). The purpose of cruise control is to allow the driver to set the vehicle speed and let the system maintain it automatically. The closed-loop control system reacts to the measured speed of the vehicle and adjusts

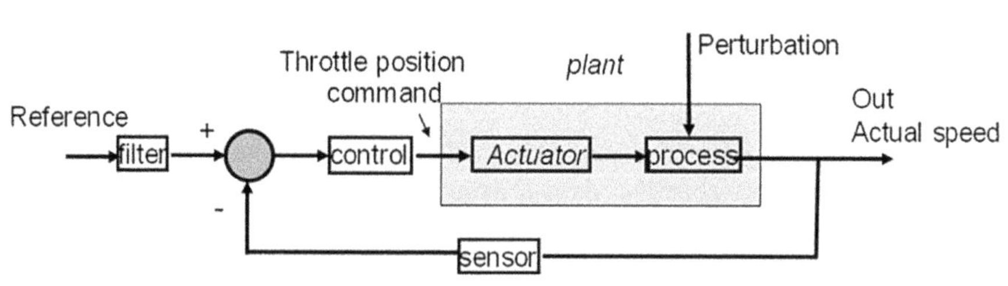

FIGURE 4.10 Principle of the feedback control system for a cruise control function.

the throttle accordingly. The reaction time is important so that the vehicle's speed does not feel to be surging up and down.

4.9.1 Modeling the Cruise Control

The design of the feedback control system follows a frequency domain analysis able to define system dynamics with a frequency model, also called the Laplace transfer function for the "process" module previously shown in Figure 4.10. In this respect, dynamic differential equations in the time-domain are converted into frequency domain equations. The time-domain differential equations are investigated first.

The mathematical description of the movement is based on a set of simplified assumptions: neglect the inertial movement of wheels, consider friction as proportional to the vehicle speed, neglect the car's shape, and consider a "body-free diagram" for the vehicle.

The analysis starts with the selection of a coordinate system for positioning along the movement direction. The force balance time-domain differential equation yields:

$$u - b \cdot \dot{x} = m \cdot \ddot{x} \Rightarrow \ddot{x} + b \cdot \dot{x} = \frac{u}{m} \tag{4.1}$$

where applied force is $u(t)$, "m" represents the vehicle weight, and "b" represents the friction coefficient. Dot marks above variables represent derivatives. Equation (4.1) can be re-written as a relationship for the speed "v," which is also the control variable.

$$\dot{v} + \frac{b}{m} \cdot v = \frac{u}{m} \tag{4.2}$$

Consider a solution in the form

$$v(t) = V \cdot e^{s \cdot t} \tag{4.3}$$

when the applied force is

$$u(t) = U \cdot e^{s \cdot t} \tag{4.4}$$

It yields

$$\left(s + \frac{b}{m}\right) \cdot V_o \cdot e^{st} = \frac{1}{m} \cdot U_0 \cdot e^{st} \Rightarrow \frac{V_0}{U_0} = \frac{\frac{1}{m}}{s + \frac{b}{m}} \Rightarrow \frac{V(s)}{U(s)} = \frac{\frac{1}{m}}{s + \frac{b}{m}} \tag{4.5}$$

This is the transfer function in Laplace form for the plant, from the force input to vehicle speed. It is further used to define the compensation law able to satisfy certain design dynamic requirements.

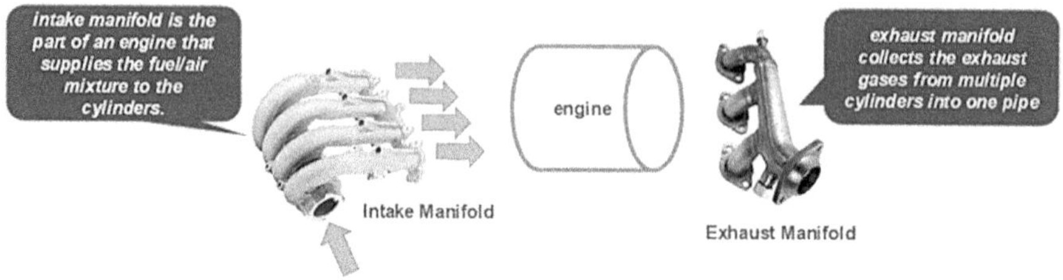

FIGURE 4.11 Using manifold pressure as an ancillary source of energy in cruise control.

Besides the implementation of the feedback control (also called closed-loop control), other requirements are also desired for a good cruise control system. Such requirements are possible with modern digital control, using software systems. These requirements include holding the vehicle speed at the selected value, holding the speed with minimum surging, allowing the vehicle to change speed, relinquishing the control immediately after the brakes are applied, storing the last set speed, and including built-in safety features.

4.9.2 Actuator for Cruise Control

The most difficult is to realize the actuator for the actual throttle control. Over the years, many solutions have been proposed. For instance, an electric motor can drive the throttle cable, or a vacuum-operated diaphragm is used, which is controlled by three simple valves.

A solution is suggested herein based on input manifold pressure as an ancillary source of energy. This source of energy is first explained with a sketch shown in Figure 4.11.

Due to the downward movement of the pistons and the restriction caused by the throttle valve, in a reciprocating spark ignition piston engine, a partial vacuum exists in the intake manifold. A partial vacuum is defined as a system with a pressure lower than atmospheric pressure. This manifold vacuum can be substantial, and can be used as a source of automobile ancillary power to drive auxiliary systems: power-assisted brakes, emission control devices, cruise control, windshield wipers, power windows, ventilation system valves, etc.

Through electrification, the usage of manifold pressure would be replaced with pure electrical solutions, involving power electronics. Until then, many cruise control systems use this ancillary source of energy.

The manifold pressure energy is used along a vacuum-operated diaphragm, which is controlled by three simple valves. This is shown in Figure 4.12. More details about the operation and construction of a solenoid valve are provided in Chapter 11.

When the speed needs to be increased, valve "*x*" is opened. This allows low pressure from the inlet manifold to one side of the diaphragm. The atmospheric pressure on the other side will move the diaphragm and throttle. To slow-down, valve "*x*" is closed and valve "*y*" is opened, allowing atmospheric pressure to enter the chamber and to vent it. The spring moves the diaphragm back.

If both valves are closed then the throttle position is held. Valve "*x*" is normally closed and valve "*y*" normally open. Thus, in the event of electrical failure, the cruise

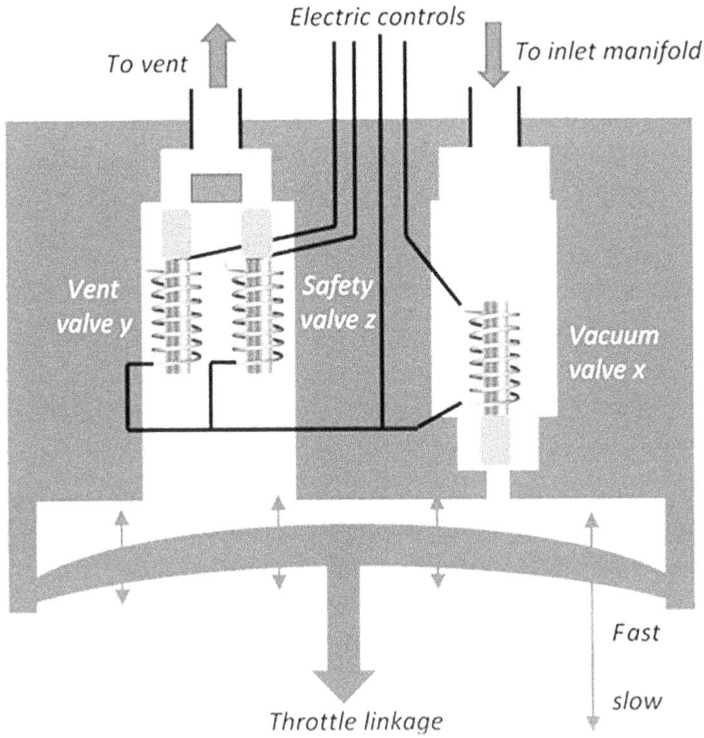

FIGURE 4.12 Cruise control actuator based on manifold pressure.

control will not remain engaged and the manifold vacuum is not disturbed. Valve "z" provides extra safety and is controlled by the brake and clutch pedals.

Control of all valves is through relay type solenoids. These are controlled with power semiconductor devices, usually in the range of a couple of Amperes. Such systems are discussed in detail, later on, in Chapter 11.

4.9.3 Drive-by-Wire

This solution represents an extension within the body systems of the electrification process started in chassis systems. The goal of the transportation' electrification is to replace all ancillary sources of energy with electrical energy. Therefore, all mechanical or pneumatical actuation systems are replaced with electrical motor drives. In this case, the usage of a manifold vacuum is replaced with an electrical drive able to set up an electronic control system.

The electronic cruise control is based on *electronic throttle control* (ETC). The electronic throttle control is an automobile technology that electronically "connects" the accelerator pedal to the throttle, replacing a mechanical linkage. A typical ETC system consists of three major components.

- an accelerator pedal module that is ideally accompanied with two or more independent sensors,

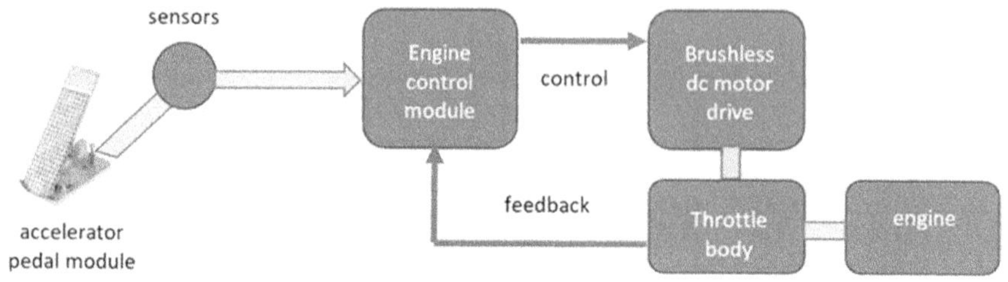

FIGURE 4.13 Drive-by-wire actuation of the cruise control system.

- a throttle valve that can be opened and closed with an electric motor and it is sometimes referred to as an electric or electronic throttle body,
- a microcontroller-based engine control module (mostly abbreviated as ECM).

The latter determines the required throttle position using calculations from data measured by other sensors, including the accelerator pedal position sensors, engine speed sensor, vehicle speed sensor, and cruise control switches. The electric motor is then used to open the throttle valve to the desired angle via a closed-loop control algorithm within the ECM. The design of the feedback control system follows the previously described modeling of the vehicle dynamics.

A sketch for the drive-by-wire system is shown in Figure 4.13, which is a proceeding from the Delphi Corporation, early in the year 2000.

4.10 CLIMATE-CONTROL

4.10.1 Heater

Design concepts are used to regulate the thermal output when heat is taken from the engine. Heat is taken from the water-cooled engine with the use of a heat exchanger, also called the *heater matrix*. The heat from the engine is used for heating the passenger cabin. A blending technique is used with a control flap, which determines how much of the air being passed into the vehicle is directed over the heater matrix. Next, a suitable arrangement of flaps makes it possible to direct air of the chosen temperature to selected areas of the vehicle interior. The motors used to increase airflow are simple permanent magnet two-brush motors, used as fans. Using dropping resistors helps to vary the voltage supplied to control the motor speed. The blower fan is often the centrifugal type and, in many cases, the blades are positioned asymmetrically to reduce resonant noise.

4.10.2 Electronic Heater Control

The role of the heater controller is to maintain the temperature at a desired automatic level, with a different temperature in the footwell above that of the upper body. A more complex system requires control of the blower motor, blend flap, direction flaps, and the fresh or recirculated air flap.

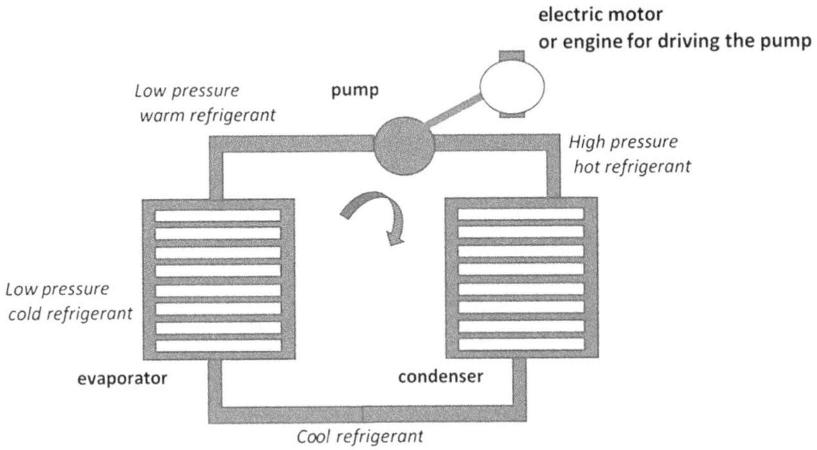

FIGURE 4.14 A/C system.

4.10.3 A/C Compressor

The system utilizes the standard heating and ventilation components, but with the important addition of an evaporator, which both cools and dehumidifies the air. It is basically a refrigerator using a refrigerant known as R134A. That is hydrofluorocarbon (HFC), not the CFC used in the past which was decidedly not environmentally OK. This substance changes state from liquid to gas at 26.3 °C (Figure 4.14).

An engine-driven pump compresses the vaporous refrigerant generating heat in the process. The refrigerant is then pumped to the condenser and returns to a liquid state releasing energy. An expansion valve sprays the cooled liquid into the evaporator, where the evaporation process is used to extract heat from the incoming fresh air. The operation is easier to follow with the application of the *Combined Gas Law* or *General Gas Equation* that is combining Boyle's Law, Charles's Law, and Gay-Lussac's Law, into the following form for a closed system.

$$\frac{\text{pressure} \cdot \text{temperature}}{\text{volume}} = \text{constant} \tag{4.6}$$

The pump needs to be energized by the engine or an electric motor. If the engine drives the A/C compressor, the automotive compressor is not very efficient because it is designed to work over a very large RPM range without having a dramatic impact on the available output in a very wide range of operating temperatures and conditions. A series of expenses are made on efficiency ratings in the name of consistency and durability. A simple comparison with a home A/C compressor highlights the advantage of electric drive, which is designed to run at a constant RPM, in a fairly stable environment, and operation is far more efficient.

While using an electric motor for the compressor is efficient, it sets a series of challenges for a 14 V battery system. Most modern automotive air conditioning systems range somewhere around 2–3 tons for a small car, and sometimes greater than 6 tons for larger SUVs. A *ton of refrigeration* is a unit of power used to describe the heat-extraction capacity of refrigeration and air conditioning equipment. It is defined as the rate of heat transfer that results in the freezing of 1 short ton of pure ice at 0°C in 24 hours.

As per manufacturers of electric compressors, a contemporary solution for electric drive provides 1.5 tons while running at full power, under optimal conditions, and it draws up to 2 kW of power. On a 14 V battery distribution system, this means huge currents of abut 100–200 Amp. The solution is not very practical, and an improvement is expected with a power distribution system at 48 Vdc.

4.11 SHAPE MEMORY ALLOY ACTUATORS

Most mechatronics applications presented briefly in this chapter are traditionally using low-power dc motor drives and solenoids as actuators, either as individual on/off motors or as complex robotic three-dimensional configurations. The increasing count of such motor drives can impose some weight limitations since electromagnetic devices are somewhat heavy. This encouraged engineers to explore new and different solutions.

A solution that has recently proved to have merit consists of *shape memory alloys*, also called *smart materials*. The shape memory alloy is an alloy that can be deformed when cold but easily returns to its predefined shape when heated. The *shape memory alloy* components are also actuated electrically when an electric current produces Joule heating. This usually produces a fast actuation and a slower de-actuation time. The two most important shape memory alloys are copper–aluminum–nickel and nickel–titanium. Components made of these materials are lightweight and hence a good replacement for the heavier motor drive actuators.

Figure 4.15 provides an example for the possible use of the shape memory alloy within an actuator. An electric current is applied in Figure 4.15a to produce the displacement of a spring made of a shape memory alloy. This is producing the outward movement of the slider and the compression of a conventional spring. When the current stops the heating effect, the shape memory alloy cools off, and the conventional spring helps the recovery of the initial state. This is shown in Figure 4.15b.

The first automotive application has consisted of an automotive valve used to control pneumatic bladders in a car seat that adjusts the contour of the lumbar support. In another application, the 2014 Chevrolet Corvette used a shape memory alloy within the trunk lock. It can be seen that slow-actuated and not critical applications are preferred herein. Both car seat and pneumatic valve assembly can benefit from a lighter actuator while the response time is not very critical.

Finally, more recent developments used wires made of these materials for actuation of a mirror assembly. The two dc motors of the conventional actuator are replaced by two pairs of shape memory alloy wires. The shape memory alloy wires used for mirror positioning are supplied with a voltage-controlled power supply with a maximum current of 1 A. An accurate and robust positioning of the mirror is achieved with a *variable structure controller*, designed after state-space variables. In a certain embodiment, the control considers two-axis for positioning and the same power converter actuates both axes. The controller receives a joystick signal from

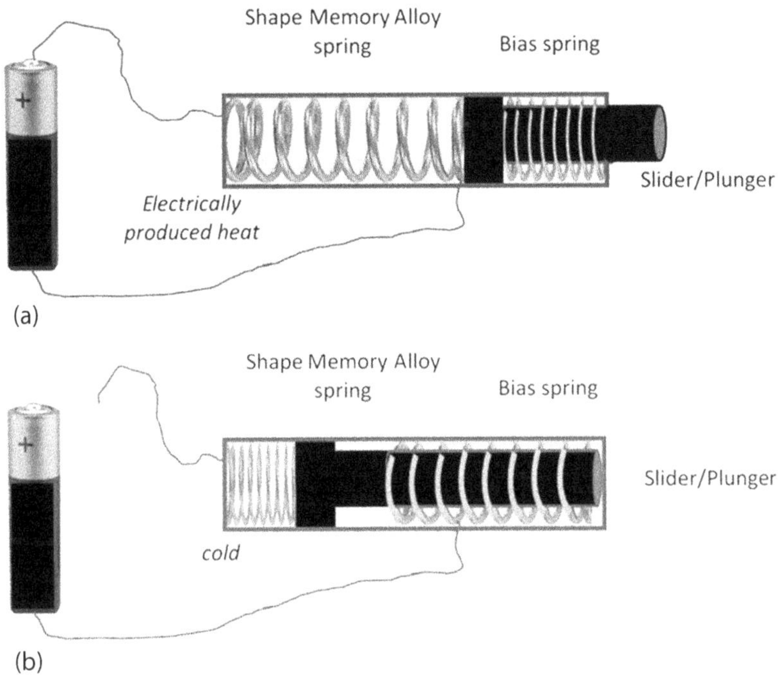

FIGURE 4.15 An example for the use of a shape memory alloy device.

the driver and routes an electrical current to the proper shape memory alloy wire in order to reduce the error in one of the two angular position coordinates. The two degrees of freedom are controlled one at a time, and the controller is able to depict the largest position error and to act for reducing it. At a certain sampling rate, it moves to the other. This maintains a fast and robust control in spite of the inherent discontinuity in the implementation. The supply voltage is accordingly routed to the proper shape memory alloy wire through a logic switching circuit based on power MOSFET transistors.

4.12 CONCLUSION

This chapter presents several motor drive applications for automotive body systems. This is a very vast field of activity, as many mechatronics applications were developed over the years. Current trends focus on adding new features that can be implemented in software. Therefore, all these body systems benefit from local microcontroller-based control. The installed power is very low and either dc or brushless dc motors are used. The latter offers good promise for the future due to improved efficiency and size.

The focus was on the principle behind the convenience feature, and the next chapter will focus on the power circuits, integrated control circuits, and power semiconductor devices.

REFERENCES

Chen, S., Lequesne, B., Henry, R., Xue, Y., Ronning, J. 2002. Design and testing of a belt-driven induction starter-generator. *IEEE Transactions on Industry Applications*, 38(6):1525–1533.

Comnac, V., Cernat, M., Mailat, A. 2008. *Optimal Use of the 14 V Alternator in 42 V Automotive Supply Systems.* Paper Presented at 13th International Power Electronics and Motion Control Conference, Poznan, Poland, pp. 1748–1754.

Denton, T. 2017. *Automobile Electrical and Electronic Systems.* 5th edition, Abingdon-on-Thames: Routledge.

Lorilla, L.M., Keim, T.A., Lang, J.H., Perreault, D.J. 2005. Topologies for future automotive generators. Part I. Modeling and analytics. Paper Presented at IEEE Vehicle Power and Propulsion Conference, Chicago, IL, USA, pp. 74–85.

Lorilla, L.M., Keim, T.A., Lang, J.H., Perreault, D.J. 2005. *Topologies for Future Automotive Generators - Part II: Optimization.* Paper Presented at IEEE Vehicle Power and Propulsion Conference, Chicago, IL, USA, pp. 831–837.

Murray, A., Wood, P., Keskar, N., Chen, J., Guerra, A. 2001. A 42V *Inverter/Rectifier for ISA Using Discrete Semiconductor Components.* Paper Presented at Future Transportation Technology Conference, Costa Mesa, CA, USA.

Naidu, M., Walters, J. 2003. A 4-kW 42V induction machine based automotive power generation system with a diode bridge rectifier and a PWM inverter. *IEEE Transactions on Industry Applications*, 39(5):1287–1293.

Neacşu, D. 2004. *Power Semiconductor and Control for Automotive Applications.* Tutorial Presented at IEEE APEC, Anaheim, CA, USA.

Stoia, D., Cernat, M. 2009. *Design of a 42 V Automotive Alternator with Integrated Switched-Mode Rectifier.* Paper Presented at 8th International Symposium on Advanced Electromechanical Motion Systems & Electric Drives Joint Symposium, Lille, France, pp. 1–8.

Whaley, D.M., Soong, W.L., Ertugrul, N. 2004. *Extracting More Power Din Lundell Alternator.* Paper Presented at Australasian Universities Power Engineering Conference, Brisbane, Australia.

5 Power Converters Used in Body Systems

HISTORICAL MILESTONES

1959–1971: Power MOSFET and microprocessor

The MOSFET (MOS field-effect transistor, or MOS transistor), invented by Mohamed M. Atalla and Dawon Kahng at Bell Labs in 1959, led to the development of the power MOSFET by Hitachi in 1969, and the single-chip microprocessor by Federico Faggin, Marcian Hoff, Masatoshi Shima, and Stanley Mazor at Intel in 1971.

1960: Brushless dc motors were made possible by the development of solid-state electronics

1990–2000: Advent of embedded units.

Multiple microprocessor-based systems are used to control the operation of small motors, under the generic name of an electronic control unit (ECU). Possible ECU are door control unit, engine control unit, electric power steering control unit, human-machine interface, powertrain control module, transmission control unit, or powertrain control module, seat control unit, speed control unit, telematic control unit, transmission control unit, brake control module (ABS or ESC), battery management system.

5.1 ELECTRICAL MOTORS USED IN BODY SYSTEMS

All mechatronics systems used in automotive body systems are based on motor drives, confirming an old saying, "*Whatever you touch in a car, there is an electric motor close by.*" The multitude of applications is illustrated in Figure 5.1.

The small electric motors used in automotive applications can be classified as either *brushed dc motors* or *brushless dc motors*. Brushed dc motors are the simplest solution, easy to control, and they can serve as actuators with an overall short operating time. Torque can be controlled down to zero speed. Brushless dc (BLDC) motors are a newer solution, due to the electronic circuits required to accompany the motor. Their advantages consist of less wear, while drawbacks include a complex control algorithm, and a higher computing power. They are most suitable for applications demanding long-term continuous duty like fuel pumps. Also, the higher cost is justified in applications with certain installed power. This means lower power drives are based on brushed dc motors.

It is also noteworthy that many low power applications involve a solenoid for actuation of a mechanical load. The operation of solenoid is explained in Chapter 11, and it may qualify as dc motor since it moves a mechanical load. Some applications are shown in Figure 5.1 actually involve a solenoid (identified therein as a motor) rather than a conventional brushed dc motor with a rotational magnetic field.

Application	motor
sliding door	DC
sunroof	BLDC, DC
seat adjustment	DC
front wiper	BLDC, DC
compressor	DC, BLDC
ventilation	DC
engine cooling	DC, BLDC

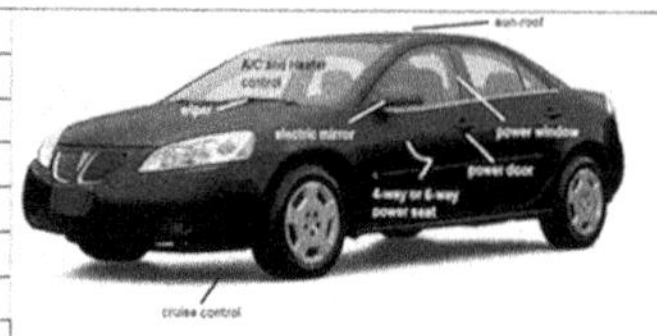

Application	motor
rear wiper	DC
trunk tailgate	DC
seatbelt pretensioner	DC
fuel pump	DC, BLDC
water pump	DC
oil ump	DC

FIGURE 5.1 Small electric motor usage in modern automotive applications shown within an Infineon classification along with the type of motor.

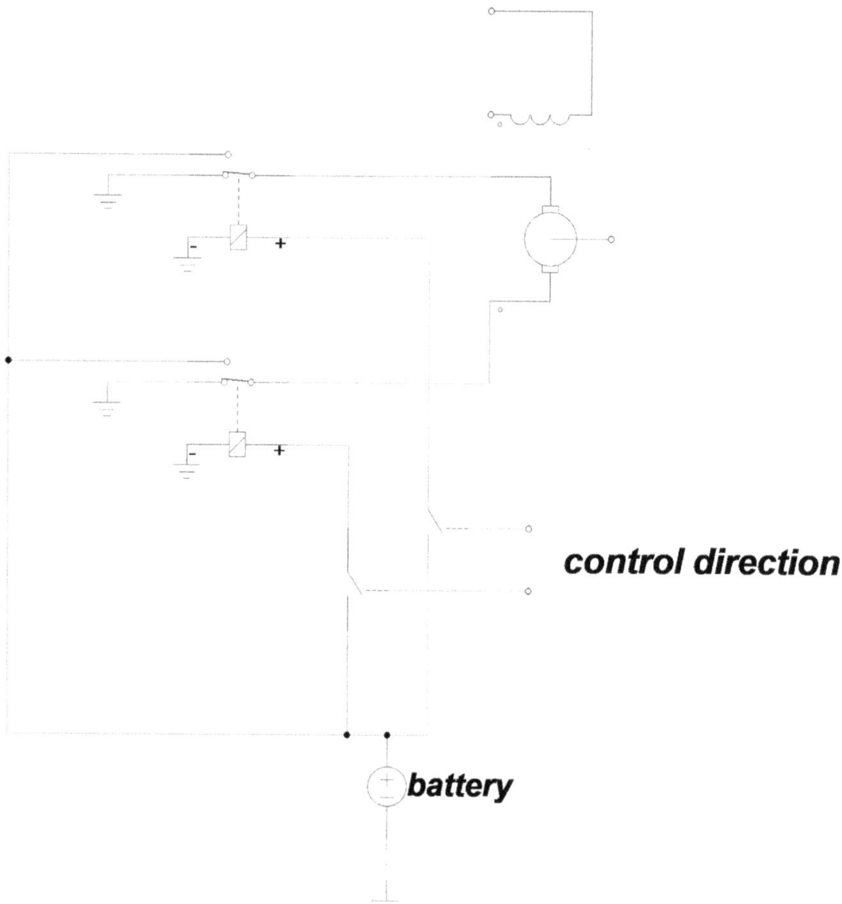

FIGURE 5.2 Electrical circuit for a bidirectional motor drive with a dc motor.

The simplest electric circuit used for supplying electrical motors within mechatronics applications is shown in Figure 5.2. The control is shown herein with relays and switches, even if modern solutions may be based on power semiconductor devices instead of relays and switches. Chapter 11 presents details of the construction and operation of relays, while Chapter 10 describes the operation of semiconductor power switches.

The two relays in the Figure 5.2 have a normal connection to the ground, at rest. When the control switch is moved to one side, one of the relays will operate and this applies a polarity of the supply to the motor. When the switch is moved the other way, the second relay will operate and this applies a different polarity of the supply to the motor. When at rest, both sides of the motor are at the same potential. The use of relays and switches allows a smaller current through the operator panel and a higher current through the actuated motor.

Due to the numerous requirements that need to be implemented in the software, the advanced comfort and safety features ask for a power electronic converter and a microcontroller for the software control of the machine drive. Power electronics at a current load level of several Amperes can be set within integrated circuits, while higher currents impose the use of integrated power staged away from the control integrated circuit.

5.2 INTEGRATION OF POWER ELECTRONICS

Depending on the power level of the small motor drive, the final power stage (power MOSFET and freewheeling diode) can be inside or outside the integrated circuit. A classification depending on the power level and complexity of control into three levels of function integration is provided herein.

- *Low level of integration*: when the microcontroller, bridge driver, external MOSFETs, and certain sensors are all used as individual, discrete components.
- *Medium level of integration*: when driver and power MOSFETs are within a fully integrated bridge, while the microcontroller and sensors are different components.
- *High level of integration*: when all components are integrated on the same platform, as a systems-on-chip or embedded power concept.

Each level of integration is next explained.

5.2.1 High Level of Integration

Figure 5.3 describes systems with a high level of integration.

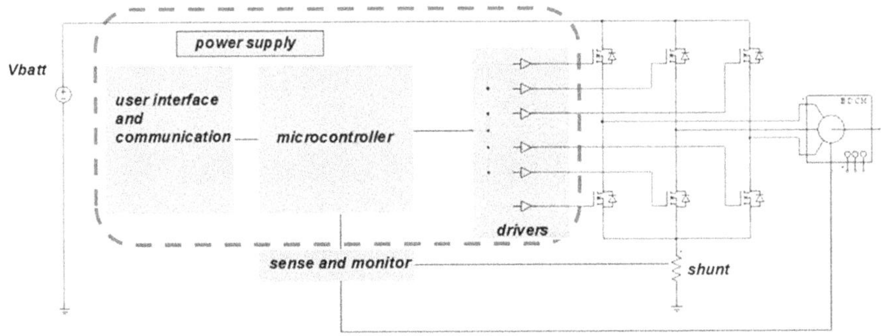

FIGURE 5.3 High-level integration platforms for power and control electronics.

As the requirements for each Body System are increased, more and more systems feature a *system-on-chip* device that represents the highest level of integration. A system-on-chip (SoC) combines all required electronic circuits of various industrial computer components onto a single, integrated chip. SoC is a complete microelectronic substrate system that may contain analog, digital, mixed-signal, or radio-frequency functions. It usually includes a central processing unit (CPU) that may be multi-core, system memory (RAM), and peripheral circuits. Because SoC includes both the hardware and software necessary for the application, it uses less power, it is faster, requires less space, and is more reliable than multi-chip systems. Most system-on-chips today come inside mobile devices like smartphones and tablets.

The system-on-chip motor control devices (like Infineon® Embedded Power ICs) are specifically designed to enable motor control solutions. In this respect, they have a special set of features as explained.

- A small package form factor and a minimum number of external components are essential.
- Applications using a higher level of integration as support for power electronics and control, include window lift, sunroof, wiper, fuel pump, HVAC fans, engine cooling fan and water pumps, and so on.
- Integration of all functions required to sense, control, and actuate a motor, additional to numerous decisions in software able to improve convenience and safety.

An automotive *system-on-chip* device integrates on a single die, the microcontroller, a non-volatile flash memory, analog and mixed-signal peripherals, communication interfaces, driving stages for either relay, half-bridge or full-bridge DC and BLDC motor applications. In most cases, the power stage is implemented outside in order to accommodate higher power levels.

5.2.2 Medium Level of Integration

This is the most common solution, reflected with the largest market share. Figure 5.4 illustrates the typical architecture for a medium level of integration, taking

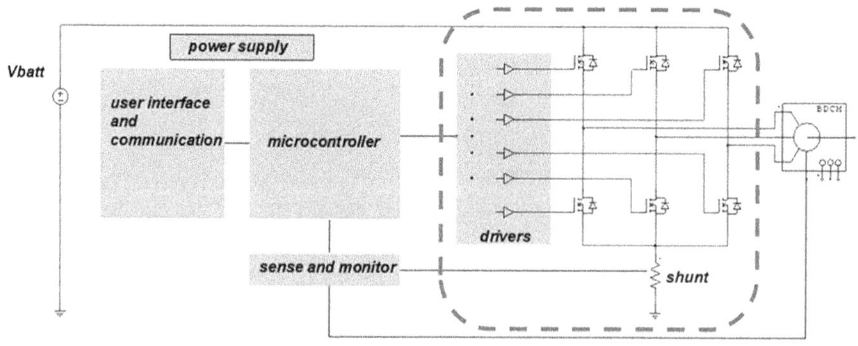

FIGURE 5.4 Medium-level integration platforms for power and control electronics.

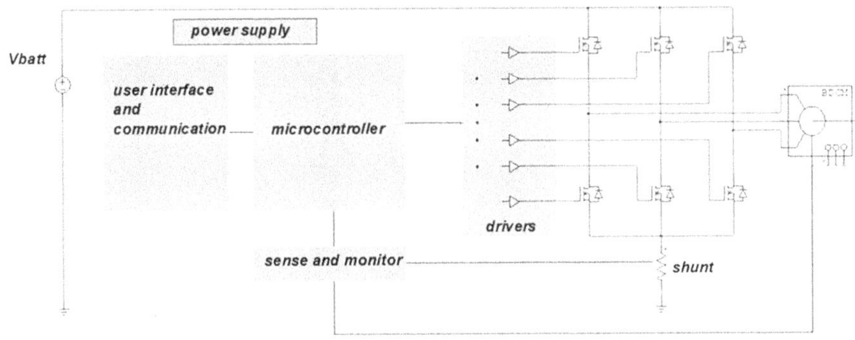

FIGURE 5.5 Low-level integration platforms for power and control electronics.

advantage herein of a single-phase power stage. The major difference from a system with a higher level of integration consists of the power MOSFETs and gate drivers being grouped together, but away from the control system.

5.2.3 Low Level of Integration

The main characteristics of this architecture consist of the usage of separate devices for control, gate drivers, and the power stage. Figure 5.5 illustrates the base architecture for the low level of integration systems.

5.3 POWER CONVERTERS

Advanced comfort and safety features require a power electronic converter and a microcontroller for software control of the machine drive. Alternatively, mixed-mode integrated circuits can be used for local motor drive control. The power converter stages can be classified depending on the application, in unidirectional or bidirectional motor drives.

5.3.1 Unidirectional dc Motor Drives

Unidirectional motor drives are used for *permanent magnet* or *wound field commutator motors*, that can be controlled by a switch in series with the power supply. The switch is herein implemented with a power semiconductor device, mostly a MOSFET. Figure 5.6 shows this simple power converter.

5.3.2 Bidirectional dc Motor Drives

Bidirectional (or reversible) motor drives require an H-bridge converter. Figure 5.7 shows the circuit usage of the H-bridge converter.

A *H-bridge converter* is made up of four MOSFET transistors and it can supply the dc motor with both polarities of the dc voltage. The H-bridge is the most-used topology for the power stage because it is supplied with dc voltage from the battery

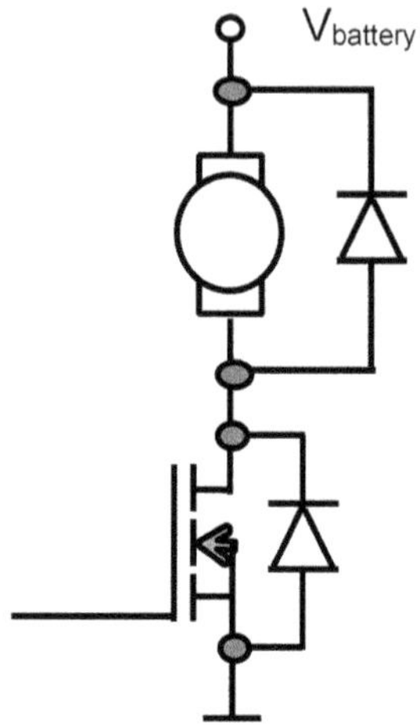

FIGURE 5.6 Unidirectional dc motor drive with power MOSFET.

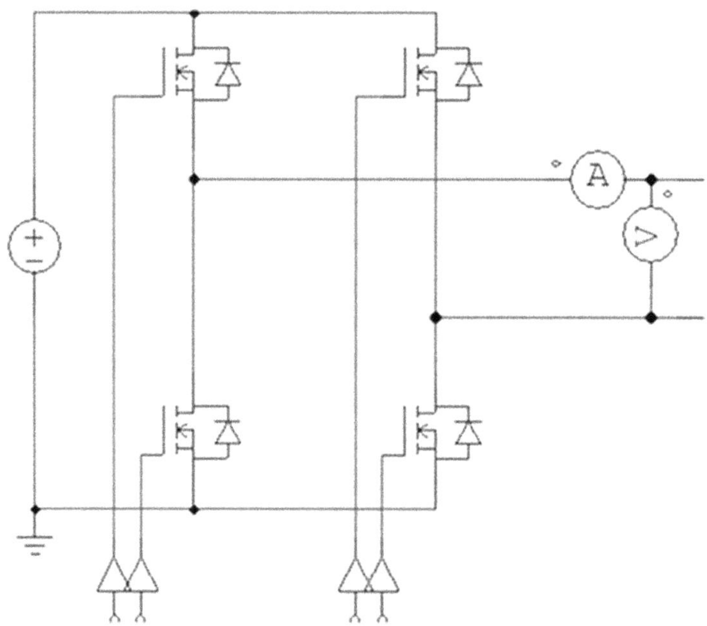

FIGURE 5.7 Circuit with H-bridge converter.

and delivers either positive or negative voltage. Numerous hybrid integrated solutions are available for MOSFET-based H-bridge power stages.

The H-bridge converter is mostly used as an on/off switch that is able to supply or not the motor. However, it has the advantage of a continuous control of the output voltage, as the output voltage always passes through a controlled device. When the motor is supplied with electrical energy, the H-bridge converter features two devices in series leading to an increase of the voltage drop and loss. This is a small drawback.

A diagonal of the H-bridge determines the application of a positive voltage on the load, while the other diagonal determines a negative voltage during the second alternance. The duty cycle defined with the operation of one or the other diagonal determines the average voltage on the load.

Using the duty cycle, the voltage applied to the load can be modified. For instance, considering a 12 V battery allows the production of the following voltages:

$D = 0.1 \rightarrow V_{out} = -9.6$ V
$D = 0.2 \rightarrow V_{out} = -7.2$ V
$D = 0.3 \rightarrow V_{out} = -4.8$ V
$D = 0.4 \rightarrow V_{out} = -2.4$ V
$D = 0.5 \rightarrow V_{out} = 0$ V
$D = 0.6 \rightarrow V_{out} = 2.4$ V
$D = 0.7 \rightarrow V_{out} = 4.8$ V
$D = 0.8 \rightarrow V_{out} = 7.2$ V
$D = 0.9 \rightarrow V_{out} = 9.6$ V

5.3.3 Single-Phase Power Converters

The emergence of ac motor drives in automotive applications requires a closer look at ac generating power converters. The most significant ac motor drive is based on a brushless dc machine, which is presented in detail later on.

The previously introduced H-bridge can also deliver a square-wave ac voltage, with magnitude equal to half the dc supply voltage. The H-bridge is configured in Figure 5.8 as a circuit used to deliver an ac voltage on the load, from an automotive battery. This converter is also called a *single-phase inverter* or *full-bridge inverter.*

Transistors located on a diagonal of the full-bridge inverter are controlled simultaneously, producing a polarity of the load voltage, followed by the control of the two transistors located on the other diagonal of the full-bridge inverter and able to deliver the second polarity of the load voltage. The duration of the conduction state for transistors on one diagonal must be equal to the duration of the on-state for the transistors on the other diagonal in order to have zero dc component of the load voltage.

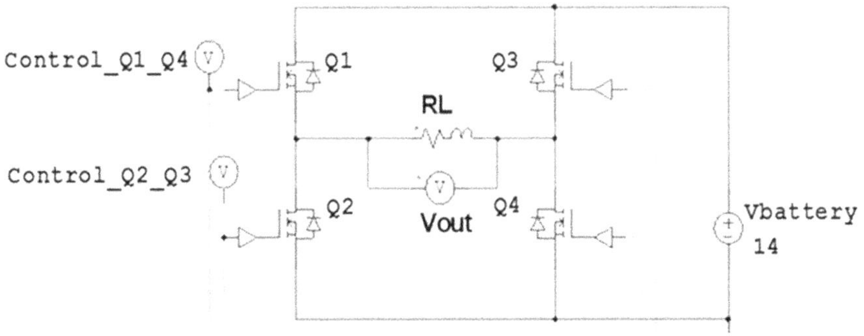

FIGURE 5.8 Single-phase inverter connected to a 14 Vdc battery.

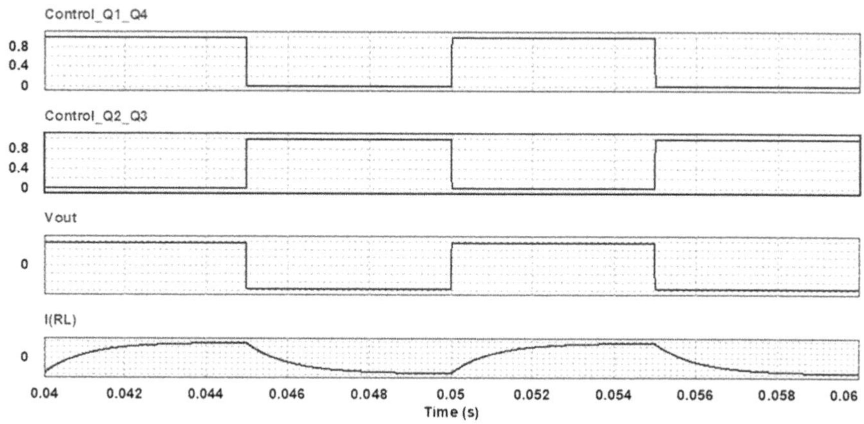

FIGURE 5.9 Typical theoretical waveforms for a single-phase inverter with R-L load.

The method used for control of the power semiconductor switches denotes an operation with *reversed control*, which means *180° out of phase* control. The load is always connected in between the two branches of the H-bridge inverter, as shown in Figure 5.8. The frequency used within the control of the transistors becomes equal to the frequency of the load voltage.

Understanding the operation of the full-bridge inverter is often done with a resistive-inductive load before analyzing an actual motor drive. Typical waveforms are shown in Figure 5.9.

Presence of an inductance in the load circuit determines short conduction intervals for diodes connected anti-parallel to transistors in Figure 5.8. The inductor character of the load introduces a phase shift between load current and voltage. This produces an effect that can be seen when associated with the conduction intervals for the power semiconductor devices. Switching processes through transistors can be observed along the polarity of the currents through the transistors Q1, Q2, Q3, and Q4. Control signals are shown in Figure 5.9. A detail on the commutation processes is also shown in Figure 5.10, for a resistive-inductive load.

Waveforms shown in Figure 5.9 for a H-bridge (full-bridge) single-phase inverter are square-waves and have a strong content of harmonics. If such a harmonic inverter output voltage would be applied to a motor winding, it would produce losses and vibration in the motor. While this may be acceptable for a low power motor, it becomes bothersome in higher power applications where a decrease in efficiency is costly and an extended vibration is damaging to parts or lifetime.

In order to reduce motor current harmonics and improve performance, a *pulse width modulation* (PWM) method is employed. The control circuit for PWM is very simple. A comparator is used to compare a sinusoidal reference with a high-frequency triangular waveform, called *carrier*. The output of the comparator defines the switching instants in the inverter. A variation of the magnitude of the sinusoidal reference determines variation in the fundamental of the output voltage. This suggests the definition of the *modulation index* (usually denoted with "*m*") as a ratio

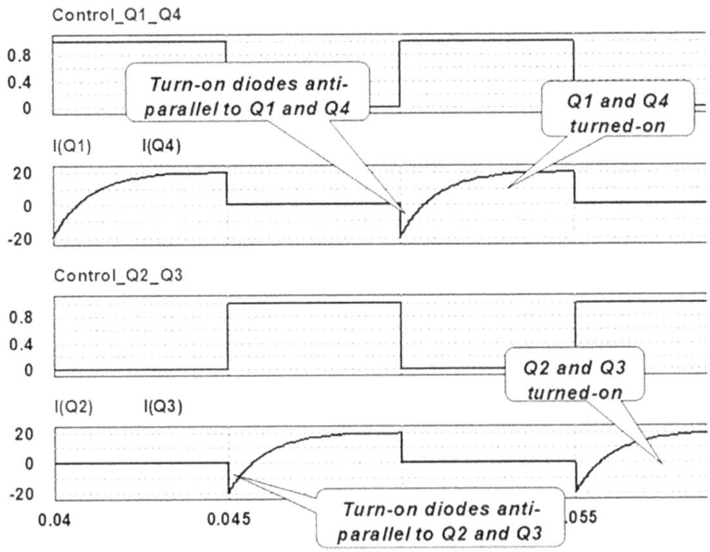

FIGURE 5.10 Details of the commutation processes within the single-phase inverter.

between the magnitude of the sinusoidal reference waveform and magnitude of the triangular carrier, which is usually considered unitary. Therefore, the modulation index $m \in (0,1)$.

The PWM control circuit produces a train of pulses on the load, as shown in Figure 5.11. It is important to observe the currents through the power switches (transistors with anti-parallel diodes). This is the most important difference from the operation of the same single-phase inverter hardware, with and without modulation. Using modulation in the control of an inverter with inductive load determines a single polarity of the current for the entire fundamental half-cycle, and this means the current is repetitively commutated either from top transistor to bottom diode (positive load current) or bottom transistor to upper diode (negative load current). This is shown with a zoom vignette in Figure 5.11.

Figure 5.12 illustrates an example of a frequency analysis of the voltage waveform generated as a PWM signal. It can be seen that the low-frequency harmonics present with a square-wave are now moved at higher frequencies. The first important frequency component occurs near the carrier frequency, in this case at 10 kHz.

When the load comes from an ac motor, the *content in fundamental* is more important than the quality of the output voltage waveform. It becomes a performance index. The content in fundamental is defined as the rms of the component at the fundamental frequency divided to the rms of the entire voltage waveform.

5.3.4 Three-Phase Inverters

The previous chapter has illustrated the use of both brushed dc and brushless dc motors in body systems. Most brushless dc (BLDC) drives benefit from a three-phase

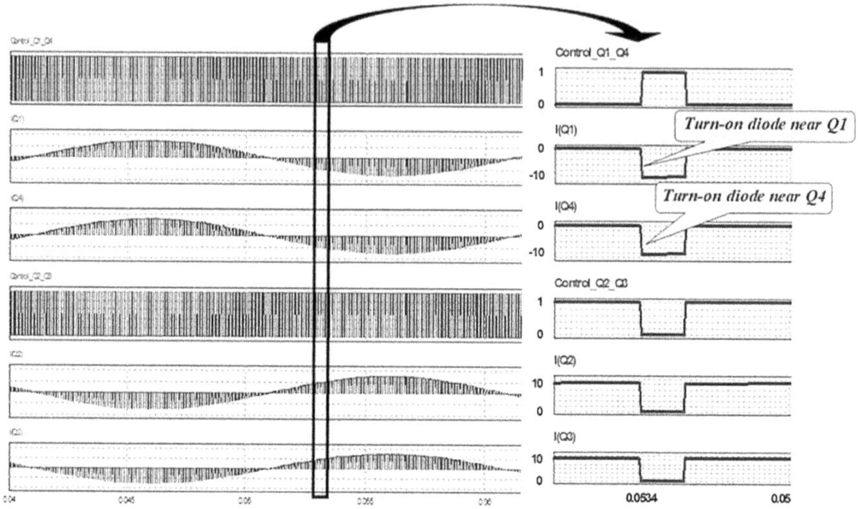

FIGURE 5.11 Waveform details for the generation of PWM for the single-phase inverter with an inductive load.

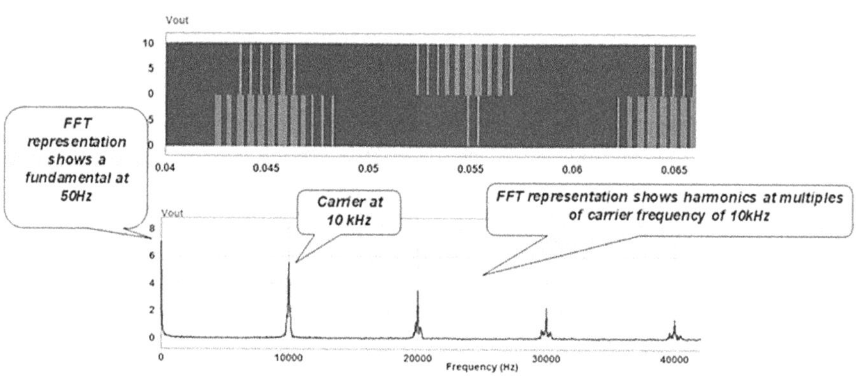

FIGURE 5.12 FFT analysis of the output voltage for a single-phase inverter.

inverter for a set of three-phase ac voltages. While the construction of the motors is detailed in Chapter 12, a brief introduction to the electronic hardware is presented herein.

There are several ways the three-phase load (or motor windings) can be connected. Figure 5.13 draws the hardware used for a three-phase inverter with the most-used load connection. The load is therein connected in a star, without a neutral connection.

The three-phase inverter is controlled with a PWM technique, not unlike the one presented for the single-phase inverters. A set of three sinusoidal references (V_{refa}, V_{refb}, V_{refc}) are compared to the same triangular carrier signal (V_{carr}) in order to produce the gate control signals for all six power switches.

Due to the way the load is connected, the waveform of the output phase voltage yields very differently from in the single-phase inverter case. Figure 5.14 illustrates

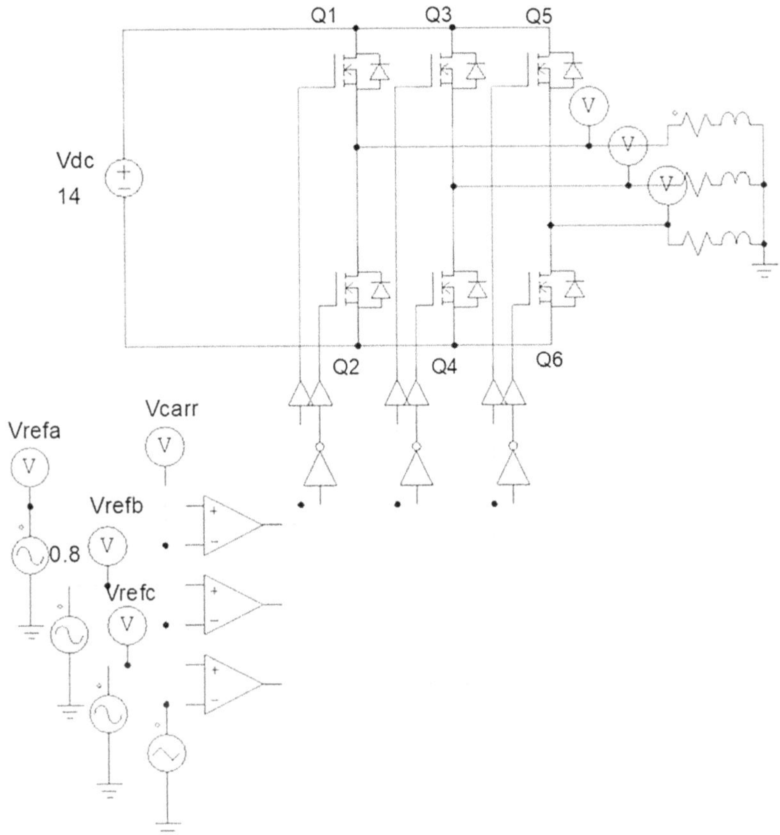

FIGURE 5.13 Three-phase inverter and control circuit.

the voltage waveforms produced by the PWM generator when used with a three-phase inverter with a star-connected load and without a neutral connection. Each phase voltage yields on three levels 0 V, 0.33 V, and 0.66 of Vdc. The dc bus voltage has been selected at the battery level of 14 Vdc, and the carrier frequency of the carrier at 1 kHz. This is a rather low value for the carrier frequency and it is used herein to allow more detail into the waveforms. Usually, the carrier frequency ranges from 1 kHz to 100 kHz.

Details of this waveform can be understood from analysis of the possible operation states. Each inverter branch has two possible states when we neglect the effect of the dead-time. Dead-time is the interval introduced in between the turn-off of a power MOSFET and the turn-on of the other MOSFET on the same inverter branch. It is used to prevent a short-circuit of the dc bus through two commutating transistors due to the inherent switching delays, which is detailed at the end of this chapter.

Each inverter branch has two possible states: either the upper-side switch or the lower-side switch is in conduction. An impedance divider is formed in the load, with the connection of each individual phase impedance to either a positive or negative dc bus. For instance, transistor Q1 determines the connection of the first phase ("*a*") to the positive dc bus, while conduction of transistors Q4 and Q6 determine the

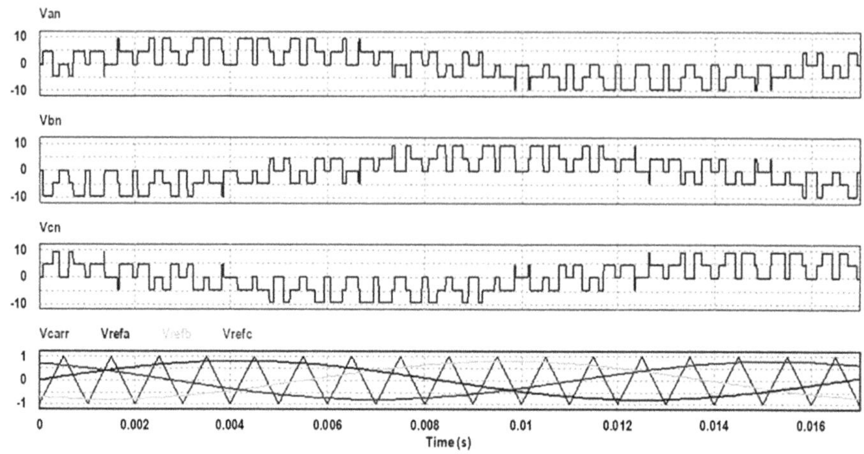

FIGURE 5.14 Waveforms at PWM operation of the three-phase inverter: phase voltages V_{an}, V_{bn}, V_{cn}; three sinusoidal references (V_{refa}, V_{refb}, V_{refc}) are compared to the same triangular carrier signal (V_{carr}).

connection of the second ("*b*") and third ("*c*") impedances to the negative dc bus. A divider from the dc bus voltage is created between the three load impedances and the result is seen with a voltage of 0.66 times the dc bus on the first phase ("*a*"), and a negative voltage of –0.33 times the dc bus on the other two phases ("*b*" and "*c*").

All possible states are depicted for combinations of conduction states for the switches, so that a short-circuit is not produced on the dc side. There are four combinations of conduction states of switches on the second and third branches when Q1 is kept on, and four combinations when Q1 is off. All eight possible states are shown in Table 5.1. For each case an impedance divider defines the output voltages as multiples of 0.33 Vdc.

Since the final application for this three-phase inverter consists of the ac motor drive, the possibility to adjust the amount of voltage required across the load

TABLE 5.1
All Possible States for the Operation of the Three-Phase Inverter, without Neutral Connection

	Q1 / Q2	Q3 / Q4	Q5 / Q6	Vph_A	Vph_B	Vph_C
2	1 / 0	0 / 1	0 / 1	[2/3] ·Vdc	[–1/3] ·Vdc	[–1/3] ·Vdc
3	1 / 0	1 / 0	0 / 1	[1/3] ·Vdc	[1/3] ·Vdc	[–2/3] ·Vdc
4	0 / 1	1 / 0	0 / 1	[–1/3] ·Vdc	[2/3] ·Vdc	[–1/3] ·Vdc
5	0 / 1	1 / 0	1 / 0	[–2/3] ·Vdc	[1/3] ·Vdc	[1/3] ·Vdc
6	0 / 1	0 / 1	1 / 0	[–1/3] ·Vdc	[–1/3] ·Vdc	[2/3] ·Vdc
1	1 / 0	0 / 1	1 / 0	[1/3] ·Vdc	[–2/3] ·Vdc	[1/3] ·Vdc
Z1	0 / 1	0 / 1	0 / 1	0	0	0
Z2	1 / 0	1 / 0	1 / 0	0	0	0

becomes mandatory. In this respect, the rms value of the output voltage is defined as follows.

$$V\left[rms\right]=\sqrt{\frac{1}{\omega T}\cdot v(\omega\cdot t)^2\cdot d\omega t} \tag{5.1}$$

The peculiar case of operation without modulation (with square-wave output phase voltage) leads to a rms value of

$$V\left[rms\right]=\sqrt{\frac{1}{\pi}\cdot\left[\int_{-\frac{\pi}{2}}^{-\frac{\pi}{6}}\left[\frac{1}{3}\cdot V_{dc}\right]^2 d\omega t+\int_{-\frac{\pi}{6}}^{\frac{\pi}{6}}\left[\frac{2}{3}\cdot V_{dc}\right]^2 d\omega t+\int_{\frac{\pi}{6}}^{\frac{\pi}{2}}\left[\frac{1}{3}\cdot V_{dc}\right]^2 d\omega t\right]}$$
$$=\sqrt{\frac{V_{dc}^2}{9\pi}\cdot\frac{\pi}{3}\cdot 6}=\frac{V_{dc}}{3}\cdot\sqrt{2}=0.4714\cdot V_{dc} \tag{5.2}$$

The drawback of an unmodulated inverter consists of the high harmonic content at low frequencies, with the undesired effects of energy loss increase and possible mechanical vibrations or oscillations.

Operation with *pulse width modulation* changes the definition of the rms due to the multitude of pulses in the waveform, as shown in Figure 5.14. The previously defined *modulation index* has now to be reported to the result in (5.2), which can be adjusted from the magnitude of the sinusoidal references, as suggested in Figure 5.13.

Furthermore, similar to the single-phase inverter, the harmonic spectrum is very important. A sample result for FFT is shown in Figure 5.15, and it can clearly be seen that harmonics have been moved from lower frequencies to the carrier's high frequency (1 kHz) and its multiples.

5.3.5 PWM Generators

The actual control of the gate of the MOSFET transistors, composing the single-phase or three-phase inverters, is discussed later on in the chapter, referring to the integrated circuits of the motor drives, after understanding the peculiars of using these power semiconductor devices. Herein, the top-level discussion continues with the *pulse width modulation algorithm* behind the control of the power transistors.

Since the usage of PWM became mandatory for both single-phase and three-phase inverters, all control integrated circuits for motor drives include one or more PWM generators. The principle of generating PWM for a single-phase inverter is presented in Figure 5.16, which follows on from Figure 5.13.

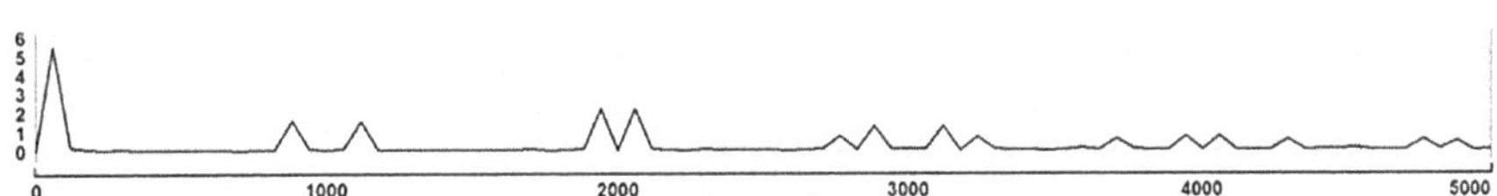

FIGURE 5.15 FFT of the output voltage for the examples in previous figures.

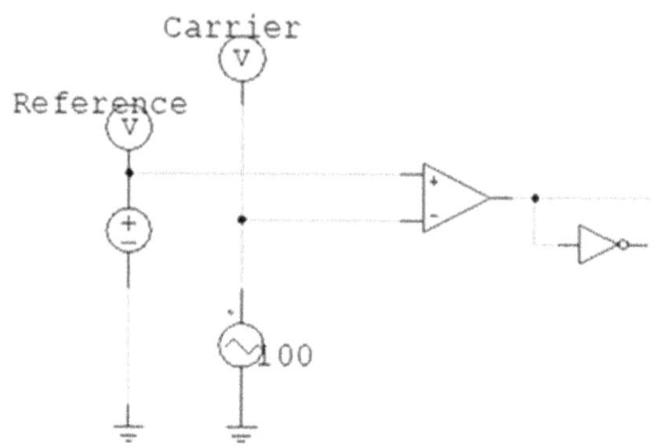

FIGURE 5.16 PWM Controller for a single-phase inverter.

A low-frequency sine-wave reference is compared with a high-frequency carrier waveform. When the sine-wave is above the triangular signal, the upper-side transistor is turned on. Conversely, when the reference is under the carrier, the low side transistor is turned on. The two transistors are complementary.

Analogously, the three-phase PWM generator has been explained based on a set of three references and the same high-frequency carrier waveform. Several details for the PWM generator are discussed next.

The triangular signal can be rising, decreasing, or symmetrical, as shown in Figure 5.17.

Mixed-mode ICs or microcontrollers generate the high-frequency carrier with an external capacitor. An internal current source charges the external capacitor and its voltage is continuously compared to a threshold voltage. When the capacitor is charged up to the level indicated by the threshold, a shorting transistor is turned on and the capacitor is suddenly discharged toward zero. The cycle repeats, creating the triangular waveform.

5.3.6 Dead-Time

The PWM generator in Figure 5.18 considers ideal switching within the power converter. Any change of state for the actual power transistors requires a finite interval of time (delay) that should be considered in the design of the control circuitry. A delay is intentionally introduced in the control of the turning-on device after the other device is ordered to turn off in order to avoid a short-circuit of the dc supply bus. This delay is called dead-time and it is designed to provide enough time for the turn-off process to finish. If the dead-time interval is too short, the short-circuit can absorb a large current and the heat produced may damage the power semiconductor. If the dead-time is too long, the pulse shapes are more compromised and the current waveform more altered.

Introducing this delay in the switching sequence modifies the width of the pulses applied to the load and their average value. Accordingly, the waveform of the load current and its harmonic spectrum are somewhat altered.

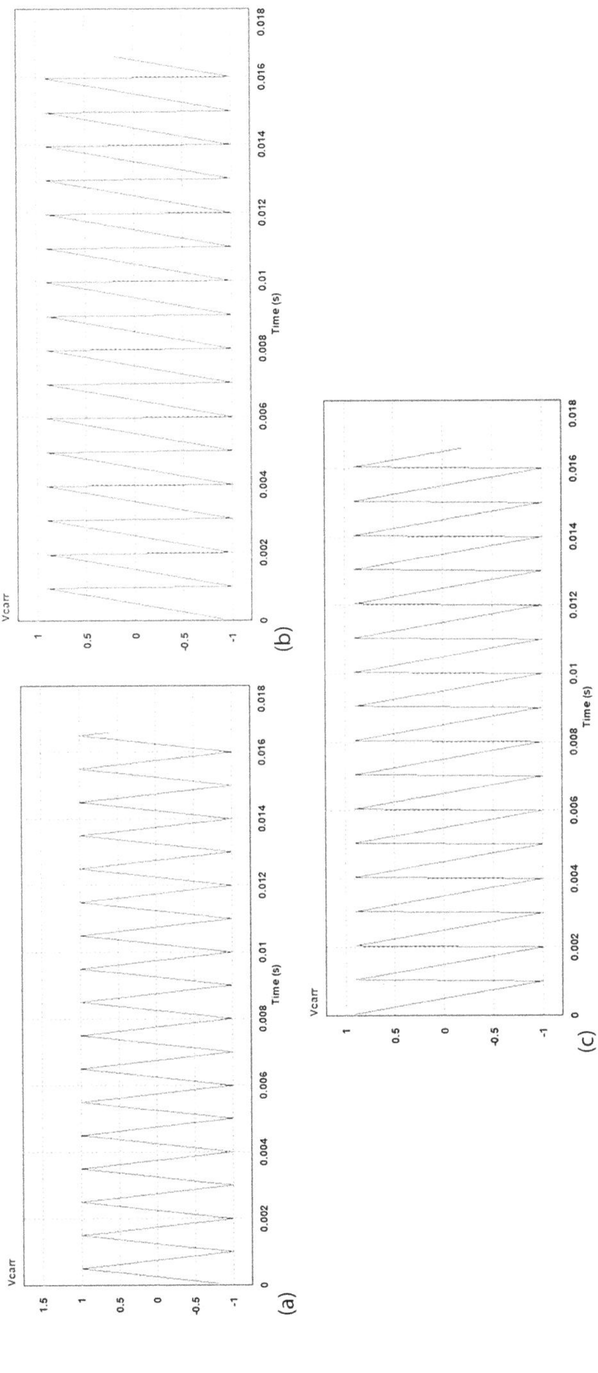

FIGURE 5.17 Various shapes for the carrier waveform.

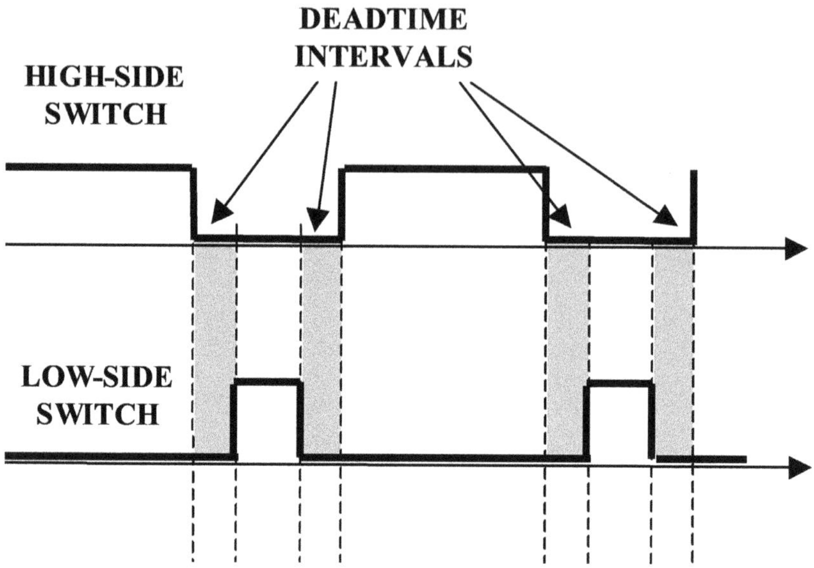

FIGURE 5.18 Dead-time use with control waveforms.

5.4 INTEGRATED CIRCUITS FOR MOTOR CONTROL

It has been shown that integrated circuits for automotive mechatronics are designed at various levels of integration. This section discusses several practical examples.

The simplest class of integrated circuits include power MOSFETs, protection circuits, and gate drivers for transistors. Power MOSFETs can be configured as a single-phase inverter. An example of this class of integrated circuits is provided by Allegro A3909, that is a *Dual Full-Bridge Motor Driver.*

More complex three-phase inverters are fully integrated into circuits, such as

- Texas Instruments DRV8313, a 2.5-A Triple half-bridge driver.
- ST Microelectronics ST6234, a complete three-phase motor driver.
- Allegro Microsystems A4931.

For the next level of integration, the PWM is included in the same integrated circuit. An example of this is offered by Allegro A4930, a single-phase fan pre-driver with external transistors or IXYS bus compatible digital PWM controller, IXDP 610.

Due to the emerging market for brushless dc machines used in automotive mechatronics, integrated solutions for complete motor drive control are required and offered. An example can be demonstrated by Allegro A4962, a *sensorless BLDC controller.* This circuit is used with external complementary P-channel and N-channel power MOSFETs. It is used as a stand-alone controller communicating directly with an electronic control unit or used in a close-coupled system with a local microcontroller (MCU). The circuit uses block commutation (trapezoidal drive) where phase commutation is determined without the need for position sensors, by monitoring the motor back-EMF of the BLDC machine. The motor can operate over speeds of less than 100 rpm up to and in excess of 30,000 rpm.

5.5 SENSORS

A series of sensors are associated with low power motor drives used in automotive body systems.

5.5.1 Thermistors

Thermistors are the most common device used for temperature measurement on a motor vehicle. A change in temperature will cause a change in resistance of the thermistor, and an electrical signal proportional to the measured resistance can be obtained. Mostly, thermistors are *negative temperature coefficient* (NTC) thermistors (Figure 5.19).

Typical values of automotive thermistors vary from several kilo-Ohms at 0°C to a few hundred Ohms at 100°C, following a typical nonlinear characteristic. This requires a linearization achievable with digital systems made of a memory-based look-up table, or analog systems that produce non-linearity with a suitable H-bridge that will partially compensate for the thermistor's non-linearity. The large change in resistance for a small change in temperature makes the thermistor an ideal sensor.

5.5.2 Hall-Effect Position Sensor

The Hall effect was first noted by Dr. E. H. Hall, hence the name. If a certain type of crystal is carrying a current in a transverse magnetic field, then a voltage will be produced at right angles to the supply current (Figure 5.20). The magnitude of the

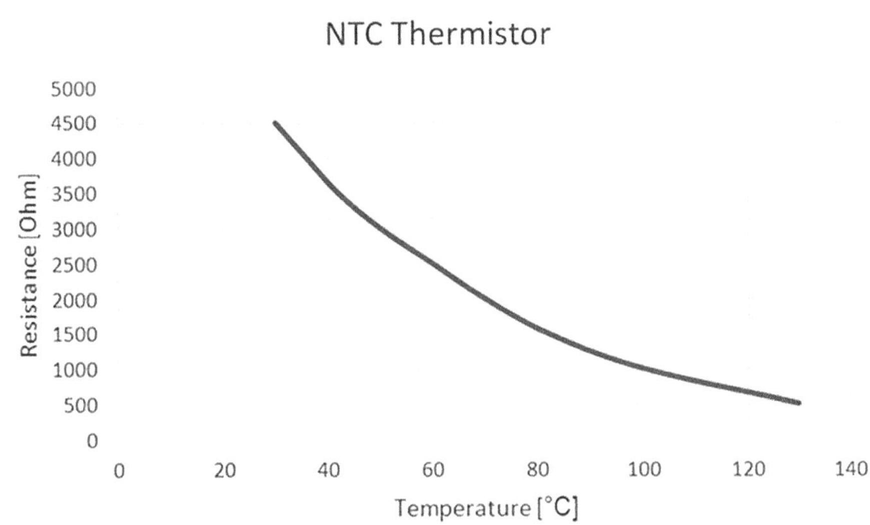

FIGURE 5.19 Example of a thermistor's resistance variation with temperature.

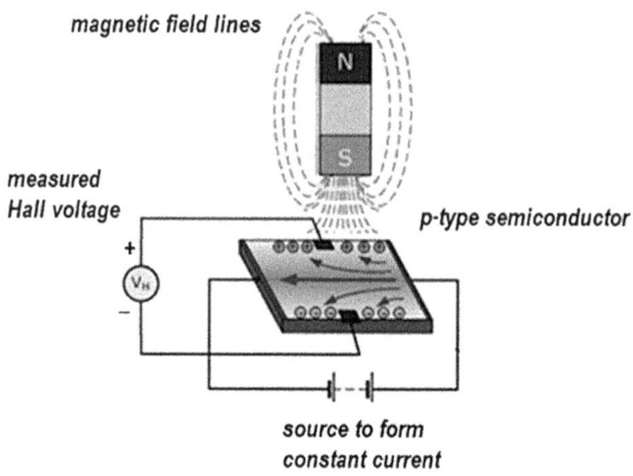

FIGURE 5.20 Hall-effect position sensor.

voltage is proportional to the supply current and to the magnetic field strength. The Hall based position sensor is typically made from semiconductors such as silicon and germanium. The Hall-effect sensors work by measuring the Hall voltage across two of their faces when you place them in a magnetic field. Some Hall sensors are packaged into convenient integrated circuit chips with control circuitry and can be plugged directly into bigger electronic circuits.

The simplest application for Hall-effect devices is to detect position. This can be within an electric window, vehicle door, soft- or hard-top roof, trunk closing, and so on. For example, a Hall sensor can be placed on a door frame and a magnet on the door, so the sensor detects whether the door is open or closed from the presence of the magnetic field, generating a sensed voltage. A device like this is also called a proximity sensor.

For a more complex application, the Hall sensor is associated with a BLDC motor drive (Figure 5.21). A BLDC motor control system has to provide a clean startup operation, maintain continuous commutation, achieve the highest possible efficiency, and extract maximum torque from the available electrical power. The key to achieving all of these goals is knowledge of the position of the rotor relative to the stator, information which enables the motor control system designer to implement a robust electrical drive management solution. Discrete Hall switch systems typically consist of three, five, or more Hall sensors fixed in position during the production of the motor (herein HA, HB, HC = see squares). An alternative solution may be to use a single device with multiple sensors within the same integrated circuit.

5.5.3 Current Sensors

The most important element in the current closed-loop control relates to current measurement. The main requirements for the sensor used for current measurement relate to their capability to work in the presence of electrical noise, temperature variation, and electromagnetic interference (EMI) radiation in the measurement system. A series of dedicated sensors have been developed to overcome these difficulties.

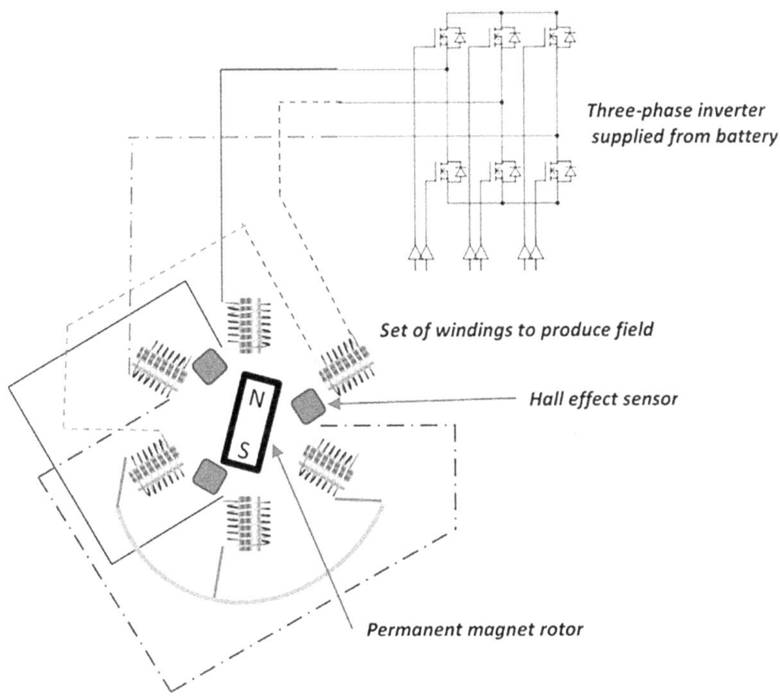

FIGURE 5.21 Brushless dc drive.

The most conventional solution for current measurement uses a low-value resistor (shunt resistor) in the current path and measures the voltage drop across it. One advantage of the shunt resistor is its practically infinite bandwidth. A shunt resistance of milliohms or microohms determines a very small voltage drop, even at full current. However, this resistance should be very stable with the current and temperature, and the sensor should introduce a small inductance. Selection and rating of the shunt resistor yields from the trade-off between the desire to have a larger voltage drop for easier signal processing and the allowable power dissipation.

The milliohms or microohms value of the shunt resistor is comparable to the resistance of the wires or connections. The measurement of voltage avoids detecting the voltage drop across the current-carrying wire connections with four connection terminals. The voltmeter measures only the voltage dropped by the shunt resistance itself, without any stray voltages, and the method is called the Kelvin or four-wire method. The current passing through the measurement circuit does not go through the power path and there is no common path for the measurement and power currents. Therefore, shunt resistors with Kelvin contacts have four connections.

Shunt resistors are usually made of a low-temperature-coefficient metal foil on an anodized aluminum substrate. For instance, *manganin wire*, an alloy of copper, manganese, and nickel, has a low-temperature coefficient within 15 ppm/°C from 0 to 80°C, and a resistivity of 44 mV/cm. Another commonly used low-temperature-coefficient material is nickel–chromium, or nichrome. This has a higher resistivity of about 110 mV/cm, which helps reduce the inductance for very low-value resistors by using shorter wires than manganin. Conversely, manganin is superior to nichrome in its temperature coefficient and long-term stability of resistance value. Another similar alloy is constantane (Eureka) with a resistivity of 49 mV/cm.

With the advent of *integrated power modules*, current sensing is carried out with integrated thin-film power resistors. More extreme contemporary designs for low voltage dc/dc converters use copper traces when building the sensing resistors, while the same principles of Kelvin connections are followed. However, as copper has a high-temperature coefficient, additional compensation may be required.

The signal from a current-sensing resistor is usually processed with an Operational Amplifier with a high common-mode rejection, as the useful signal is usually floating from the ground under a large common-mode voltage. The signal processing usually continues with a low-pass filter implemented around another operational amplifier. In lower power integrated circuits, these amplifier stages are inside the same mixed-mode integrated circuit.

Another solution for signal processing senses the current through an external shunt resistor and modulates a fixed frequency train of a pulse with the sensing information. These pulses are transferred to the low side. The output format is a discrete pulse width modulation that eliminates the need for an A/D input interface and can be directly connected to a timer circuit within any digital signal processor (DSP) or microcontroller.

Shunt resistors are less used today in high-current applications due to the inherent voltage drop. The alternative lies in the use of Hall-effect sensors. Hall-effect current sensors are available in hundreds of Amperes and provide highly accurate measurements for a large class of power electronic applications. Their bandwidth is usually around 100 kHz, enough for high-power converter applications.

An open-loop Hall-effect current sensor has a block of semiconductors as the sensing element, supplied by a constant current source, and an amplifier to raise the millivolt output to a reasonable value (Figure 5.22). A current proportional to the measured current is produced in a sensing resistor through the Hall effect. Better

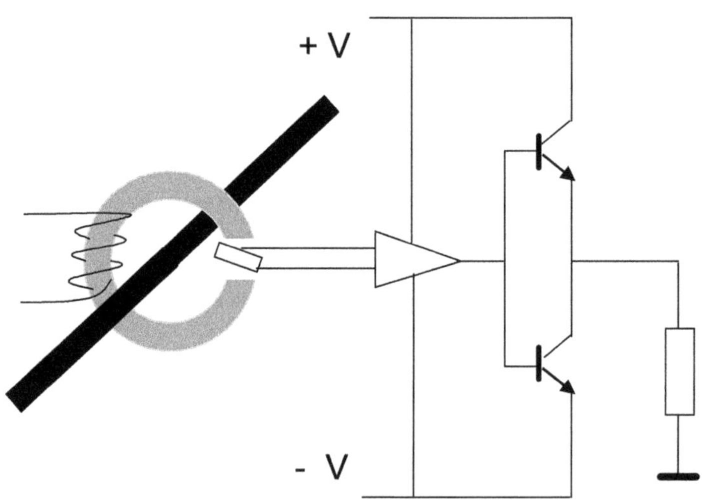

FIGURE 5.22 Open-loop Hall sensor.

performance can be achieved with closed-loop current sensors. These sensors represent a different class of Hall-effect current sensors that include an application-specific integrated circuit (ASIC) to provide an extremely low offset drift with temperature, resulting in stable, repeatable, accurate measurements.

5.5.4 Voltage Measurement

The voltage measurement is used around power converters, following a high-precision resistive voltage divider. This means the voltage measurement is referenced to the ground, and a local measurement with isolation to the main controller also needs to be employed.

5.6 CONCLUSION

Mechatronics represents an engineering field highly developed in automotive applications. It requires extensive use of electronic control and power electronic converters for the actuation of solenoids, relays, or motors.

Chapter 4 provided several examples of mechatronics applications in automotive body systems. This chapter illustrates the principles of the low power motor drives used to actuate these applications. The power electronics implications are discussed along with the requirements of electronic equipment. It is shown that both brushed dc motors and brushless dc motors are used in body systems.

The main power electronic converter used in this application class is the three-phase inverter controlling a brushless motor drive. Details about control and construction are discussed.

Several examples of its main use in automotive mechatronic applications are explained along with the possible integration of functions within integrated circuits (ICs). This is possible mainly due to the low power required by the motor. Sensors are also a constant presence in mechatronic systems and their operation is herein also briefly illustrated.

REFERENCES

Anon. 2017. *Automotive Motor Drives System Solutions*, Infineon Corporation, Document order number: B000-H0000-X-X-7600.

Anon. 2019. *The 12V Climate Cooling Solution*. Internet reading at http://www.revoltev.com/2013/08/02/the-12v-climate-cooling-solution/ Last reading December 13, 2019.Denton, T. 2017. *Automobile Electrical and Electronic Systems*. 5th edition, Abingdon-on-Thames: Routledge.

McKay, D., Nichols, G., Schreurs, B. 2000. *Delphi Electronic Throttle Control Systems for Model Year 2000; Driver Features, System Security, and OEM Benefits*, SAE Technical Paper Series, No. 2000-01-0556.

Neacşu, D. 2004. *Power Semiconductor and Control for Automotive Applications*. Tutorial Presented at IEEE APEC, Anaheim, CA, USA.

Neacşu, D. 2018. *Telecom Power Systems*, Boca Raton, MA: CRC Press.

Zulkifli, A.A., Dahlan, A.A., Zulkifli, A.H., Nasution, H., Aziz, A.A., Perang, M.R.M., Jamil, H.M., Misseri M.N. 2015. *Impact of the Electric Compressor for Automotive Air Conditioning System on Fuel Consumption and Performance Analysis*. Paper Presented at 3rd International Conference of Mechanical Engineering Research, Kuantan, Malaysia, pp. 1–6.

6 Chassis Systems

HISTORICAL MILESTONES

1867: The first power steering system on an automobile was apparently installed in 1876 by a man with the surname of Fitts.

1900: Robert E. Twyford, a resident of Pittsburgh, Pennsylvania, included a mechanical power steering mechanism in a four-wheel-drive system.

1903: The next power steering system was put on a Columbia 5-ton truck where a separate electric motor was used to assist the driver in turning the front wheels.

1927: Albert Dewandre, from Liege, Belgium, was the inventor of the servo-brake or brake booster system "Dewandre," which used a manifold vacuum as an ancillary energy source. His invention was manufactured and sold through the Robert Bosch company.

1954: Hydro-pneumatic suspension was developed by Paul Magès at Citroën.

1980: Colin Chapman developed the original concept of electronic actuation for hydraulic suspension to improve cornering in racing cars. Lotus fitted and developed a prototype system to a 1985 Excel racing car.

1988: Subaru XT6 was fitted with a unique Cybrid adaptive electro-hydraulic steering system that changed the level of assistance based on the vehicle's speed.

1990: Toyota introduced its second-generation MR2 with electro-hydraulic power steering, which was able to avoid running hydraulic lines from the engine up to the steering rack.

1992: Williams Grand Prix Engineering prepared an active suspension for F1 cars, creating such successful cars that the Fédération Internationale de l'Automobile decided to ban the technology.

1994: Volkswagen used an electric pump, allowing the power steering to operate while the engine was stopped.

1999: Mercedes Benz CL-Class (C215) introduced Active Body Control, where high-pressure hydraulic servos are controlled by electronic computing, and this feature is still available today.

2004: Michelin's Active Wheel incorporates an in-wheel electrical suspension motor that controls torque distribution, traction, turning maneuvers, pitch, roll, and suspension damping to the wheel, in addition to an in-wheel electric traction motor.

2017: Audi active electromechanical suspension system drives each wheel individually and adapts to the prevailing road conditions. Each wheel has an electric motor that is powered by the 48-V main electrical system.

2018: Audi A8, Audi SQ7 TDI, Porsche Panamera, and Bentley Bentayga, employ both 12-V and 48-V systems inter-connected by a dc/dc converter. The 48 V is used for motor drives in chassis systems.

6.1 ELECTRIFICATION OF TRANSPORTATION

6.1.1 Introduction

This chapter discusses applications of power electronics and motor drives in chassis systems, and it distinguishes three major classes of applications: *power converters used in brake systems*, *power converters used in steering systems*, and *power converters used in suspension systems*. The term "chassis" is used to designate the complete car less the body. It therefore consists of the engine, power-transmission system, and suspension system. These systems are attached to, or suspended from, a structurally independent frame. with the separation of the body and chassis as mechanical parts, various power electronics systems are associated with each. It is worthwhile to note that in rare cases both the body and chassis are welded together, forming a single unit.

If the buzzword for body systems was "mechatronics," here we are excited by "all-electric vehicle" or "more electric vehicle," which is a replacement for pneumatics/hydraulics with electrical drives. This contemporary electrification of the transportation process also takes place in marine and aviation applications.

The evolutionary technological steps for all three main applications related to chassis systems subsequently include:

- Direct driver/operator action.
- Assisting operator action with an *auxiliary energy source* derived from the engine through either manifold or belts and a hydraulic system of sorts.
- Replacing the engine-based energy source with electrical energy through an electromagnetic device, still within the hydraulic system.
- All-electric solution (or more electric vehicle), with direct action of the motor drive instead of hydraulics or mechanical systems.

Even if this is the evolution, contemporary designs can reside in any of these categories, depending on the particular complexity of the motor vehicle.

6.1.2 Ancillary Energy Sources Derived from the Engine

The above-mentioned chassis applications of motor drives require a higher energy level than each of the body systems introduced in the previous chapter. Since the need and the desire to implement these precluded the larger alternators in operation, designers had to find other energy sources inside the car, preferably derived from the operation of the internal combustion engine.

The conventional car burns gas or diesel to produce mechanical power. Various equipment systems—that need energy for their displacement—had to derive energy from the internal combustion engine's operation. The engine's energy may be directly tapped through a mechanical connection with belts or gears, with a notable example of the *air conditioning compressor.*

Similarly, an alternator offers the possibility of using a higher energy generator to supply various electrically controlled accessories. While the construction and operation of a conventional alternator are explained in Chapter 3, newer designs may increase the installed power within the generator and supply more electrical loads. This is the trend taken up with the *electrification in transportation* process, and it can be sustained with energy stored in battery banks. In the future, the high-energy batteries used in hybrid and all-electrical vehicles can shift the balance of technologies even further in the direction of electrically powered accessories.

Other sources of energy consist of the hydraulic links from a pump, mechanically driven by the engine, with the most notable example of the power steering. Several interesting expansions to body systems were also considered. For instance, the French company Citroën devised a high-pressure hydraulics system for cars, which was used for all manner of systems, even power-adjustable seats. Also, the 1999–2004 Jeep Grand Cherokee had a hydraulically driven radiator fan, powered by the SUV's (*Sport Utility Vehicle*) power steering pump. Finally, it is worthwhile to mention that most hard-top convertible cars use engine-driven hydraulics.

A distinct source of energy was briefly mentioned with the cruise control body system. The partial vacuum available at the *intake manifold* is used, with the most notable application in chassis systems being the booster for the power brake system.

Compressed air is rarely used in passenger cars and widely used in large vehicles (*remember the hissing bus door?*).

6.1.3 Road to All-Electric Vehicle Concept

The electrification process aims at replacing other sources of energy with electrical energy. The goal is justified with better energy savings, smaller and simpler systems, better controllability, including herein safety, and convenience functions impossible to imagine before.

Conventionally, only 2–8% of a vehicle's total power output has gone toward powering accessories through other energy sources. Advanced accessories require more energy that may not be possible to be driven out of the engine with current means. There is a justified need for high voltage batteries with more storage capacity. Such batteries are also charged from an alternator, and can level off load peaks so that better utilization of the power distribution system occurs.

The second trend required by electrification is the increase in the dc bus voltage. Since the load power is evidently increased with more and more loads, the current through wires and associated loss can be kept at bay with an increase of the dc bus voltage. Until a full conversion is implemented, modern automotive systems employ dual systems, with both 12 V and 48 V distribution dc buses. For instance, 2018 Audi A8, Audi SQ7 TDI, Porsche Panamera, and Bentley Bentayga, employ both 12-V and 48-V systems inter-connected with a dc/dc converter.

Due to the increase of the dc bus voltage, motors used within the chassis applications are rated at higher power levels than the body systems (previously grouped as mechatronics). Higher installed power means more possible loss, and it is justified to have advanced control through microcontrollers and PWM converters. A good control means understanding the dynamics of the overall system and applying principles of modern digital control, maybe with state-space based equations. Hence, this chapter also includes some examples of the dynamic modeling of electromechanical systems used in chassis applications.

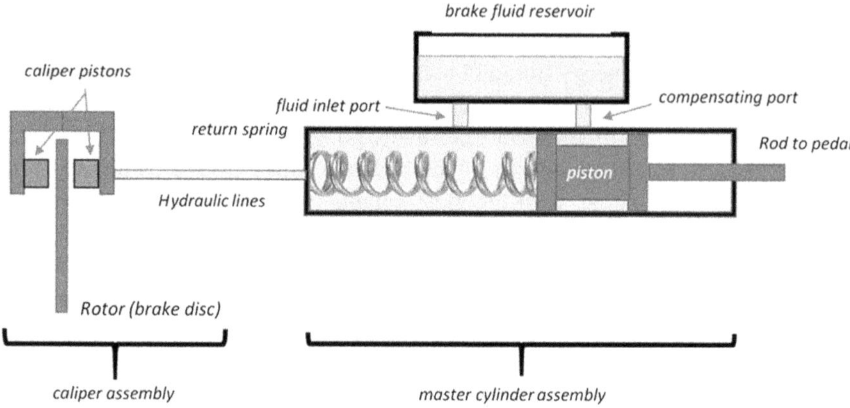

FIGURE 6.1 Drum brakes.

6.2 BRAKE SYSTEMS

6.2.1 Drum Brakes

From the beginning, motor vehicles were equipped with brakes. Traditionally, a brake is a mechanical device that inhibits motion by absorbing energy from the motor vehicle. It is used for slowing or stopping a moving vehicle, most often accomplished with friction, converting the vehicle's kinetic energy into heat energy. There are two major types of brakes, *drum brakes* and *disk brakes*. Improvements are possible with the assistance of electric power systems, and their operation is studied herein.

Old cars with *drum brakes* do not need an electric power assist. A drum brake is a brake that uses friction caused by a set of shoes or pads that press outward against a rotating cylinder-shaped part called *a brake drum*. Drum brakes are still often applied to the rear wheels since most of the stopping force is generated by the front brakes of the vehicle and therefore the heat generated in the rear is significantly less. Figure 6.1 illustrates a drum brake system.

6.2.2 Disk Brakes

A disk brake is a type of brake that uses the calipers to squeeze pairs of pads against a disk or "rotor" to create friction. Figure 6.2 illustrates a disk brake system. Modern *disk brakes* need a power assist, which is achieved with a device called a *booster.* A highly effective braking system is needed in the front wheels of a car; hence *hydraulic disc brakes* are used in the front.

The conventional solution calls for a *vacuum booster* to increase the power applied to the disk brakes following pedal action. Until recently, the majority of vehicle's brake boosters used the vacuum generated from the intake manifold of the internal combustion engine. As explained for the cruise control system, a *manifold vacuum*, or engine vacuum in an internal combustion engine, is the difference in air pressure between the engine's intake manifold and the Earth's atmosphere caused by

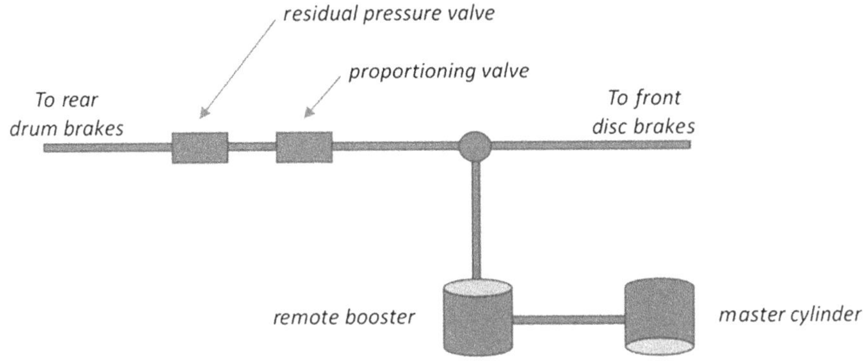

FIGURE 6.2 Disk brakes with a booster.

the piston's movement on the induction stroke and the choked flow through a throttle in the intake manifold of an engine. The manifold vacuum is used as an auxiliary power source to drive accessories, like the disk brake.

However, with the contemporary decrease in the physical size of the combustion engine, with the advanced engine technologies, or under certain operating conditions like a cold start or engine warm-up phase, the vacuum provided by the engine is no longer sufficient. Furthermore, advanced engine technologies, such as electric or hybrid vehicles, are not capable of generating any vacuum pressure at all.

The modern solution for the *brake booster* consists of an electric vacuum pump that provides an alternative or additional vacuum for these advancements in technology. Power electronics help therein.

6.2.3 Electric Vacuum Power with a MOSFET Power Converter

This is a relatively simple application, with dc supply currents under 10 A. The *electric vacuum pump* consists of an electric motor and a vane pump. There are open-loop (with examples in motor vehicles like Volkswagen Golf 1998, Bora, Audi A3) and closed-loop (with examples in motor vehicles like Volkswagen Passat 2001, Audi A4, Audi A8) solutions to actuating the electrical motor.

The electric motor is controlled "on" and "off" through a set of relays in order to keep a certain pressure range, without advanced speed or a torque control algorithm. If an advanced speed control is rarely employed, it may be a hysteresis control. Solid-state MOSFET based relays are easily available and these will be discussed in Chapter 11.

An example of an electrical motor used as a booster pump is included herein. The system considers a nominal voltage of 12 Vdc, which is considered constant during an "on" state; a rated motor speed 6,750 RPM = 707 rad/sec; a vacuum [operational] rated current of 6.0–8.0 A; a stall torque of 5,800 g-cm = 0.56 N·m; and power consumption of around 100 W.

This numerical example leads to the characteristics shown in Figure 6.3. For torque, note that 1 newton-meter is equal to 10197.162129779 g-cm or 1 g-cm is equal to 9.80665E-5 newton-meter. The "g-cm" unit is used for torque in small motors

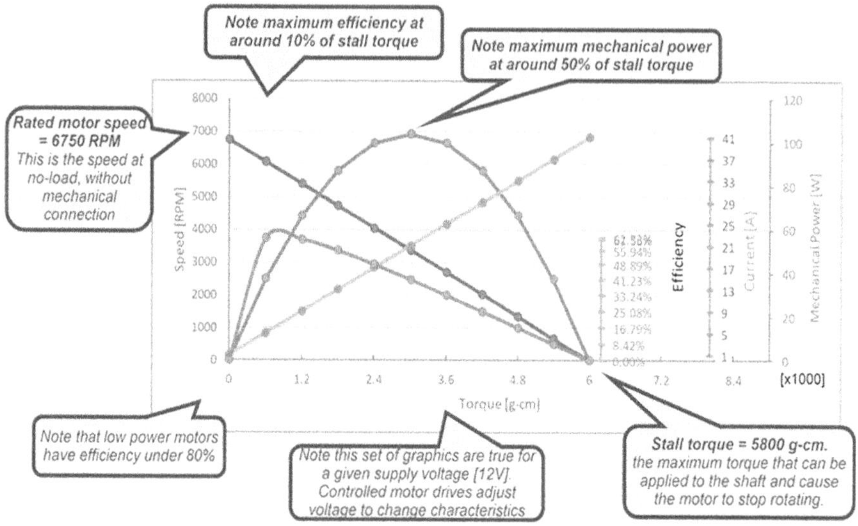

FIGURE 6.3 dc motor characteristics for the numerical data from example.

to avoid working with small numbers. The shapes of the characteristics shown in Figure 6.3 are typical for dc motors and they should be similar to any dataset.

It is worthwhile to note that most applications on 12 V battery use a permanent magnet dc (brush or brushless) motor. The main advantage of PMDC motor over conventional shunt wound dc motor is its linear speed-torque characteristics due to less interaction of the armature field (back-EMF) with the main field. More content related to small motors is provided in Chapter 11.

Equations used to derive the characteristics in Figure 6.3 follow the ideal equations for any dc motor. In dc motors, electrical power (P_{el}) is converted to mechanical power (P_{rot}). In addition to frictional loss, there are power loss (P_{loss}) components such as Joules/sec. The iron loss in low power coreless dc motors is usually negligible.

Therefore

$$P_{el} = P_{rot} + P_{loss} \tag{6.1}$$

In the case of rotational motion, mechanical power is the product of torque (T) and rotational speed (ω), defined as a rotational distance per unit time.

$$P_{rot} = T \cdot \omega \tag{6.2}$$

Attention needs to be paid to the units of measure. Speed (ω) is measured in rad/sec, and equals speed in revolutions/minute (RPM), multiplied by $(2\pi)/60$. That yields a conversion gain of 0.1047 from RPM to rad/sec. As explained, torque is often given in [g-cm] for low power motors. This means that 1 newton-meter is equal to 10197.162129779 g-cm or 1 g-cm is equal to 9.80665E-5 newton-meter.

The equation for the electrical circuit is depicted from the stator circuit. The nominal supply voltage from the power source must be equal in magnitude to the sum of the voltage drop across the resistance of the armature windings and the back-EMF generated by the motor.

$$V_o = (I \cdot R) + e \tag{6.3}$$

where the back-EMF generated by the motor (*e*) is directly proportional to the angular velocity of the motor. The proportionality constant (K_e) is the back-EMF constant of the motor.

$$e = \omega \cdot K_e \tag{6.4}$$

where:

ω = angular velocity of the motor
K_e = back-EMF constant of the motor

The *back-EMF constant* of the motor (K_e) is usually specified by the motor manufacturer in volts/RPM or mV/RPM. The torque produced by the rotor is directly proportional to the current in the armature windings. The proportionality constant is the *torque constant* (k_M) of the motor.

$$T_o = I \cdot K_m \tag{6.5}$$

where:

T_n = torque developed at rotor (output), and k_M = motor torque constant.
Substituting this relationship leads to the result as follows.

$$V_o = \frac{(T_o \cdot R)}{K_m} + (\omega \cdot K_e) \tag{6.6}$$

Assuming that a constant dc voltage is applied to the motor terminals, the motor velocity will be directly proportional to the sum of the friction torque and the load torque. The constant of proportionality is the slope of the torque-speed curve and can be calculated by:

$$\frac{\Delta n}{\Delta T} = \frac{n_0}{T_H} \tag{6.7}$$

where: T_H = stall torque (5,800 g-cm) and n_0 = no-load speed (6,750 RPM).
For instance, calculation of speed at 1,200 g-cm leads to

$$6750 - ((6750 / 5800) \cdot 1200) = 5353 \text{ RPM} \sim 5400 \text{ RPM} \tag{6.8}$$

The motor current under load is the sum of the no-load current and the current resulting from the load. The proportionality constant relating the current to the torque load is the torque constant (K_M), in this case, 6.67 kg-cm/A (practical determination or from given characteristics).

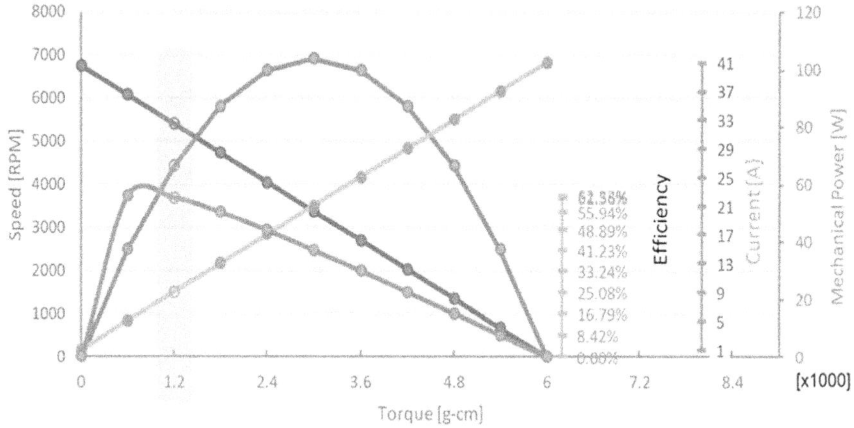

FIGURE 6.4 Particular design requirement on motor graphical characteristics shown as hatched.

In this case, the load torque is 1,200 g-cm, and the current yields

$$1200 \text{ g-cm} \times 6.67 \text{ kg-cm/A} \times 0.001 = 8 \text{ A} \tag{6.9}$$

The total motor current must be the sum of this value and the motor no-load current.

The data sheet lists the motor no-load current as 1 A. Therefore, the total current is 9 A.

Finalize the calculation example with a design requirement for the load torque in Figure 6.4. The system needs to provide for a load torque of 1.2 kg-cm = 0.1176 N·m. From characteristics, it is seen that the motor speed will drop to 5,400 rpm. The rotational mechanical power yields as follows.

$$P_{rot} = 0.1176\left[\text{Nm}\right] \cdot 5400\left[\text{RPM}\right] \cdot 0.1047 = 66.5 \text{ W} \tag{6.10}$$

Following this on the graph or, recalculating, leads to the electrical current required for this application, $I = 9.0$ A. The electrical power required, $P = 9.0$ A * 12 V = 108 W and the efficiency equals $\eta = P_{rot}/P = 66.5$ W / 108.0 W = 61.57 %.

6.3 ELECTRONIC CONTROL OF POWER STEERING

6.3.1 Application

First generations of vehicles used a direct link between the steering wheel and the front axle. As motor vehicles became heavier, it became more and more difficult to turn the car using a direct link. Various solutions have been proposed despite the complexity of the system. Today, all production cars have power steering, and the technology has been transferred to smaller vehicles like a snow blower or yard mower.

The difficulty of producing a good design resides in the different trajectories of the front wheels when the car turns because this is a complex mechanical system. Figure 6.5a illustrates the two trajectories of the wheels.

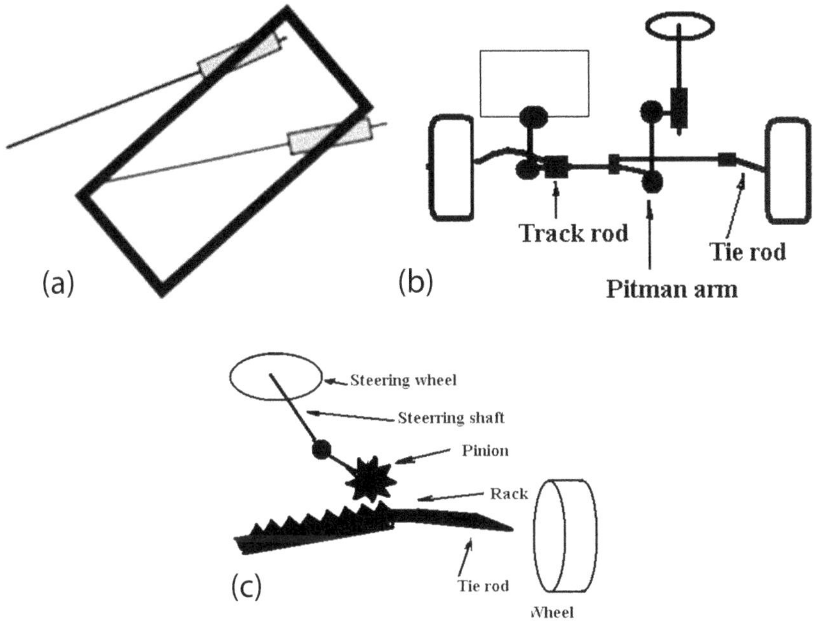

FIGURE 6.5 Power steering: (a) different trajectory for wheels; (b) Rack-and-pinion solution; (c) recirculating ball solution.

Different mechanical solutions are used in practice. The recirculating ball is illustrated in Figure 6.5b and it is used in large SUV vehicles (Figure 6.5d). Rack-and-pinion power steering is very common in small vehicles (Figure 6.5c).
With the advent of vehicle electrification, previous hydraulic systems tend to be replaced with electrical drives. Hence, an electric drive-based solution can be applied to the power steering application to augment the power applied to the steering rack. Modern solutions involve:

- Solution 1: A variable-speed electric motor is used to power the steering pump. Hence, this eliminates the hydraulics by using sensors to sense the displacement of the steering wheel.
- Solution 2: A variable-speed electric motor acts directly on the axle.
- Solution 3: Using the *steer-by-wire* concept totally eliminates the mechanical connection between the steering wheel and the steering mechanism. A sensor system detects the displacement of the steering wheel and electronically transmits the movement to the steering electrical motor drive.

A combination of conventional front-wheel steering with electronic controlled rear wheels steering is used in certain heavy four-wheel-drive vehicles.

6.3.2 Solution 1: An Electrical Motor Drives the Pump

This solution uses a different rack design, including a cylinder with a piston, which can be moved by high-pressure fluid. The hydraulic power is provided by a *pump*

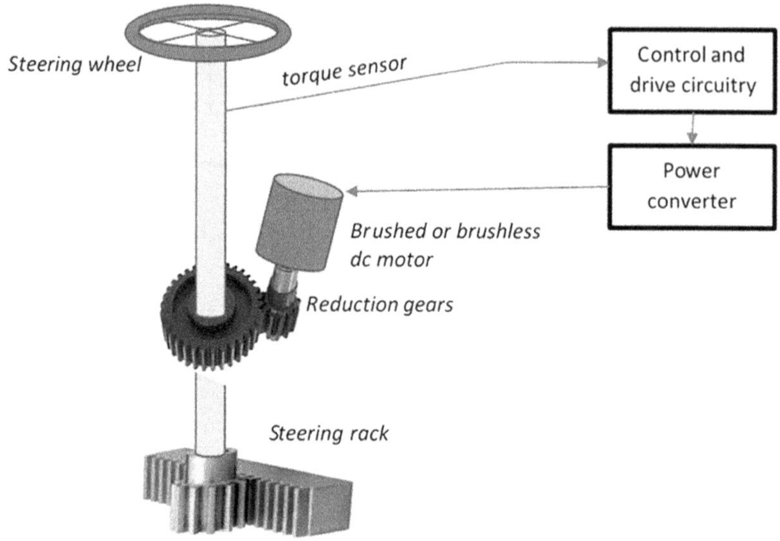

FIGURE 6.6 Power steering.

driven by a belt from the car engine. The noticeable drawback relates to the pump moving more fluid than necessary at high speeds. The second important element is a *rotary valve*, which is able to sense the force applied on the steering wheel and to transform this force into a rotation of the valve to provide high-pressure fluid in the appropriate direction. The drawback to this solution relates to pumping fluid all the time, which wastes power.

6.3.3 Solution 2: Electrically Assisted Power Steering

An electrical motor is used to supply additional torque to a moving steering rack. A system with two torque sources on the same axle occurs and this requires complex dynamic modeling. The complete system is shown in Figure 6.6.

6.3.4 Solution 3: Principle of Electronic Power Steering

This is shown in Figure 6.7. A typical application uses a bidirectional brushless dc motor (BLDC) with electronic switching. A set of sensors are located in the steering column to measure two primary driver inputs: torque (steering effort) and steering wheel speed or position. Contactless Hall-type sensors monitor the twist of the torsion bar by measuring the change in magnetic flux generated by its position to the vanes located on the sensor stator rings. Contact-type torque sensors use a wiper attached to the torsion bar and voltage divider attached to the rotating bridge attached to the motor shaft to measure the twist of the torsion bar. An electronic controller to provide steering assist based on a series of complex algorithms, also using information from the stability controller and ABS braking. A worm gear is used that can be connected to the steering column shaft or the steering rack to convert the motor action into steering rack movement (Figure 6.8).

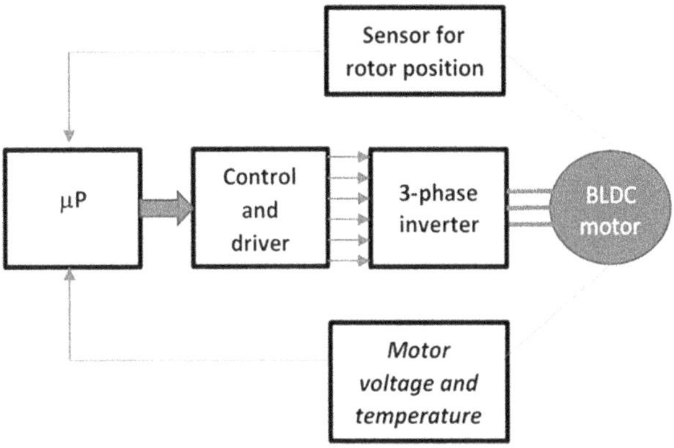

FIGURE 6.7 Electronic control of power steering.

FIGURE 6.8 Electronic power steering principle illustrated with a generic worm gear.

6.3.5 Dynamic Modeling of the Power Steering

An example of modeling the power steering system is provided in Figure 6.9, with the note that this is not the only mechanical solution for the problem. This model is aimed at modeling the cumulative effect of the two torque sources. The main torque is provided by the steering column and it is herein characterized with torque T_l. An auxiliary torque T_2 is applied through the spring K_2 and a gear to the main axle.

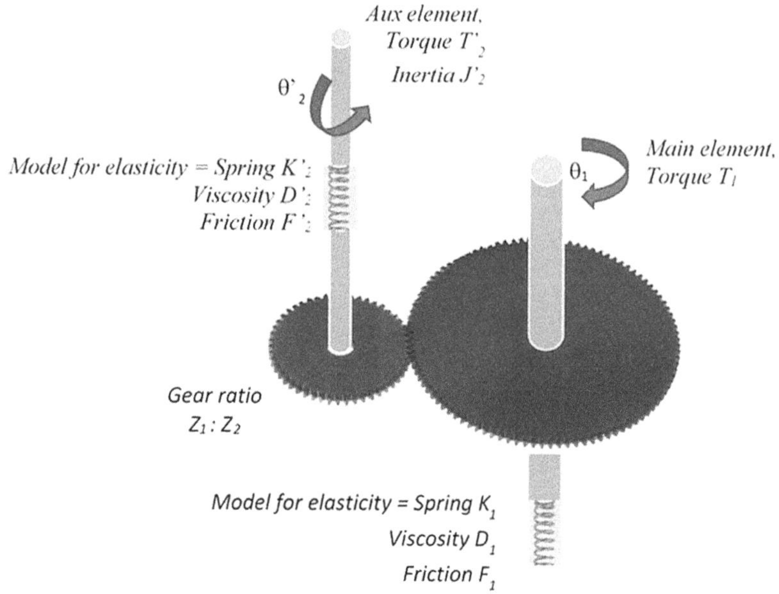

FIGURE 6.9 Modeling the power steering assembly.

The inertia moment of the auxiliary system is denoted with J_2 while the model also includes viscosity D_2 and friction coefficient F_2.

The gear $g = Z_1/Z_2$ produces a speed reduction from an electrical dc motor dependent on the needs of the steering rack. This also increases the moment of inertia from the motor's moment of inertia multiplied with the squared gear ratio (may be in the range of 10). Such a large moment of inertia worsens the dynamic response as well as creates the possibility of low-frequency resonance. Therefore, a mechanism to disconnect the motor linkage when not in use is necessary. A clutch may do this task. This is a general problem when using an electric motor.

After the gears, the notations change to T_2, D_2, J_2, K_2, through gear equations.

$$\begin{aligned} &T_2 = g \times T_2' \quad J_2 = g^2 \times J_2' \quad g = \frac{Z_1}{Z_2} \\ &D_2 = g \times D_2' \quad K_2 = g \times K_2' \end{aligned} \tag{6.11}$$

Both sources of torque act on J_1. The auxiliary torque acts through the spring K_2, contributing with a component defined as follows.

$$\Delta T = K_2 \cdot (\theta_1 - \theta_2) - T_{j2} - T_{f2} \tag{6.12}$$

where T_{j2} represents the torque required to accelerate J_2 and T_{f2} represents the torque required to overcome friction. Finally, an equivalent diagram can be drawn for this system with two sources of torque, as it is shown in Figure 6.10.

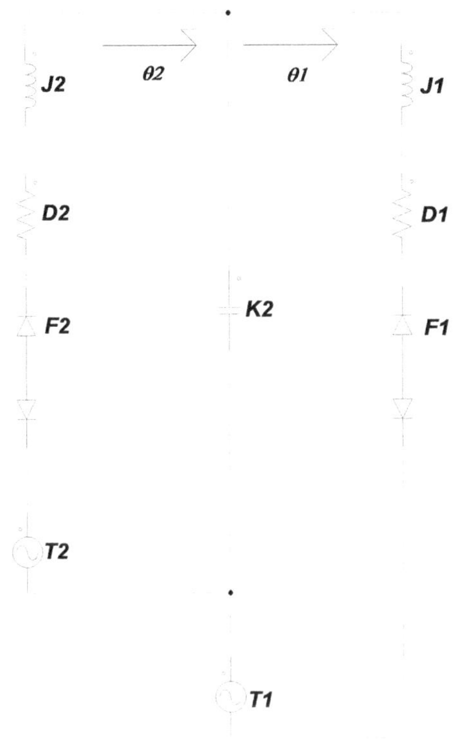

FIGURE 6.10 Equivalent diagram of the two torque sources.

6.3.6 Design With a BLDC Motor

A torque of 2 … 8 N·m, at a rated speed of around 1,200 RPM is required. This is before the worm gear is used for speed reduction. That is a power requirement of 400 … 800 W or 30 … 70 A at 12 V. The motor can be used with an inverter switched at 20–25 kHz to limit ripple, but only from motor's leakage inductance. A reduced size is further achieved with the integration of electronics and a motor. An example of this size and type of motor is shown in Figure 6.11. This is a brushless dc motor, for 500 W, and a design using an external rotating set of magnets in a so-called Halbach array. Other design choices are obviously possible.

Electromagnetic (brush dc or brushless dc motors) work at relatively high speeds, thousands of RPM, and a considerable speed reduction should be achieved with a gear set. The gear is used to create the low speed and high torque necessary to the steering rack. The usual reduction ratio is around 100 for this application. This is the major drawback of this technology since it also increases the effective moment of inertia to the steering wheel. The moment of inertia is equal to the motor moment of inertia times the square of the reduction ratio (in a numerical example: 10,000!). This worsens control performance, with a possible low-frequency resonance. It implies an additional requirement for a mechanism able to disconnect the electrical motor's mechanical linkage when not in use. This function is achieved with a clutch.

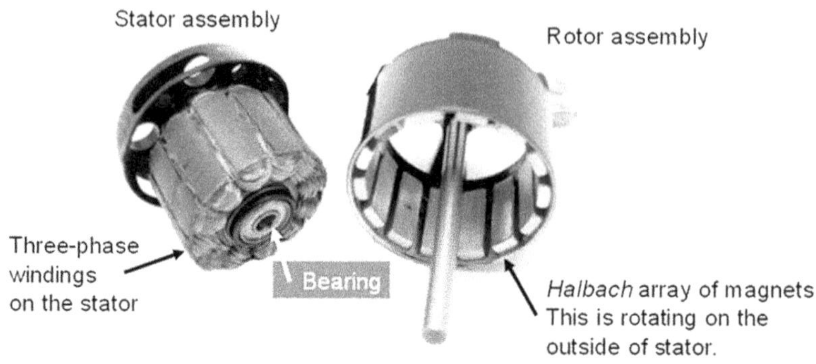

FIGURE 6.11 Example of brushless dc motor at around 500 W, usable in power steering applications.

This limitation related to the speed difference between an electric motor and a driven axle is inherent to any electromagnetic motor (dc or otherwise) and cannot be solved with improvements in motor design or structure.

An alternate solution is possible with an ultrasonic motor since it is able to achieve low speed coupled to high torque without a speed reduction gearbox. Hence the equivalent inertia is small. The ultrasonic motor is also reviewed in Chapter 11, which is dedicated to small automotive motors.

6.3.7 Progressive Steering Wheel

While the previous sections describe the electrification process—which in essence means to do the same things with electricity—this section takes a quick look into a technological advance that benefits from an already existing electrified system.

A conventional steering system uses a constant steering ratio, usually allowing a conversion within the range of –479° to 479° (for example, a 2020 Volkswagen vehicle, or similar Mercedes Benz, Audi, Ford, and so on). In most passenger cars, the constant steering ratio is between 12:1 and 20:1. For example, a complete turn of the steering wheel with 480 degrees causes the wheels to turn 30 degrees; that is a ratio of 480:30 = 16. The new technology proposes a progressive steering ratio which is able to compress the range from –378° to 378°. Furthermore, this interval is achieved with a progressive ratio, which conserves somewhat small turns and compresses wide turns. The advantage is obvious when maneuvering the vehicle for parking with lots of turns in small spaces and at low speed. Another advantage may be noticed only on a sporty driving on twisting roads, when the vehicle dynamics yields improved since it benefits from the effect of a nonlinear controller/actuator (not unlike the theory of asymmetrical membership functions for fuzzy control systems).

The progressive steering feature is usually achieved mechanically, with a variable tooth spacing on the rack-and-pinion gear and a more powerful electric motor than the standard electromechanical steering system. The electric motor should be able to tackle the variable gear ratio and to provide enough torque at the most disadvantageous ratio. Figure 6.3 of Chapter 4 illustrates the dependency of output torque on the gear ratio and input torque, justifying the need for more power at a different gear ratio.

Other variants of the progressive steering technology may use other mechanical arrangements for progressive or logarithmic gear ratios, like the variable slope of the teeth.

6.4 AUTOMOTIVE SUSPENSION

Suspension is the system of tires, tire air, springs, shock absorbers, or struts that connects a motor vehicle to its wheels and allows a relative motion in between. Suspension systems must support ride quality, which comes in contradiction with road handling.

Relatively smaller vehicles use coil springs while large trucks may use leaf springs, not unlike the rail car suspension systems (see Figure 6.16 later on). Figure 6.12 illustrates two major types of coil suspension systems: *MacPherson* and *double wishbone*. Along with the actual suspension springs, Figure 6.12 also illustrates the other joints and connections. The main advantage of the *double wishbone* system is that it maintains a flat tire contact patch with the road while turning or navigating a bump. The main disadvantage of the *double wishbone* system is that it is more complex and wears out much faster. The double wishbone system has the lower ball joint always loaded with the weight of the vehicle, whereas in a *MacPherson suspension* the lower ball joint is merely a follower. Our dynamic modeling and analysis disregard the effect of the joints and the force applied to the lower ball joint.

A third possible suspension design is called *multi-link suspension* and is not detailed in this book. The multi-link suspension uses three or more lateral arms, and one or more longitudinal arms. This makes the suspension system costly, complex, and it is difficult to tune the geometry without a complete tri-dimensional computer-aided design analysis. Compliance under load can have an important effect and must be checked using simulation software. Conversely, the multi-link suspension is recommended for certain off-road vehicles since it allows the vehicle to flex more; this

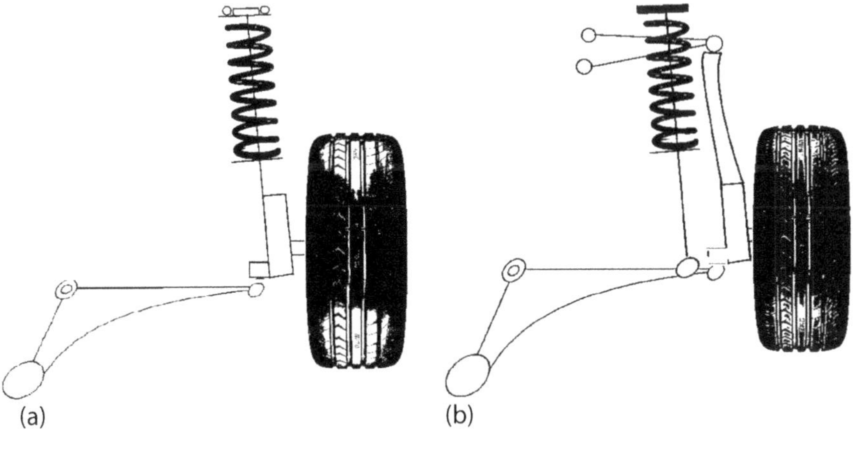

(a) (b)

FIGURE 6.12 Illustration of the spring-based suspension: (a) MacPherson suspension principle; (b) double wishbone suspension principle.

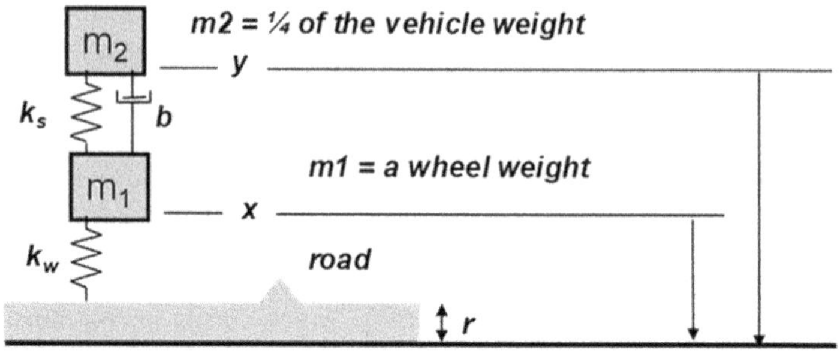

FIGURE 6.13 Dynamic model for the automotive suspension with generic springs.

means simply that the suspension is able to move more easily to conform to the varying angles of off-road driving.

While most of the traditional suspension systems are mechanical, modern solutions augment the effect of mechanical systems with an electrically assisted one. Before explaining the usage of electrical drives for automotive suspension, the design of the control system is advanced by a frequency analysis of the plant. In this respect, a numerical example is considered along with the dynamic modeling of the suspension system equipped with just springs.

The weight is considered equally distributed among the four wheels (Figure 6.13). The spring constant for the suspension system is $k_s = 130{,}000$ N/m. The spring constant for each wheel/tire due to the deformation under weigh, $k_w = 1{,}000{,}000$ N/m. Additionally, a shock damping device is considered, and modeled similar to the friction. This is denoted with b. The force applied to the shock absorber is directly proportional to the difference between the displacement of the two weights (suspension spring and wheel).

The key modeling equation comes from physics and it is called Hooke's Law for springs.

$$F = Kx + bv \tag{6.13}$$

where F = force, K = spring constant, x = displacement, v = speed, and b = friction coefficient.

It yields:

$$\begin{cases} m_1 \cdot \ddot{x} = b \cdot (\dot{y} - \dot{x}) + k_s \cdot (y - x) - k_w \cdot (x - r) \\ m_2 \cdot \ddot{y} = -k_s \cdot (y - x) - b \cdot (\dot{y} - \dot{x}) \end{cases}$$
$$\Rightarrow \begin{cases} \ddot{x} + \dfrac{b}{m_1} \cdot (\dot{x} - \dot{y}) + \dfrac{k_s}{m_1} \cdot (x - y) = \dfrac{k_w}{m_1} \cdot (x - r) \\ \ddot{y} + \dfrac{b}{m_1} \cdot (\dot{y} - \dot{x}) + \dfrac{k_s}{m_1} \cdot (y - x) = 0 \end{cases} \tag{6.14}$$

The differential equations are next transformed in the Laplace equivalent. This helps to define the Laplace transform that is later on used to design the feedback control system for the electrically assisted suspension.

$$\begin{cases} \ddot{x}+\dfrac{b}{m_1}\cdot(\dot{x}-\dot{y})+\dfrac{k_s}{m_1}\cdot(x-y)=\dfrac{k_w}{m_1}\cdot(x-r) \\ \ddot{y}+\dfrac{b}{m_1}\cdot(\dot{y}-\dot{x})+\dfrac{k_s}{m_1}\cdot(y-x)=0 \end{cases}$$

$$\Rightarrow \begin{cases} s^2\cdot X(s)+s\cdot\dfrac{b}{m_1}\cdot\big(X(s)-Y(s)\big)+\dfrac{k_s}{m_1}\cdot\big(X(s)-Y(s)\big)=\dfrac{k_w}{m_1}\cdot\big(X(s)-R(s)\big) \\ s^2\cdot Y(s)+s\cdot\dfrac{b}{m_1}\cdot\big(Y(s)-X(s)\big)+\dfrac{k_s}{m_1}\cdot\big(Y(s)-X(s)\big)=0 \end{cases}$$

$$\Rightarrow \frac{Y(s)}{R(s)}=\frac{\dfrac{k_w\cdot b}{m_1\cdot m_2}\cdot\left(s+\dfrac{k_s}{b}\right)}{s^4+\left(\dfrac{b}{m_1}+\dfrac{b}{m_2}\right)\cdot s^3+\left(\dfrac{k_s}{m_1}+\dfrac{k_s}{m_2}+\dfrac{k_w}{m_1}\right)\cdot s^2+\left(\dfrac{k_w\cdot b}{m_1\cdot m_2}\right)\cdot s+\left(\dfrac{k_w\cdot k_s}{m_1\cdot m_2}\right)} \tag{6.15}$$

This relationship represents the vehicle/passenger displacement on a perpendicular axis to the road when a road hazard is met. In order to reduce the oscillation after displacement, the "*b*" constant should be improved with an electrically assisted shock absorber.

Conventional suspension systems are always a compromise between soft springs for comfort and harder springs for better cornering ability. In order to avoid a rough ride when cornering there is a need to continuously change the constant with the road conditions. A large constant would help with large bumps in the road, while a small constant would help during cornering on a flat surface.

A MATLAB™ analysis of the dynamic system uses the previous Laplace transfer function. The model is herein finalized with numerical data: wheel weight $m_1 = 20$ [kg]; motor vehicle weight without wheels, $m^2 = (1{,}580–80)/4$ [kg]; spring constant for suspension, $ks = 130{,}000$ N/m; spring constant for each wheel, $kw = 1{,}000{,}000$ N/m; generic shock absorber, $b = 9{,}800$.

A step load is applied to the dynamic model, without considering any electrically assisted suspension therein. The effect of doubling "*b*" is illustrated in Figure 6.14, as the effect of the shock absorber for the suspension system under the generic step response at a step road hazard of 10 cm. This can be the edge of the sidewalk beside the road. Note a full response in less than 0.4 seconds for the original system and two times faster for the improvement. Both cases feature an overshoot of around 13 %.

Technical solutions for improvement are called *active suspension*, or *damper* (Figure 6.15). Interestingly enough, despite being invented in Formula 1 and Grand Prix racing, the same forum has banned them since the 1993–1994 season. The

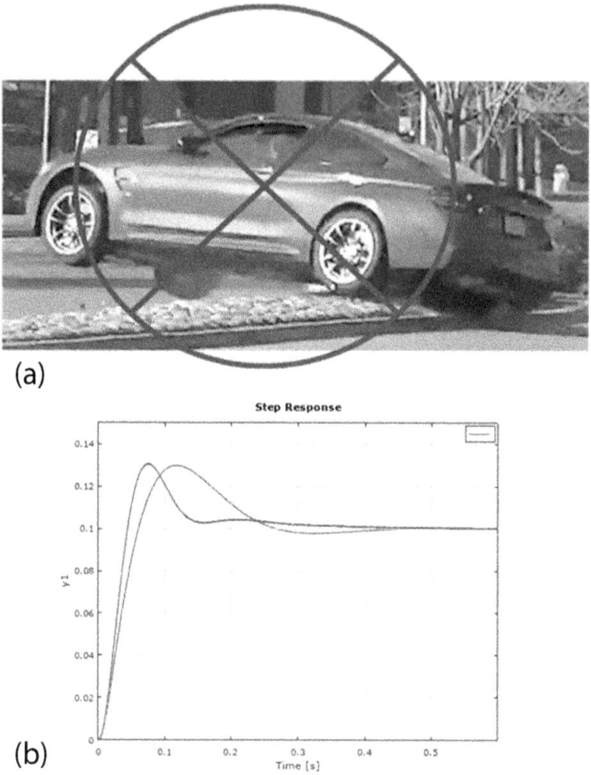

FIGURE 6.14 Step response for the suspension system: (a) danger of an unmitigated step, (b) ideal step response by MATLAB™ analysis.

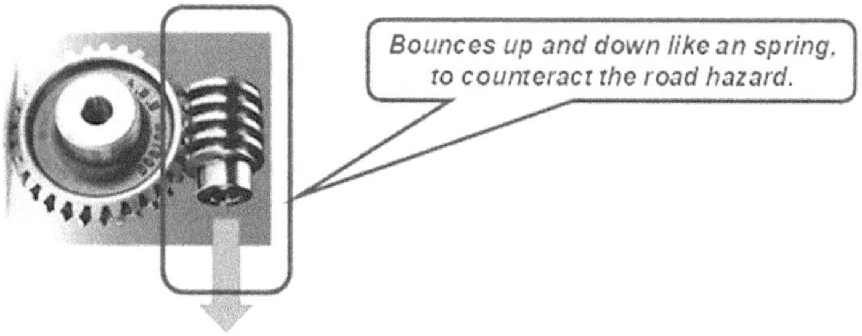

FIGURE 6.15 A possible principle solution of gears' displacement within electrically assisted suspension.

most-used solution consists of double-acting hydraulic units controlled by a microprocessor which receives signals from various sensors. A servo-valve controls the oil pressure in excess of 150 bar into the hydraulic system.

The all-electric-vehicle concept encourages the replacement of this hydraulic system with an electromagnetic system. A contemporary system was developed for the 2018 Audi A8. This all-new active suspension is a fully active electromechanical suspension system, which drives each wheel individually and adapts to the prevailing road conditions. In this respect, each wheel has an electric motor that is powered by the 48 Vdc main electrical system, available in a dual 12/48 Vdc bus voltage system. Additional components include gears, a rotary tube together with an internal titanium torsion bar and a lever which exerts up to 1,100 N·m (811.3 lb-ft) on the suspension via a coupling rod.

The electrical motor produces a large torque of around 1,000 N·m to a gear that converts rotation into linear displacement on the vertical direction of the suspension. A fast control algorithm is required to respond to road excitation (Figure 6.14). The design considers a maximum mechanical frequency of 30 Hz, which is also demonstrated in the previous transfer function.

Other hydraulic solutions use a solenoid/valve actuator. This is the most economic and basic type of semi-active suspensions. It consists of a solenoid valve which alters the flow of the hydraulic medium inside the shock absorber, therefore changing the damping characteristics of the suspension setup. Chapter 11 explains the details of the construction and operation of a solenoid valve.

Additionally, the *spool-type solenoid valve* is featured within the Dynamic Suspensions Spool Valve (DSSV). This is a chassis suspension technology that delivers high levels of suspension damping performance. It is used on various high-performance General Motors vehicles, including the Chevrolet Camaro Z/28, Camaro ZL1 1LE, and Colorado ZR2 (Figure 6.16).

FIGURE 6.16 DSSV suspension along a leaf spring (instead of coil springs).

Since the frequency model in Equation (6.13) is directly used for the design of the electronic control system, a new technology has been proposed by Volkswagen and is called *Adaptive Chassis Control* (also known as *Dynamic Chassis Control*). The spring constant corresponding to an active suspension component in the system is changeable by the driver, adjusting the actual level of firmness. Adjustments can be made in real-time in order to suit the road conditions, acceleration, steering, or braking. While most cars go for three levels of firmness, the 2020 Volkswagen Arteon allows adjustment in 15 degrees of firmness.

6.5 CONCLUSION

The chapter reviewed several chassis systems that can benefit from electric drives. Power converters are used in brake systems, in steering systems, and in suspension systems. Some examples were presented to provide the reader with basic knowledge before designing or using electrical drive systems.

Since the actuation of the mechanical system defines the size and control of the electrical drive, constructive details help with dynamic modeling. Conversely, dynamic modeling allows for a better knowledge of power converter stress during the dynamic regime.

REFERENCES

Anon. 2001. *Electric Vacuum Pump for Brake Servo Unit.* Published as Volkswagen Self-Study Programme 257, pp.1–20.

Anon. 2018. *Introducing the Breakthrough SMVP 0500 Electric Vacuum Pump.* Marketing documentation from LPR Global Inc., Canada. Internet reading at www.uskoreahotlink.com/wp-content/uploads/Electric-Vacuum-Pump.pdf.

Badawy, A., Zuraski, J., Bolourchi, K., Chandy, A. 1999. *Modeling and Analysis of an Electric Power Steering System.* Published in SAE Technical Paper series, no.1999-01-0399.

Denton, T. 2017. *Automobile Electrical and Electronic Systems.* 5th edition, Abingdon-on-Thames: Routledge.

Neacşu, D. 2004. *Power Semiconductor and Control for Automotive Applications.* Tutorial Presented at IEEE APEC, Anaheim, CA, USA.

Rajashekara, K. 2013. Present status and future trends in electric vehicle propulsion technologies. *IEEE Journal of Emerging and Selected Topics in Power Electronics*, 1(1):3–10.

Rajashekara, K. 2014. Power conversion technologies for automotive and aircraft systems. *IEEE Electrification Magazine*, June 2014, pp. 50–60.

Zhang, H., Zhang, Y., Liu, J., Ren, J., Gao, Y. 2009. *Modeling and Characteristic Curves of Electric Power Steering System.* Presented at IEEE International Conference on Power Electronics and Drive Systems (PEDS), Taipei, Taiwan, pp. 1390–1393.

7 Lighting

HISTORICAL MILESTONES

1878: Joseph Swan in the UK demonstrated the first light bulb.

1892: Thomas Wilson discovered an economically efficient process for creating calcium carbide in an electric arc furnace. This also facilitated the emergence of acetylene lights.

1900s: Acetylene lights were commonly used in automobiles.

1913: Irving Langmuir and Lewi Tonks discovered that the lifetime of a tungsten filament could be greatly lengthened by filling the bulb with an inert gas, such as argon.

1940s: Vehicle alternators were perfected and first used by the military in WWII, to power radio equipment on specialist vehicles.

1950s: Usage of alternators on production cars allowed the use of the conventional light bulb.

1962: Nick Holonyak at GE developed the first visible [red] light LED, which was followed by George Craford with the yellow light LED.

7.1 AUTOMOTIVE LIGHT SOURCES

Motor vehicle lights must allow the driver to both see and be seen in the dark. While the sidelights, tail lights, brake lights, and so on are relatively straightforward, the headlights present the most challenging problems. On a dipped beam, the headlights must provide adequate light for the driver but without dazzling other road users. A "*Dip light*" or "*dip beam*" is projected downwards so that the incoming traffic is not distracted by the glare.

Many techniques have been tried over the years to solve this problem and great advances have been made, but the conflict between seeing and dazzling is very difficult to overcome.

The most important parameter for motor vehicle lighting is light intensity, which is measured in units of *candela* or *lumens*. The "candela" (cd) represents the base unit of luminous intensity in the international system of units (SI), which is the luminous power per unit solid angle emitted by a pointed light source in a particular direction. For instance, a common wax candle emits light with a luminous intensity of roughly one candela.

The conversion factor between lumens and candelas is 12.57, or 4π. One candela has a light intensity of 12.57 lumens (lm). For instance, a spotlight with a light output of 600 lumens has a light intensity of 48 cd.

Years of experience in automotive applications have provided the light intensity requirements for the function/role of each light source. Common expectancies are:

- Sidelights at up to 60 cd
- Rear lights at up to 60 cd
- Brake lights at up to 40–100 cd
- Reversing lights at up to 300–600 cd
- Day running lights at up to 800 cd

There are no clear standard requirements for fog lights and headlights, as these usually vary depending on the size of the vehicle and location of the light source.

"*Color temperature*" is a way to describe the light appearance provided by a light source like a light bulb or lamp. This parameter has been disregarded for a long time in lighting applications, while it was always a factor in photography activities. More recently, the emergence of multiple light sources with electronic control allowed for a better understanding of the opportunity to adjust the color temperature of the light source.

The color temperature is measured in *temperature degrees of Kelvin* (symbol: K) on a scale from 1,000 K to 10,000 K. For commercial and residential lighting applications, Kelvin temperatures fall somewhere on a scale from 2,000 K to 6,500 K. For a color temperature under 3,000 K, colors are called "*warm*" and look yellowish. This is simply known as "*warm light*." A light source with a color temperature of around 3,000 to 3,500 K appears less yellow and more whiteish. Above 5,000 K, the light produced appears bluish-white (sometimes improperly called xenon). This is simply known as "*cool light*." Finally, it is noteworthy that the color temperature of daylight varies, but is often in the 5,000 K to 7,000 K range.

As the light intensity level and the color temperature have been standardized based on application, a series of specialty bulbs have been defined for automotive applications. These are shown in Figure 7.1 and are briefly reviewed herein.

- *Festoon* = The glass envelope has a tubular shape, with the filament stretched between brass caps cemented to the tube ends.
- *Miniature center contact* (MCC) = This has a bayonet cap consisting of two locating pins projecting from either side of the cylindrical cap.

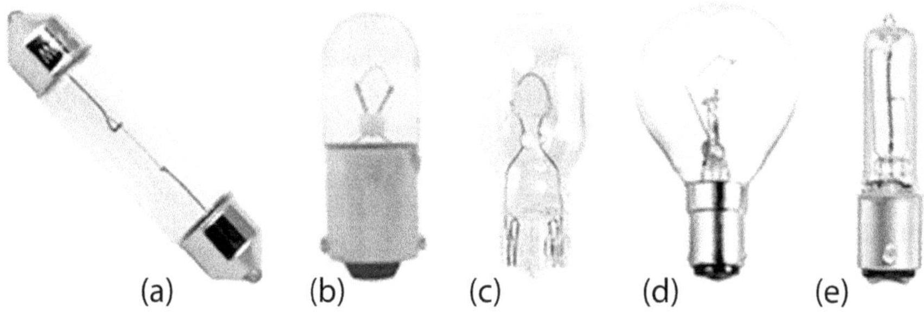

FIGURE 7.1 Specialty light bulbs for automotive applications: (a) festoon; (b) miniature center contact; (c) capless bulb; (d) single contact, small bayonet cap (SBC); (e) double contact.

- *Capless bulb* = This has a semi-tubular glass envelope with a flattened end, which provides the support for the terminal wires, which are bent over to form the two contacts.
- *Single contact, small bayonet cap* (SBC) = This has a bayonet cap with a diameter of about 15 mm with a spherical glass envelope enclosing a single filament.

Because the headlight should be very powerful and directed to the ground or away, headlight assembly is composed of reflectors and lenses. Light from a source can be projected in the form of a beam of varying patterns by using a suitable reflector and a lens system.

A reflector is basically a layer of silver, chrome, or aluminum deposited on a smooth and polished surface. Reflectors are used behind the light source, forcing a return of the spotlight to the front. The role of the reflector is to direct the light rays produced by the bulb into a beam of concentrated light by applying the laws of reflection. This means that the bulb filament position relative to the reflector is important. Reflectors used for headlights are as shown in Figure 7.2.

- Parabolic
- Bifocal (has two reflector sections with different focal points)
- Homifocal (has multiple sections, each with a common focal point)
- Other rare shapes (like poly-ellipsoidal headlight system)

Lenses are also a headlight assembly component. Beam formation can be considerably improved by passing the reflected light rays through a transparent block of

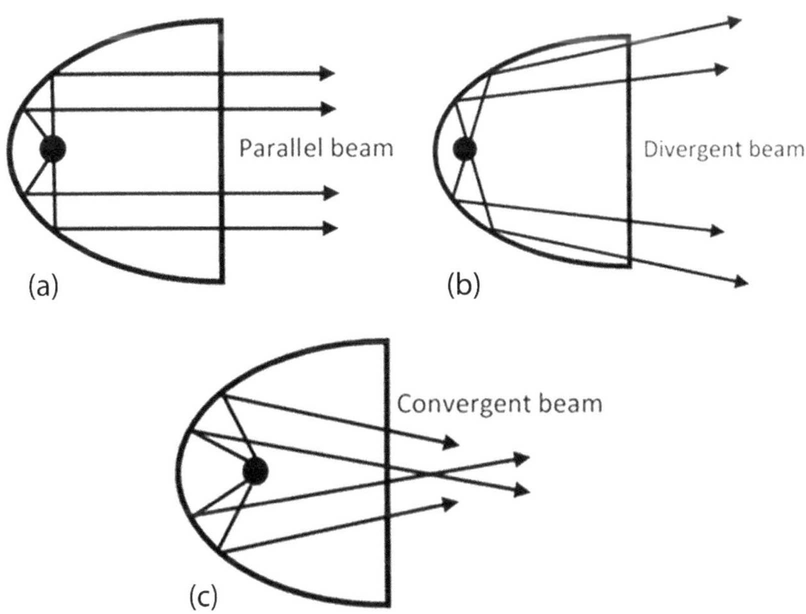

FIGURE 7.2 Headlight reflectors: (a) parallel beam, (b) divergent beam; (c) convergent beam.

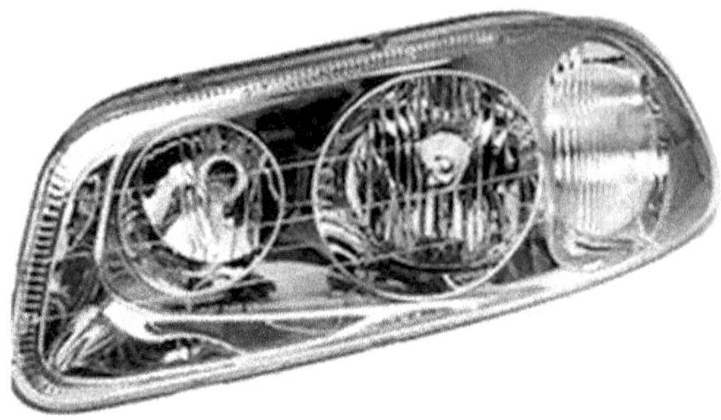

FIGURE 7.3 Lenses within a headlight assembly.

lenses. This is placed in front of the light source. Lenses work on the principle of refraction, which means the change in the direction of light rays when passing into or out of a transparent medium. Lenses can be made of glass or plastic. Figure 7.3 illustrates a sample headlight and locates lenses as apparently small cuts. The headlight's front cover and glass lens are divided up into a large number of small rectangular zones, each zone being formed optically in the shape of a concave flute or a combination of flute and prisms (Figure 7.4). The prisms sharply bend the rays downwards to give diffused local lighting just in front of the vehicle. The flutes control the horizontal spread of light.

7.2 CONVENTIONAL LIGHTING CIRCUITS

Various lighting technologies have been used by the automotive industry. Their historical evolution is sketched in Figure 7.5. It is noteworthy that transition from one technology to the next is not very sharp, and these technical solutions are in production at the same time. This chapter reviews these technical solutions.

Conventional lighting circuits do not have electronic control of the light source. Various operational states are inter-connected with relays and switches. Figure 7.6 provides an example. Switches allow various conditions between light bulbs, so that they are or not all turned on simultaneously.

It is also worthwhile observing the important standardization of connecting wires, terminals, and meaning/role, as suggested in Table 7.1. Hence, the actual lights are often specified for an application.

Figure 7.7 shows a precursor of acetylene lights, which were used a lot in automotive applications in the early 1900s. For instance, early models of the Ford Model T automobile used carbide lamps as headlamps. Acetylene lamps are also called carbide lamps. Their technology follows the discovery by Thomas Wilson in 1892 of an economically efficient process for creating calcium carbide in an electric arc furnace. Acetylene lamps are simple lamps that produce and burn acetylene (C_2H_2), which is created by the reaction of calcium carbide (CaC_2) with water (H_2O).

It took some time for the automotive world to switch to the conventional bulb due to the lack of enough electrical energy available onboard vehicles. The introduction of the electric alternator encouraged the usage of the conventional light bulb, which became a constant presence after the 1950s.

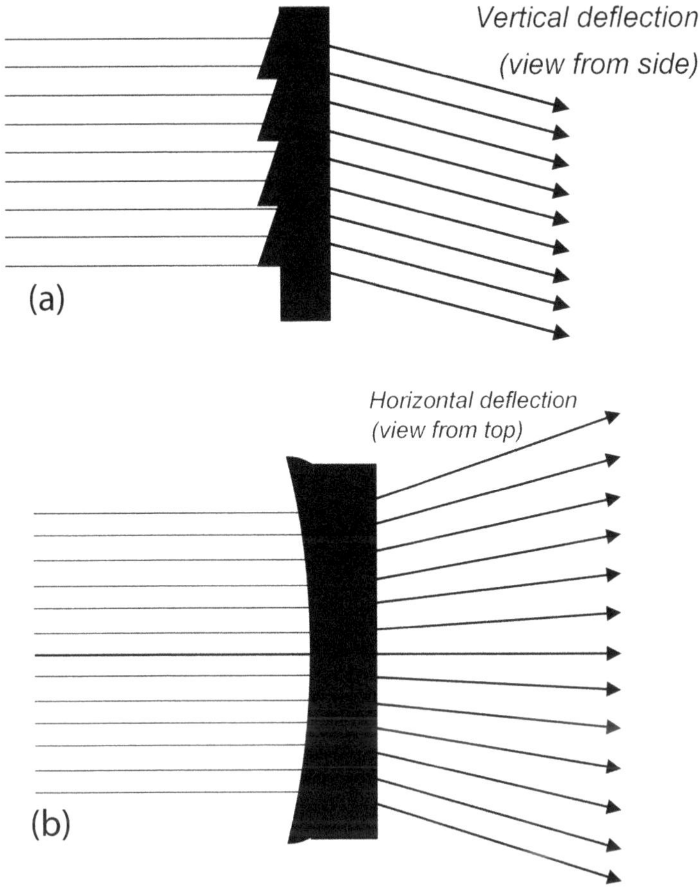

FIGURE 7.4 Effect of (a) prisms, (b) flutes.

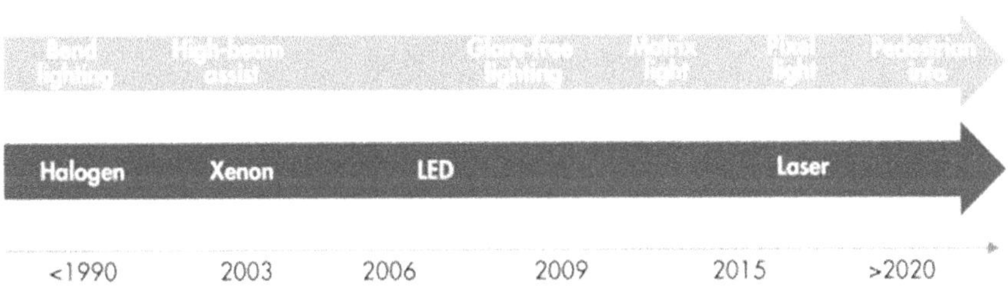

FIGURE 7.5 Evolution of lighting technologies.

Concerning its construction, a tungsten filament is placed in a vacuum and heated to incandescence by an electric current at a temperature of about 2,300°C. Tungsten is a heavy metallic element and has the symbol W, its atomic number is 74, and its melting temperature is 3,410°C. The tungsten filament is wound into a spiral to get more length of thin wire into a small space. Improved bulbs are filled with gas instead of a vacuum, most typically argon. This allows the tungsten filament to work

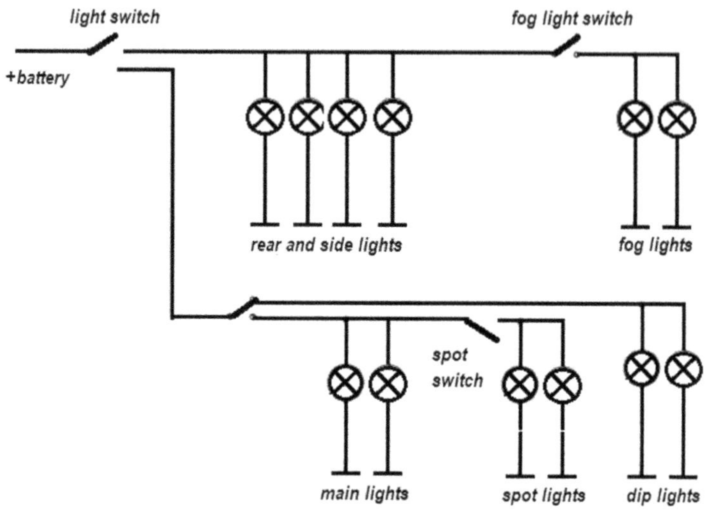

FIGURE 7.6 Possible wiring for a motor vehicle's lights.

TABLE 7.1
Standardized Connection Codes

	British Standard color codes for cables	European color codes for cables	Popular terminal designation numbers
Headlight main beam	Blue/White	White	56a
Headlight dip beam	Blue/Red	Yellow	56b
Headlight switch to dip switch	Blue	White/Black	
Sidelight main feed	Red	Grey	
Left-hand sidelights and number plate	Red/Black	Grey/Black	58L
Right-hand sidelights	Red/Orange	Grey/Red	58R
Reverse	Green/Brown	Black	
Stop lights	Green/Purple	Black/Red	54
Rear fog light	Blue/Yellow	Green/Black	55

at a higher temperature without failing and therefore produces a whiter light. Argon bulbs produce about 17 lm/W compared with a vacuum bulb, which will produce about 11 lm/W only.

Using argon gas as filler was not the final answer. A combination of using tungsten for the filament and halogen for the gas filler is the top choice today. The name halogen is used because there are four elements within the group VIIA of the periodic table, known collectively as the halogens (bromine, chlorine, fluorine, and iodine). Figure 7.8 illustrates the periodic table and the locations of these elements.

FIGURE 7.7 Acetylene light.

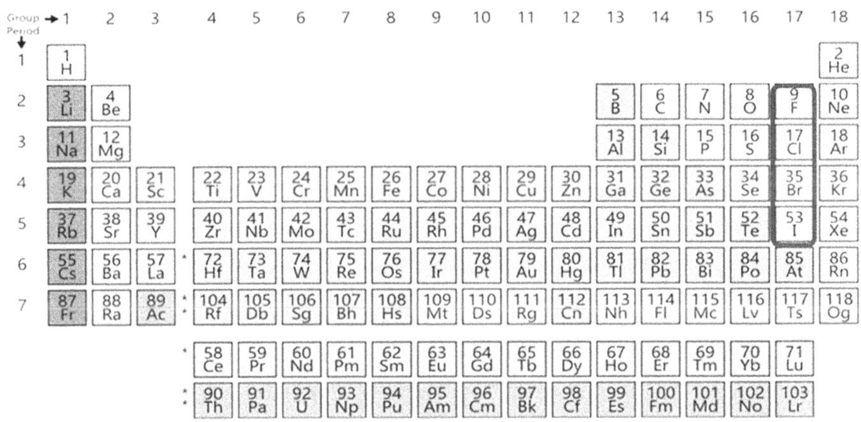

FIGURE 7.8 Periodic table and halogens in column 17.

Iodine is mostly used with a higher pressure of 7–8 atmosphere. The bulbs are able to produce over 24 lm/W, with a color temperature of 2,800–3,400 K. The drawback of vacuum- or argon-filled bulbs is that conventional bulbs blacken over time due to the evaporation of tungsten. The tungsten filament still evaporates when using halogen but combines with two or more halogen atoms forming a tungsten halide, which is a gas.

7.3 GAS DISCHARGE LAMPS AND THEIR ELECTRONIC CONTROL

Since headlights are critical in automtive design, new solutions are looked for. Among them, gas discharge lights (GDL) and LED are considered therein. The next technical solution after a conventional bulb with halogen was based on GDL and their unavoidable electronic control. The GDL lights are also known as Xenon or HID lights. The

TABLE 7.2
Comparison of Lighting Technologies

	Standard halogen bulb	HID light source
Light source	**Filament**	**Arc discharge**
Color temperature	3,000 K	4,100 K
Lumens, light output	700 .. 1,000	3,200
Light source, Watts	55 W	35 W
Lifetime	320 .. 1,000 hours	Up to 3,000 hours

source of light in the *gas discharge lamp* is an electric arc. To sustain this arc, the bulb contains a mixture of mercury, various metal salts, and xenon under pressure. When the light is switched on, the xenon illuminates suddenly and evaporates the mercury and metal salts. The high luminous efficiency is due to the metal vapor mixture. The mercury generates most of the light and the metal salts affect the color spectrum. Special filters are required for the high output of ultraviolet (UV) radiation from the GDL. The success of the GDL lights is due to its 80 lm/W intensity.

A comparison between a standard halogen bulb and a gas discharge lamp is provided in Table 7.2.

Components of a GDL system are illustrated in Figure 7.9. The lamp requires a high voltage for operation. The electronic ballast contains an ignition and control unit and converts the electrical system voltage into the operating voltage required by the lamp.

A possible solution for the power electronic circuit is shown in Figure 7.10.

First, a flyback converter (Figure 7.11a) is used for the dc/dc converter from an automotive battery to a higher regulated dc voltage bus, usually up to 300 Vdc. This dc voltage is used with a full-bridge inverter for dc/ac conversion. The full-bridge inverter (Figure 7.11b) produces a 300–400 Hz square-wave voltage. During the steady-state production of ac square-wave voltage, it needs to supply ~35 W for the lamp.

Since the same power converter cannot provide both the square-wave and the 25 kV pulses required by the GDL, a take-over phase is introduced. The igniter works as a separate circuit able to produce a high voltage on the arc (Figure 7.11c).

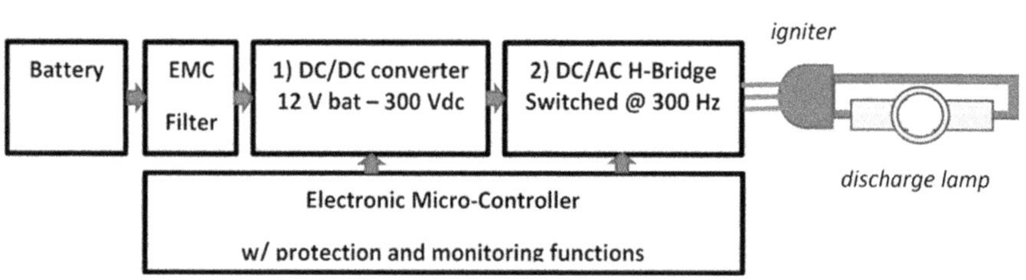

FIGURE 7.9 Components of the gas discharge lamp.

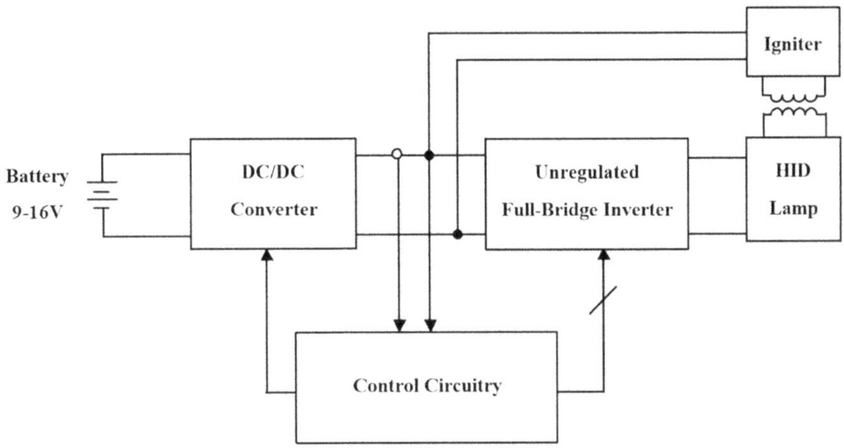

FIGURE 7.10 Power electronic circuit for GDL.

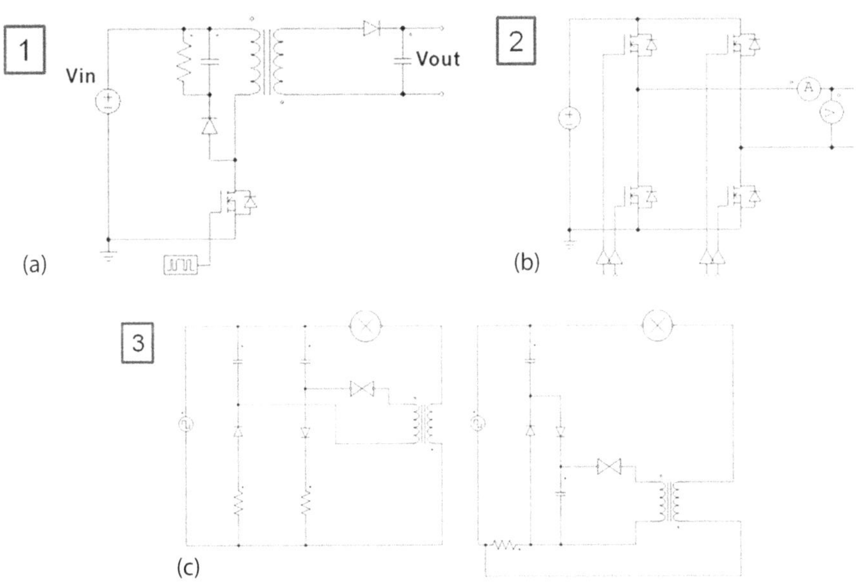

FIGURE 7.11 Power electronic circuit used as components of a GDL lighting system.

The operation of this hardware inside a GDL lighting system is fairly complex. For ignition, a high voltage pulse is repeated at 20 Hz in order to cause a spark between the electrodes, which ionizes the gap and creates the arc. Since the same power converter cannot provide both ac and 25 kV pulses, a take-over phase is introduced with electronic control. A warm-up phase follows for immediate light. The current flowing through the electric arc excites the xenon, which then emits light at about 20% of its continuous value. The temperature rises rapidly and the mercury and metal salts evaporate. A run-up phase follows and the lamp is now operated at an increased wattage. The pressure in the lamp increases as the luminous flux increases

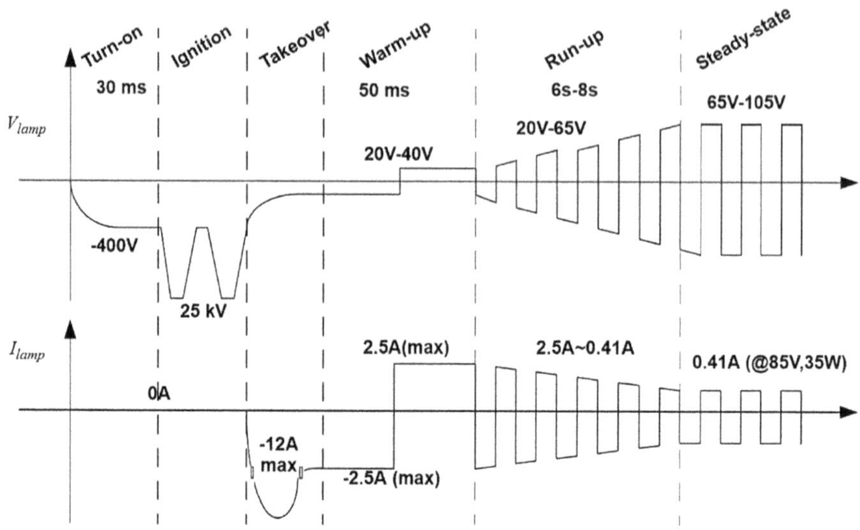

FIGURE 7.12 Various steps in operation.

and the light shifts from the blue to the white range. Finally, for a continuous action, the lamp is operated at a stabilized power rating for a still arc, showing no flicker.

Figure 7.12 sketches all the steps in the operation, while Figure 7.13 illustrates the subsequent usage of various control algorithms on the same hardware platform. The control software switches between current control and voltage control, depending on the operation stage.

- During the turn-on phase, a voltage feedback control is used to adjust the dc bus voltage (Figure 7.14) in order to delivered the dc bus voltage with no power to the lamp.
- During the warm-up phase and at the beginning of the run-up phase, a current-feedback control mode is required (Figure 7.15) in order to keep a constant current while settling the power level within the lamp.
- During the steady-state (continuous) phase, a power-feedback control mode is required in order to maintain a constant power level within the arc (Figure 7.16).

An example of the complete schematics for the power electronic circuits used within the controller is shown in Figure 7.17. The battery protection, reverse protection, EMI filter, sensors, digital (DSP) controller are not shown.

7.4 LED LIGHTS AND THEIR ELECTRONIC CONTROL

Rubin Braunstein of the *Radio Corporation of America* and Robert Biard and Gary Pittman of *Texas Instruments* contributed to the development of the infrared LED. In 1962, a GE scientist, Nick Holonyak, developed the first visible [red] light LED, which was followed by George Craford with the yellow light LED. In 1994,

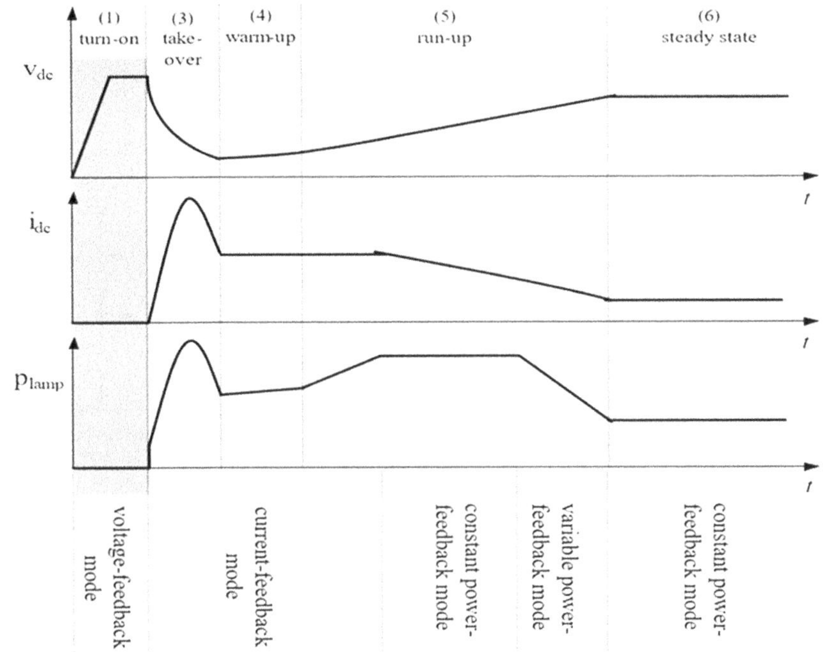

FIGURE 7.13 Various control methods adopted on the same power stage.

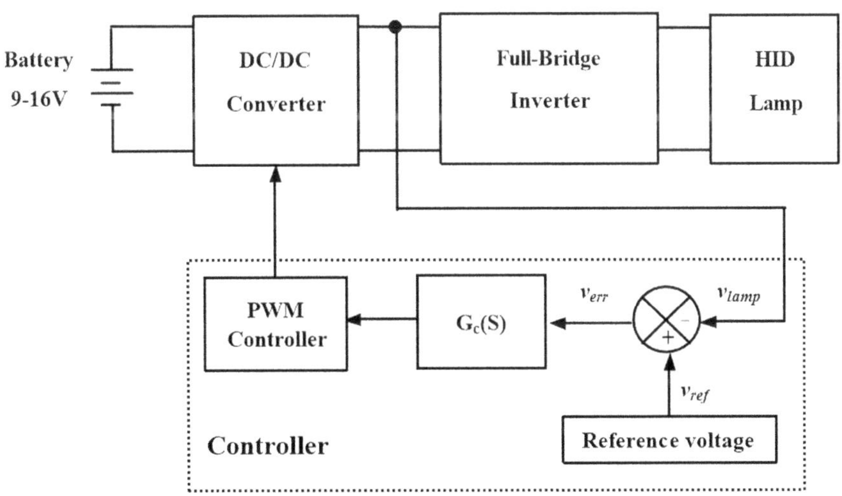

FIGURE 7.14 Voltage feedback control.

Hewlett-Packard increased the efficiency of the LED to ten times the efficiency of a red filtered light bulb.

White LED is used today for lighting applications. This was first demonstrated by Zhao et al. in 2002, by locating red, green, and blue LEDs adjacent to one another, and properly mixing the amount of their output. A poor color rendering was achieved

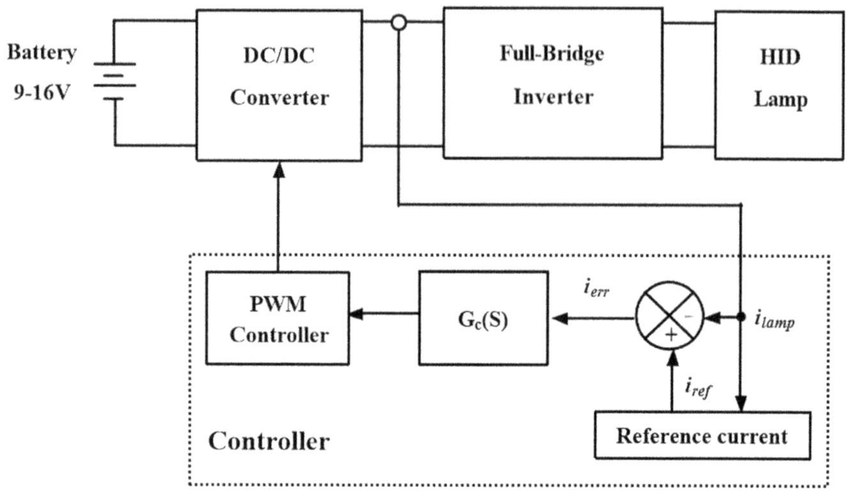

FIGURE 7.15 Current-feedback control mode.

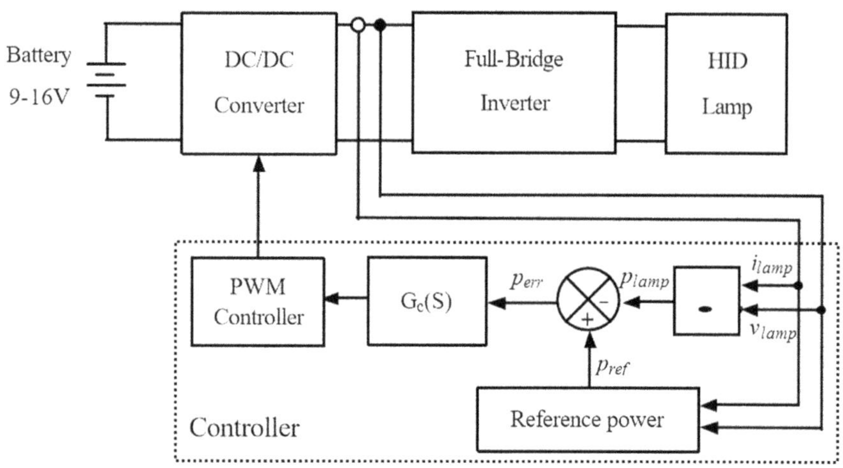

FIGURE 7.16 Power-feedback control mode.

since only three narrow bands of wavelengths of light were emitted. Later on, better results were achieved with phosphor: a $Y_3Al_5O_{12}$:Ce (known as "YAG") cerium doped phosphor coating produces yellow light through fluorescence. The combination of yellow with the remaining blue light appears white to the eye.

Using different phosphors produces green and red light through fluorescence. The mixture of red, green, and blue is perceived as white light, with improved color rendering compared to the blue LED/YAG phosphor combination.

The principle for the operation of any LED device is rather simple (Figure 7.18). When a voltage is applied across the electrodes, the current flows from anode (P) to the cathode (N), not unlike a conventional diode. When an electron meets a hole at

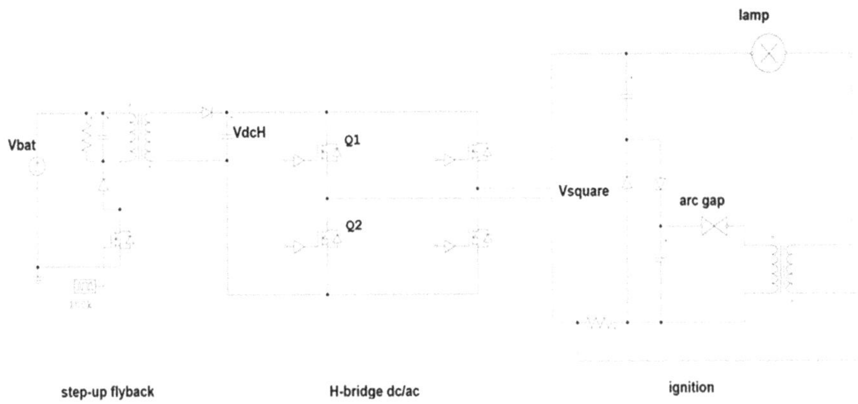

FIGURE 7.17 Power electronic circuit.

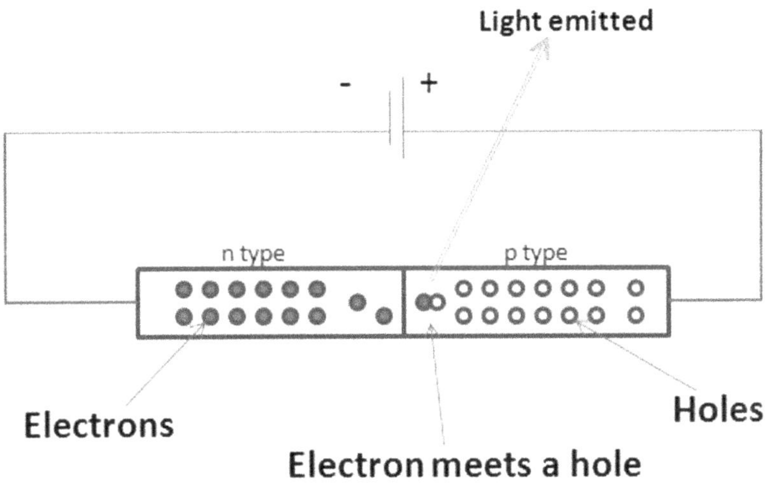

FIGURE 7.18 Principle of a LED device.

the P–N junction, it falls into a lower energy state. *The difference in energy of the two states is called the "band gap," which is a characteristic of the material comprising the P–N junction.* The excess energy of the electron is emitted as a photon. The more the "band gap," the higher is the energy difference, and the shorter the wavelength of the light emitted.

Control means producing a quasi-dc current through the LED, Figure 7.19.

LED color is determined by the energy required for electrons to cross the band gap of the semiconductor and this depends on the semiconductor used:

- Indium gallium nitride (InGaN): blue, green, and ultraviolet high-brightness LEDs;
- Aluminum gallium indium phosphide (AlGaInP): yellow, orange, and red high-brightness LEDs;

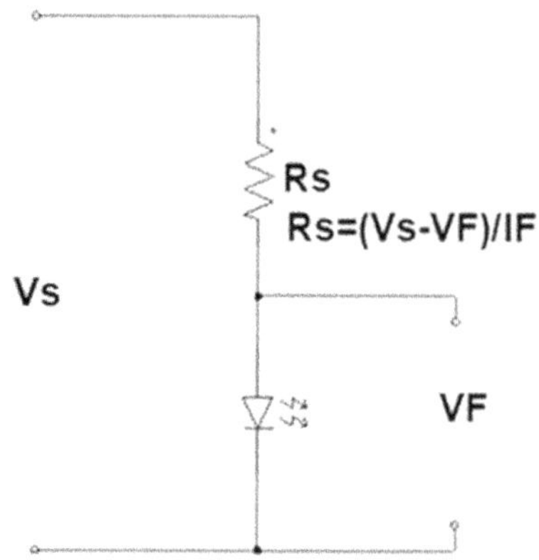

FIGURE 7.19 LED bias.

- Aluminum gallium arsenide (AlGaAs): red and infrared LEDs;
- Gallium phosphide (GaP): yellow and green LEDs.

White light is obtained by using multiple semiconductors or a layer of light-emitting phosphor on the semiconductor device. The exact choice of the semiconductor material used will determine the overall wavelength of the photon light emissions and therefore the resulting color of the light emitted.

Until recently, legislation has prevented the use of LEDs for exterior lighting.

An LED can produce approximately 100 lm/W. The current state-of-the-art (SOTA), phosphor-coated LEDs regularly achieve 140 lm/W for "warm white" LEDs and 160 lm/W for "cool white" lighting, and this technology is expected to be limited to an efficacy of 255 lm/W. The greatest advantage comes from reliability since LEDs have a typical rated life of over 50,000 hours, compared with just a few thousand for incandescent lamps.

Another advantage for automotive applications: the turn-on times are about 130 msec for the LEDs (due to minority carrier), and 200 msec for bulbs. This advantage is especially beneficial for brake lights.

Finally, there is a better thermal response than with any other technology. This presents a major advantage since all former technologies needed more space for natural cooling. Modern LED lights allow different light receptacle shapes, with smaller, slender curves (Figure 7.20).

The LED lighting technology evolution can be followed with examples from a world leader in vehicle lighting technologies:

- 2004: Audi A8 W12 with LED daytime running lights
- 2008: Audi R8 with all-LED headlights
- 2010: Audi A8 in which the headlights are networked with the navigation data
- 2012: Audi R8 with dynamic turn signal lights
- 2013: Audi A8 with Matrix LED headlights

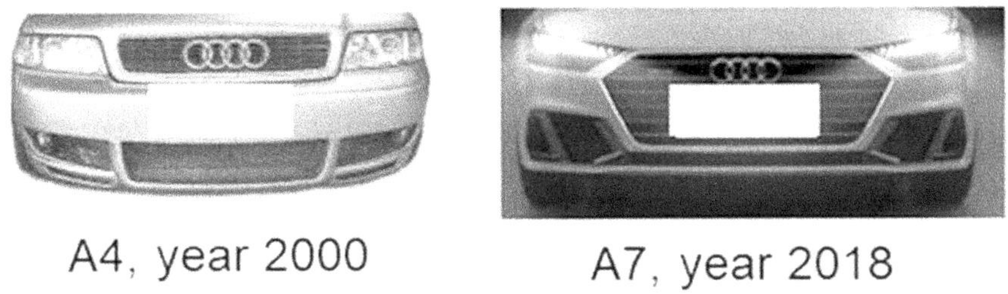

FIGURE 7.20 New shapes for the headlights equipped with LED lights.

At higher power levels, a power converter replaces the voltage source from Figure 7.19, because high-intensity LEDs are designed to work with a higher current. This is possible with the pulse-width modulation (PWM) operation of a power converter (Figure 7.21) so that a high-intensity producing current is delivered to the LED and the average current is low. This secures a good energy efficiency. If the pulse frequency is high enough, this "on–off" flashing does not affect what is seen by the human eye as it "fills" in the gaps between the "on" and "off" light pulses, making it appear as a continuous light output. Interestingly enough, pulses at a frequency of 100 Hz or more actually appear brighter to the eye than a continuous light of the same average intensity.

The brightness of an LED is approximately proportional to its average current. The brightness is adjusted through a process called dimming. Both amplitude modulation (AM) and pulse-width modulation are used for dimming that changes the average value of the current based on a reference command (Figure 7.21).

For a simple control, the duty cycle of the train of pulses is adjusted to change the average value of current through the LED. A higher performance control avoids a color temperature shift. This problem relates to the fact that current variation in an LED may cause a color temperature shift. To prevent the LED lighting from color shift, the LED lamp can be dimmed by low-frequency PWM control and usage of a single current level through an LED. This process is illustrated in Figure 7.22.

Numerous integrated modules have been developed for automotive LED control (Figure 7.23). For example, MPM6010. The MPM6010 is a high-frequency,

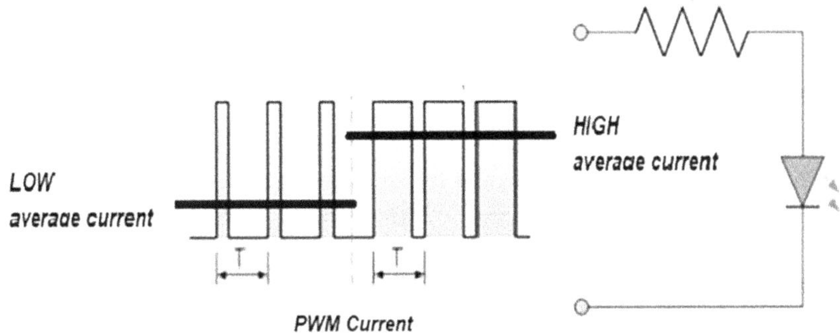

FIGURE 7.21 PWM control of the LED current.

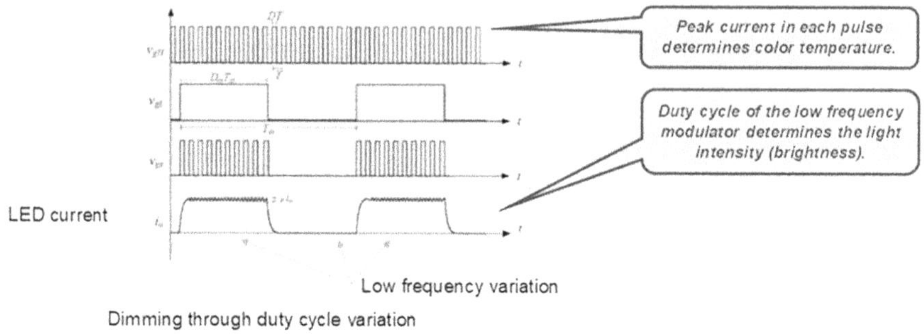

FIGURE 7.22 Advanced PWM control to avoid color temperature shift.

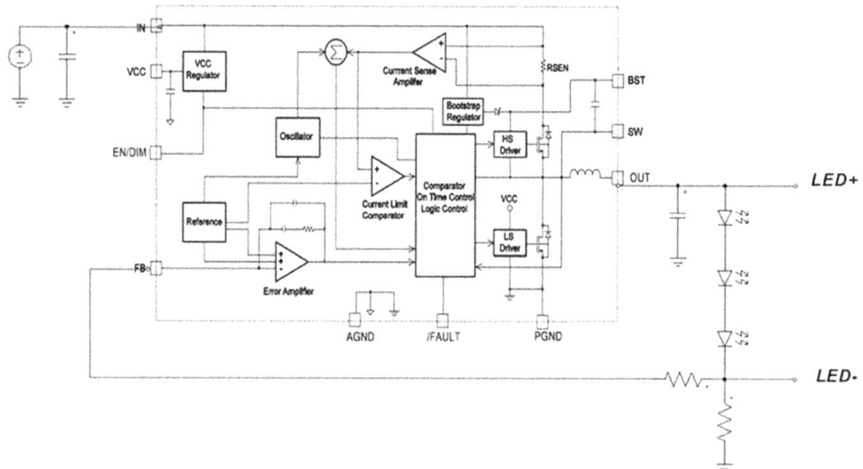

FIGURE 7.23 Integrated circuit for LED lighting control.

synchronous, rectified, step-down, switch-mode, white LED driver with built-in power MOSFETs, an integrated inductor, and two capacitors, which Also includes a control with an internal compensation network along with multiple protection features (Figure 7.23).

7.5 LASER LIGHTS

Laser lighting is very similar to LED devices and their control. The laser diode is a solid-state device, which emits light through photons. Instead of the P–N junction, the laser diode uses either a *P–I–N* structure or a *N–p–P* double heterostructure, where a narrower band gap semiconductor (for example, GaAs) is sandwiched between two semiconductors of a wider band gap (such as AlGaAs). This solution is used in most cases. The surface of the area around the junction is prepared to a mirror-like finish. Hence, the photons are bouncing around the polished resonant cavity, creating the laser effect.

TABLE 7.3
Typical LED Characteristics

Typical LED characteristics			
Material	**Wavelength**	**Color**	**Voltage drop**
GaAs	850–940 nm	Infrared	1.2 V
GaAsP	630–660 nm	Red	1.8 V
GaAsN	605–620 nm	Amber	2.0 V
GaAs P:N	585–595 nm	Yellow	2.2 V
AlGaP	550–570 nm	Green	3.5 V
SiC	430–450 nm	Blue	3.6 V
GaInN	450 nm	White	4.0 V

TABLE 7.4
Relationship between Wavelength and Color of Visible Light for a Laser Lighting System

Color	Wavelength [nm]
violet	380–450
blue	450–495
green	495–570
yellow	570–590
orange	590–620
red	620–675

The bias current needs to be a stable, low-noise, transient-free current source; otherwise, current variation changes the intensity and light color. *Laser diodes* are current-driven and current-sensitive semiconductors. A change in drive current equals a change in the devices' wavelength and output power. As a noteworthy figure of merit, the variation of wavelength with the current is around 10 pm/mA, while the variation with temperature is around 90 pm/°C. Table 7.4 illustrates the wavelength of visible light colors, and this can be analyzed in relation to Table 7.3.

A high-power current source is designed starting off a buck/boost converter transformed into a current source. A simple example is shown in Figure 7.24 for a simple current source from a voltage regulator integrated circuit.

A laser diode also has a low thermal shock tolerance. Hence, protection is desired through the way the current source is enabled and disabled. This means using slow-start circuits and overcurrent protection with current limitation circuits. It is noteworthy that a bench-top power supply cannot be used since it ramps up the voltage at turn-on, while the current is not controlled. This is not good for diodes, which require a constantly controlled current and may get damaged by the surge. Therefore, if the application demands constant laser output and a stable wavelength (color), a

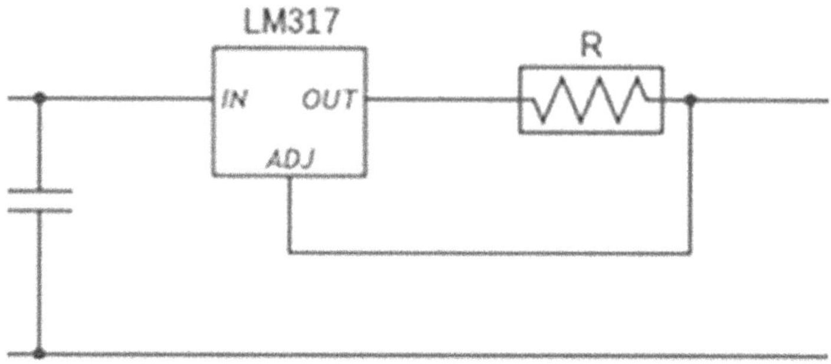

FIGURE 7.24 High-power current source produced with a voltage regulator.

voltage source will not work and may put the laser at risk from thermal shock and/or transients due to a quick change in current.

Not unlike the LED drivers, the laser diode drivers can be classified in *constant current* and *pulsed current* drivers. While the constant current solution is very intuitive, the pulsed current driver requires more explanation.

Pulsed current drivers are current sources that offer pulse repetition rates from a single shot up to 500 MHz. These constant current sources are designed and optimized to produce clean current pulses at a user-set amplitude and at a user-set pulse repetition interval and/or pulse-width. Such a clean pulse has a lack of overshoot and a low amount of noise/ripple on the pulse of overall square-wave pulses. Very rarely, a sinusoidal current wave is used. The pulsed current wave is often chosen on the shortest (most narrow) pulse width that they can deliver to the laser diode.

Furthermore, the laser drivers also have overvoltage clamping and an analog current limit to protect the laser.

As an OEM example, RLS/SF6090 represents a complete hardware solution for a high current laser driver. Such laser diode drivers are somewhat expensive, around ~$500 for a 100 A driver.

Laser diode lighting for automotive applications is a newer solution. The 2014 BMW i8 and the Audi R8 LMX were launched almost at the same time, in 2014, as the first series production vehicles with laser headlights. Since laser lighting produces a narrow beam or spot of light, the lighting system uses a laser matrix. This consists of a rapidly moving micro-mirror (about 3 mm in size), which redirects more or less of the laser beam. At low vehicle speeds, the light is distributed over a larger projection area producing a lower light intensity (diffused light), and the road yields an illumination with a very wide aperture angle. At high vehicle speeds, the aperture angle is reduced, while the intensity and range of the light are increased significantly. The advantage is obvious in highway driving at high speed.

The brightness of different lighting zones can be varied by controlling the illumination dwell times corresponding to each specific zone. This means the light can be distributed precisely and a series of new features can be introduced. For instance, two parallel beams can be accentuated along the car width to help the driver pass through narrow spaces. Or, the left side of the light spot can be blocked temporarily if a pedestrian is detected on the front-left side of the vehicle. The features derived from experimentation with the light spot are therein endless.

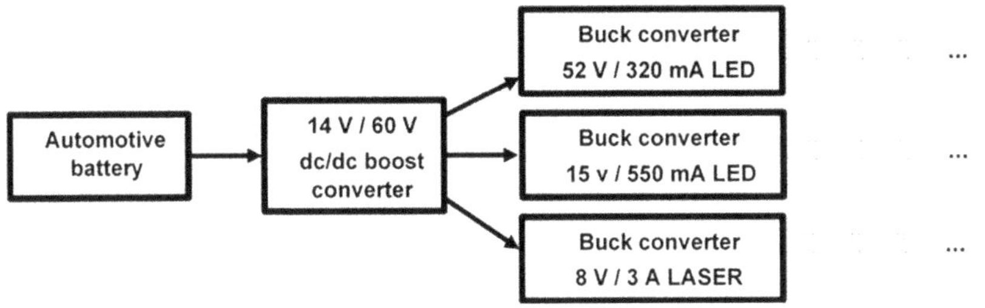

FIGURE 7.25 Example of multiple power converters for lighting systems.

Laser headlights have a series of benefits that make the technology attractive for the future. The brightness is already almost four times that of an LED, at 300 lm/W. This means that headlights can be made even smaller in the future—without compromising on light intensity.

The primary benefit for drivers is that the laser-based headlights have the longest range provided by any current headlight technology. The full-beam has a range of up to 600 meters—double the distance of the current standard LED headlights.

Due to the special properties of laser lighting, modern solutions consider the laser diode for the high beam in combination with LED lighting for the daylights or low dip beam. This follows a previous trend of using LED lighting for day use, tail lights, and low dip beam, while using stronger GDL lights for the high beam.

Unfortunately, the power converter requirements for the two diodes are different and the same converter cannot be used to supply both. This is not very efficient and the power converter incurs additional heat dissipation on the laser channel, since the voltage designed for LED-based daylighting yields a higher voltage than is needed to drive the laser diode.

For instance, combination lighting should design-in multiple buck converters. A 60 W light control module may have three load groups: 52 V/ 320 mA for day running lights, 15 V/ 0.55 mA for the smaller fog lamp, and an 8 V/3 A laser module. Figure 7.25 sketches this example.

7.6 CONCLUSION

Automotive lighting systems are very demanding to design since they need to respond to both the requirement of allowing the driver to see and the requirement for making the motor vehicle seen without dazzling.

Various solutions for automotive lighting have been reviewed in this chapter from both technological and historical perspectives. While gas discharge lights are very common, modern solutions based on LED and laser diodes are very economical and allow more control features. These are also a great field for further power electronics research and development.

REFERENCES

Anon. 2009. *AN2823 Application Note, Current Source for LED Driving Based on the L5973AD*. ST Microelectronics Corporation, Doc ID 15015 Rev 1.

Anon. 2015. *Laser Light for Headlights = Automotive Product of the Year 2015 by the Specialist Magazine Elektronik*. Published by OSRAM Sylvania.

Anon. 2017. *Automotive LED Headlamp With A Laser Channel Provides Versatile and Flexible Solution Multi Phase Driver IC Enables Flexible Topology for Headlamp LCU Design*. Paper Published in Automotive Solid State Lighting Application Note from NXP Semiconductors, Document Number: solidstlghtwp-rev-0.

Anon. 2017. *SF6090 Laser Diode Driver, Datasheet & User Manual*. Internal documentation at Maiman Electronics Corporation.

Bhattacharya, A., Lehman, B., Shteynberg, H., Rodriguez, H. 2006. *Digital Sliding Mode Pulsed Current Averaging IC Drivers for High Brightness Light Emitting Diodes*. Presented at IEEE COMPEL Workshop, Rensselaer Polytechnic Institute, Troy, NY, USA, July 16–19, 2006, pp. 136–142.

Cho, C.G., Song, S.H., Park, S.M., Park, P., Jang, S.R., Ryoo, H.J. 2019. A novel series-connected xenon lamp power supply system using a pulse trigger with simmer circuits for pulsed light sintering application. *IEEE Transactions on Power Electronics*, 66(1):233–241.

Davis, S. 2017. *LED Lighting is Penetrating the Automotive Industry. Power Electronics Magazine*, Editorial, October 26, 2017. Internet source: https://www.powerelectronics.com/automotive/led-lighting-gradually-penetratingautomotive-industry.

Denton, T. 2017. *Automobile Electrical and Electronic Systems*. 5th edition, Abingdon-on-Thames: Routledge.

Hodgson, D., Olsen, B. 2003. *Protecting Your Laser Diode*. ILX Lightwave Corporation documentation, revision Rev01.060103.

Hu, Y. 2001. *Analysis and Design of High-Intensity-Discharge Lamp Ballast for Automotive Headlamp*. Ph.D. Thesis, Virginia Tech.

Moo, C.S., Chen, Y.J., Yang, W.C. 2012. An efficient driver for dimmable LED lighting. *IEEE Transactions on Power Electronics*, 27(11):4613–4619.

Neacşu, D. 2004. *Power Semiconductor and Control for Automotive Applications*. Tutorial Presented at IEEE APEC, Anaheim, CA, USA.

Pang, H.M., Pong, M.H. 2009. *A Stability Issue with Current Mode Control Flyback Converter Driving LEDs*. Presented at IEEE Power Electronics and Motion Control IPEMC, Wuhan, China, pp.1402–1408.

Thompson, M.T., Schlecht, M.F. 1997. High power laser diode driver based on power converter technology. *IEEE Transactions on Power Electronics*, 12(1):46–52.

Wang, J. 2011. *Automotive Headlamp HID Ballast Reference Design Using the dsPIC® DSC Device*. Paper Published by Microchip Corporation, Application Note DS01372A.

8 dc/dc Converters

HISTORICAL MILESTONES

1902: Peter Cooper Hewitt invented mercury-arc rectifiers that were used to provide power for industrial motors, electric railways, streetcars, electric locomotives, radio transmitters, and for high-voltage direct current (HVDC) power transmission.

1913: Irving Langmuir demonstrated that a negative electric potential introduced between a cathode and anode prevented the current starting from the anode, with applications to vacuum radio tubes.

1923: Tests proved that a mercury-arc rectifier could be an inverter in order to convert dc into ac. This created a new industry, and after 20 years showed installations up to 20 MW.

1935: Vacuum tube car radio becomes the first electronic equipment inside a car.

1955: The first transistorized ignition system is introduced by the British Lucas Industries, also on BRM and Coventry Climax Formula One engines in 1962.

1963: Pontiac was the first automaker to offer electronic ignition on production cars.

1970: The Japanese electronics industry began producing integrated circuits and microcontrollers dedicated to the automobile industry.

1959–1971: The MOSFET (MOS field-effect transistor, or MOS transistor), invented by Mohamed M. Atalla and Dawon Kahng at Bell Labs in 1959, led to the development of the power MOSFET by Hitachi in 1969.

1977: Alex Lidow and Tom Herman co-invented the HexFET, a hexagonal type of power MOSFET, at Stanford University.

1977–1979: The insulated-gate bipolar transistor (IGBT), which combines elements of both the power MOSFET and the bipolar junction transistor (BJT), was developed by B. Jayant Baliga at General Electric.

8.1 ROLE OF dc/dc POWER CONVERTERS

Since the entire dc distribution system is built around the battery and various circuits need to also be supplied with a dc voltage, dc/dc converters are popular in automotive applications. They range from tens of Watt to tens of kW, which means working with currents from several Amperes to several hundred Amperes. While the working principle and circuitry are the same, the building of the power stage depends on the installed power.

The dc/dc converters can be classified

- After the grounding connection
    - Converters with direct transfer of energy
    - Converters with galvanic isolation when the input and output circuits do not share the same grounding
- After the operation mode
    - Switched-mode converters
    - Analog (linear) mode voltage regulators

8.2 DIRECT CONVERSION (WITHOUT GALVANIC ISOLATION)

The switched-mode dc/dc converters without galvanic isolation can produce an output voltage higher than the input voltage (boost converters) or can produce an output voltage lower than the input voltage (buck converters). In both cases, the circuit is built around a power transistor that alternates its state in between on to off. This sequence of on and off states of the transistor produces a train of pulses; wherefrom one can extract a dc component with a low-pass filter. The dc level of the output voltage can thus be adjusted through the duration of the conduction interval for the transistor. A theoretical dependency can be established between the pulse duration and the output voltage. The successive control of the on and off states of the transistor is called Pulse-Width Modulation (PWM).

8.2.1 Buck Converter

A power converter able to decrease the level of dc voltage from the input source to the output connected load is called a buck converter (Figure 8.1).

The circuit is composed of a power transistor (Q) working in a switched mode and a power diode (D) able to take over the inductive current when the transistor is in the off state. A power inductor (L) serves as a temporary storage of energy, and an output capacitor (C) filters the output voltage.

First, the transistor is brought into an on state and current flows from the input source through the inductor and toward the output capacitor bank. Since the inductor

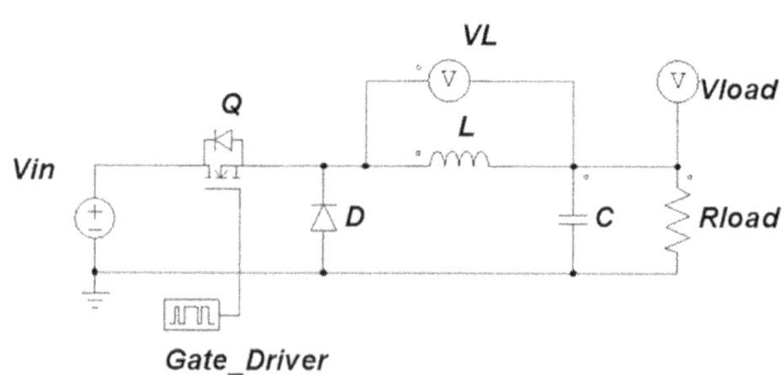

FIGURE 8.1 Buck converter.

sees an almost constant voltage drop, equal to the difference between the input and output voltages, its current increases linearly.

When the transistor is turned off, the inductor tends to maintain the current circulation with the generation of a voltage (v = L·di/dt) at the first tendency of current decrease. The induced voltage has the polarity able to favor the previous current circulation, which means with the positive end toward the output capacitor bank. This voltage is as high as necessary to forge a path for the current. Eventually, it can produce a transistor breakdown due to overvoltage. Herein, the diode turns on and offers a path for the inductive current. The inductor now sees a voltage drop equal to the output voltage.

Details of operation are easier to follow along with a numerical example, herein considered for a conversion from $V_{in} = 48.0$ V to $V_{load} = 12$ V. The main waveforms are shown in Figure 8.2 when the switching frequency is 50 kHz, and the nominal load current equals 20 A. Since the current waveform never reaches a zero Amperes value, this operation mode is called a *continuous conduction mode* (CCM), and the average value of the output voltage yields:

$$V_o = \frac{1}{T_s} \cdot \int_0^{T_s} v_o(t) = \frac{1}{T_s} \cdot \left[\int_0^{t_{ON}} V_{in} dt + \int_{t_{ON}}^{T} 0 dt \right] = \frac{t_{ON}}{T_s} \cdot V_{in} = D \cdot V_{in} \tag{8.1}$$

where D = t_{on}/T = duty cycle. Control of the output voltage can be achieved at a variation of D.

To benefit from this switched-mode operation, a low-pass filter (LPF) is required in order to eliminate the effect of switching and to extract the average value of the train of pulses. Since the filter has a typical L–C second-order structure, the magnitude characteristics will drop with 40 dB/decade. The frequency characteristic to this filter should be selected such that the harmonic components, multiples of the PWM frequency, are attenuated with more than 60–80 dB, which is that 1 Vdc goes into less than 1 mVdc ripple.

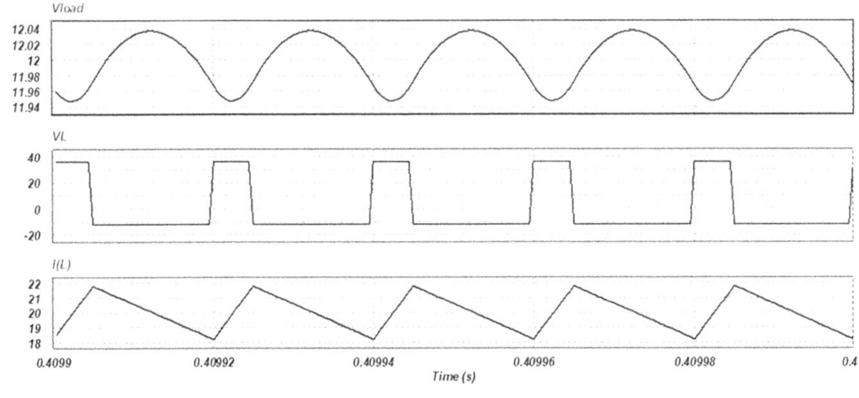

FIGURE 8.2 Waveforms for the operation of a buck converter in CCM.

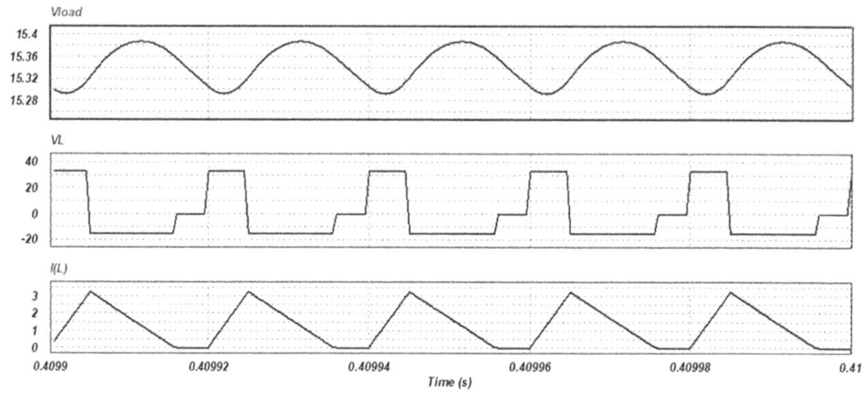

FIGURE 8.3 Waveforms for the operation of the buck converter in DCM.

While in CCM mode, the waveforms for the output voltage as well as the voltage drop on the inductance do not depend on the value of the load resistance and the Equation (8.1) stays true at any variation in the load circuit.

The inductance current is composed of a dc component equal to the average value of the load current (in this case, 20 A), that also equals the ratio between the load voltage and the load resistance (in this case 0.6 Ohm), and an ac component (ripple) approximately the same for any load resistance. The waveform for the inductor current slides up or down with the amount of load in the dc current.

Whenever the load changes and a lower current passes through the circuit, the swing of the inductor current may reach zero Amperes and create a no-conduction interval. This operation mode is illustrated in Figure 8.3 and is called a *discontinuous conduction mode* (DCM). The waveforms change more drastically at light currents and the diode eventually turns off at zero current. This case can be studied with the same practical setup with an increase in load resistance. Figure 8.3 features a load resistance of 12 Ohm, which is a 1 A load.

The value of the output voltage depends on the load resistance and the dependency relationship between the output voltage and duty cycle does not stand true. The average value of the output voltage can be calculated based on the load and duty cycle and not directly from the duty cycle. The output voltage yields higher as the load resistance is higher and the zero current interval is larger.

The waveforms in Figure 8.3 can be used to calculate the transfer characteristic. The current through the inductance can be calculated from the voltage across the inductance:

$$V \cdot \Delta t = L \cdot \Delta i \tag{8.2}$$

This will be the same for the two intervals:

$$\left(V_{in} - V_{load}\right) \cdot D \cdot T_s + \left(-V_{load}\right) \cdot \Delta_1 \cdot T_s = 0 \Rightarrow \frac{V_{load}}{V_{in}} = \frac{D}{D + \Delta_1} \tag{8.3}$$

The dc current within the load is denoted here with I_{Load} and the time interval with zero current is denoted with Δ_1.

$$\begin{cases} L\cdot\dfrac{0-i_{L,peak}}{\Delta_1\cdot T_s}=-V_{load} \\ I_{Load}=\dfrac{1}{T_s}\cdot\displaystyle\int_0^{T_s} i(t)\,dt \end{cases}$$

$$\Rightarrow\begin{cases} i_{L,peak}=\dfrac{V_{Load}}{L}\cdot\Delta_1\cdot T_s \\ I_{Load}=i_{L,peak}\cdot\dfrac{D+\Delta_1}{2} \end{cases}\Rightarrow I_{Load}=\frac{V_{Load}}{L}\cdot\Delta_1\cdot T_s\cdot\frac{D+\Delta_1}{2}=\frac{V_{in}\cdot T_s}{2\cdot L}\cdot D\cdot\Delta_1$$

$$\Rightarrow\Delta_1=\frac{2\cdot L\cdot I_{Load}}{V_{in}\cdot T_s\cdot D} \tag{8.4}$$

Transfer characteristic for the DCM operation yields

$$\frac{V_o}{V_d}=\frac{D}{D+\dfrac{2\cdot L\cdot I_0}{V_d\cdot T_s\cdot D}} \tag{8.5}$$

In conclusion, the discontinuous conduction mode can be avoided with the proper selection of components. A unique value of the load resistance can be identified to produce operation at the boundary between continuous and discontinuous conduction. This is the basis of modern control circuits with a boundary operation mode for improved efficiency of the power converter.

The average value of the critical current in boundary mode is:

$$\begin{aligned} I_{LB}&=\frac{1}{2}\cdot i_{L,peak}=\frac{t_{on}}{2\cdot L}\cdot\left[V_d-V_o\right]=\frac{D\cdot T_s}{2\cdot L}\cdot\left[V_d-V_o\right] \\ &=\frac{D\cdot T_s\cdot V_d}{2\cdot L}\cdot\left[1-D\right]=I_{OB} \end{aligned} \tag{8.6}$$

8.2.2 Boost Converter

Whenever the output voltage is required to be larger than the input voltage, a step-up converter is built, as shown in Figure 8.4, and this is also called a boost converter.

When the transistor (Q) turns on, the voltage across the inductance equals the input voltage and the current rises linearly. There is no current circulation from the input toward the load, and the load is supplied solely from the output capacitor. When the switch turns off, the inductance tends to maintain the current circulation by generating a voltage in the circuit. This voltage adds to the supply voltage to

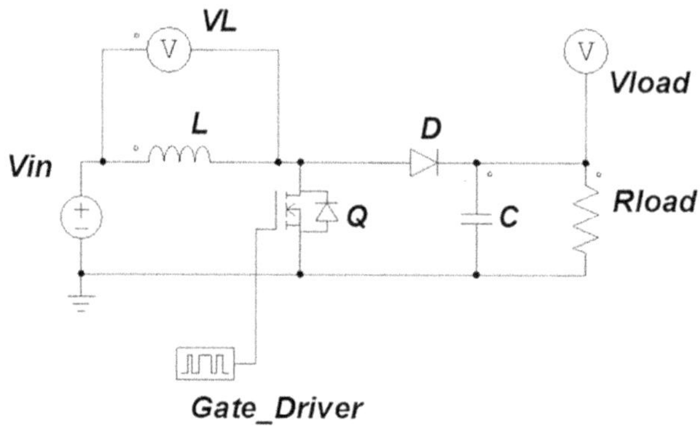

FIGURE 8.4 Boost converter.

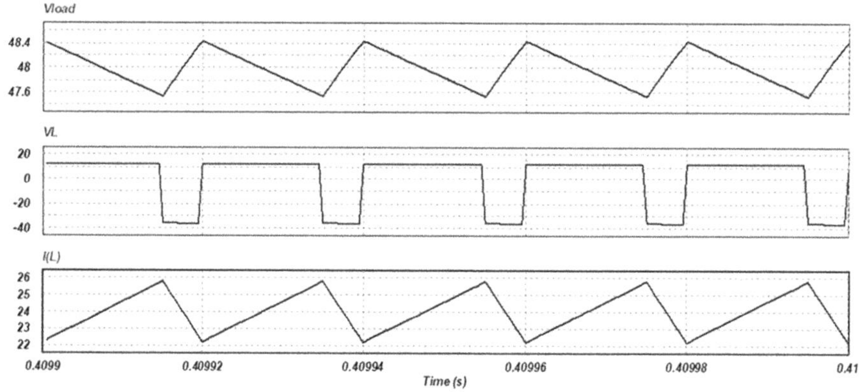

FIGURE 8.5 Waveforms for CCM operation of a boost converter.

create a larger voltage before getting to the diode (D). The diode D becomes forward-biased and turns on. The inductance delivers energy to the output circuit through the diode D and the output capacitor is charged from the input voltage plus the voltage induced in the inductance. The generic waveforms for inductance voltage and current are shown in Figure 8.5 for the case with continuous conduction mode when the inductor current never reaches zero Amperes.

The numerical example follows the example of the buck converter, this time for a reversed conversion from V_{in} = 12 V to V_{load} = 48 V. The switching frequency is 50 kHz, and the nominal load current equals 8 A, from a load of 6 Ohm.

The average value of the output voltage yields:

$$\frac{V_o}{V_d} = \frac{T_s}{t_{off}} = \frac{1}{1-D} \tag{8.7}$$

$$V_d \cdot I_d = V_o \cdot I_o \Rightarrow \frac{I_o}{I_d} = 1 - D \tag{8.8}$$

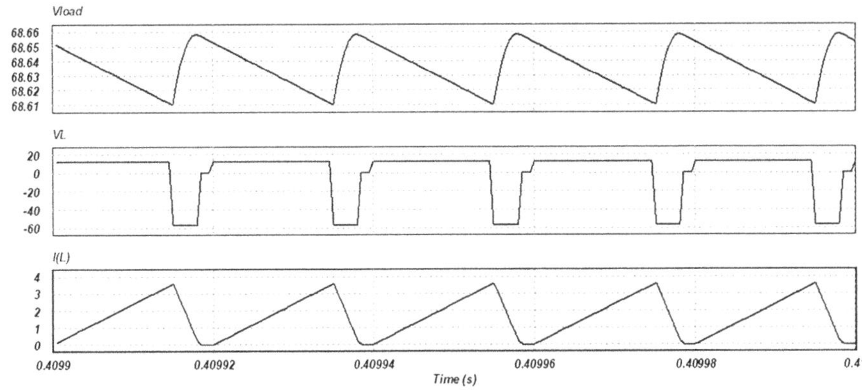

FIGURE 8.6 Waveforms for the case of a resistance of 240 Ohm, and *discontinuous conduction mode.*

The voltage drop on the inductance is approximately identical for the same duty cycle and a wide range of values for the load resistance. This produces the same ripple of the inductance current with a different average value, and the current waveform seems to slide up and down with the load current.

If the load resistance is increased and the load current decreases, an operation mode is possible when the inductor current occasionally reaches zero Amperes. This case is illustrated in Figure 8.6, where the load resistance becomes 240 Ohm. Note the appearance of zero current intervals within the inductance current which coincide with intervals of zero voltage across the inductance.

An operation mode with intervals of zero current through inductance is called a *discontinuous conduction mode.* Due to the zero current intervals, the output voltage yields higher than expected for the same duty cycle and the result from Equation (8.7) is not valid.

A different calculation of the output voltage considers the average value for the inductance voltage should be zero. It yields:

$$\left(V_d\right)\cdot D\cdot T_s+\left(V_d-V_o\right)\cdot\Delta_1\cdot T_s=0\Rightarrow\frac{V_o}{V_d}=\frac{D+\Delta_1}{\Delta_1}\Rightarrow\frac{I_o}{I_d}=\frac{\Delta_1}{D+\Delta_1} \tag{8.9}$$

The maximum current is calculated from

$$V_{in}=L\cdot\frac{i_{L,peak}}{D\cdot T_s}\Rightarrow i_{L,peak}=\frac{V_{in}}{L}\cdot D\cdot T_s \tag{8.10}$$

The input dc current equals the average value of the current through inductance and it can be calculated as

$$I_{in}=I_{L,av}=i_{L,peak}\cdot\frac{D+\Delta_1}{2} \tag{8.11}$$

The output current yields

$$I_o = I_d \cdot \frac{\Delta_1}{D+\Delta_1} = \frac{\Delta_1}{D+\Delta_1} \cdot \frac{D+\Delta_1}{2} \cdot \frac{V_{in}}{L} \cdot D \cdot T_s \Rightarrow \Delta_1 = \frac{2 \cdot L \cdot I_0}{V_{in} \cdot D \cdot T_s} \quad (8.12)$$

The transfer characteristic is expressed as

$$\frac{V_o}{V_{in}} = \frac{D+\Delta_1}{\Delta_1} = \frac{D + \dfrac{2 \cdot L \cdot I_0}{V_{in} \cdot D \cdot T_s}}{\dfrac{2 \cdot L \cdot I_0}{V_{in} \cdot D \cdot T_s}} \quad (8.13)$$

The discontinuous conduction mode can be avoided through a proper selection of passive components. Analogous to the buck converter, there is a value of the load resistance which produces an operation at the boundary between continuous and discontinuous conduction. The boundary conduction mode has advantages in reducing power loss since the diode turns off naturally, at zero current, and the transistor turns on at zero current as well.

8.2.3 Other Topologies of Non-Isolated dc/dc Converters

While the buck and boost converters have a very simple structure and are highly used in practice, several other topologies have been proposed to address specific applications. The buck/boost converter (Figure 8.7) and the Cuk converter (Figure 8.8) are other possible solutions using a single switch.

The H-bridge converter (Figure 8.9) features two devices in series leading to a bipolar voltage on the load. The conduction of transistors located on a diagonal of the H-bridge determine a positive voltage on the load. Transistors located on the other diagonal determine a negative voltage. The duty cycle associated with the operation of one diagonal or the other determines the average voltage on the load, which can yield positive or negative.

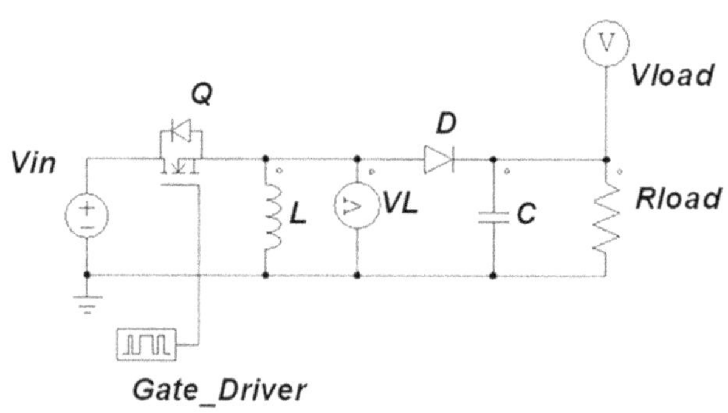

FIGURE 8.7 Buck–boost converter.

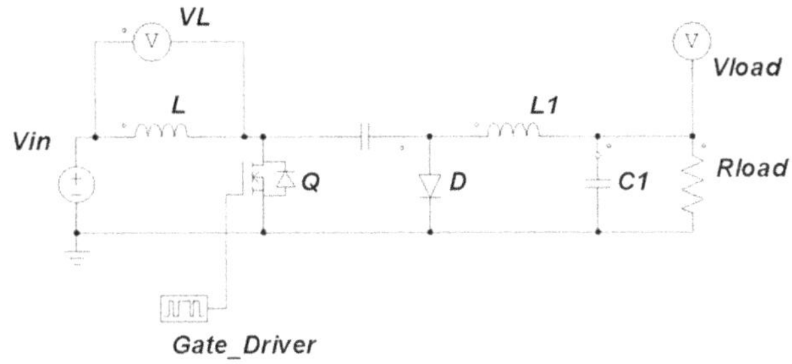

FIGURE 8.8 Cuk converter.

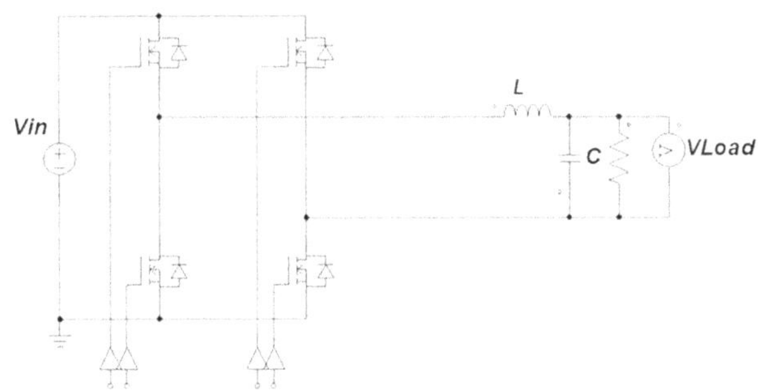

FIGURE 8.9 H-bridge converter.

8.2.4 Multi-Phase Converters

Most automotive applications require power supplies delivering large currents, which cannot be achieved with the single switch from the buck or boost converter. For instance, a typical dc/dc converter for bidirectional energy transfer between the 12 V and 48 V dc distribution buses may be built at several kW. This means around 100 A on the low-voltage side and over 20 A on the 48 V side. In order to reduce ripple at such current levels and to seek for the proper packaging of the converter, a multi-phase (usually, three to six phases) converter topology is used. Furthermore, as long as several transistors are connected in parallel anyway, the control pulses can be phase-shifted from each other. Figure 8.10 illustrates this converter, while Figure 8.11 shows the main waveforms.

8.2.5 The Synchronous Converter

Diodes used within the conventional buck or boost converters can be replaced with MOSFET transistors for reduction of loss due to a lower voltage drop in a conduction state and a turn-off operation without recovery currents. A converter featuring this replacement of switching diodes with power transistors and their appropriate control to work as diodes is called a synchronous converter.

Within a conventional buck converter, when the main transistor turns off, the diode turns on. If this diode is replaced with a power MOSFET, the MOSFET transistor sees a negative voltage before being controlled. The command of this transistor

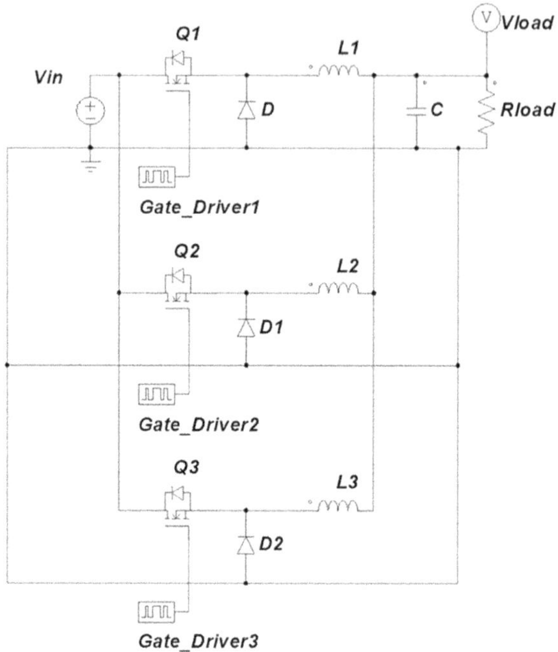

FIGURE 8.10 Three-phase buck converter.

thus controls conduction in the third quadrant, at a negative voltage and current. The drain-source voltage drop across the transistor is given by the product between the drain-source resistance and the converter current. The drain-source voltage yields are lower than the voltage across a conducting diode.

Furthermore, the MOSFET used to replace the diode turns off faster, without reverse recovery.

Since the newly-introduced power MOSFET needs to conduct current when the diode conducts current, the converter is called a *synchronous converter.* However, the diode is still used in the circuit to quickly turn on with the effect of the voltage induced by the inductance when the main transistor turns off. The synchronous command comes shortly after and the synchronous MOSFET may already see the diode in conduction. The synchronous MOSFET turns on with the command at a drain-source voltage lower than the voltage drop across the diode, which means the diode will turn off and the transistor stays turned on.

8.3 ISOLATED CONVERTERS

Certain applications require a galvanic separation between the supply and load circuits. This is achieved inside the power supply with a power transformer able to transfer pulses from a switch-mode circuit toward a secondary. The pulsed voltage from the secondary is rectified into the output dc voltage. Examples of topologies using this principle are presented next.

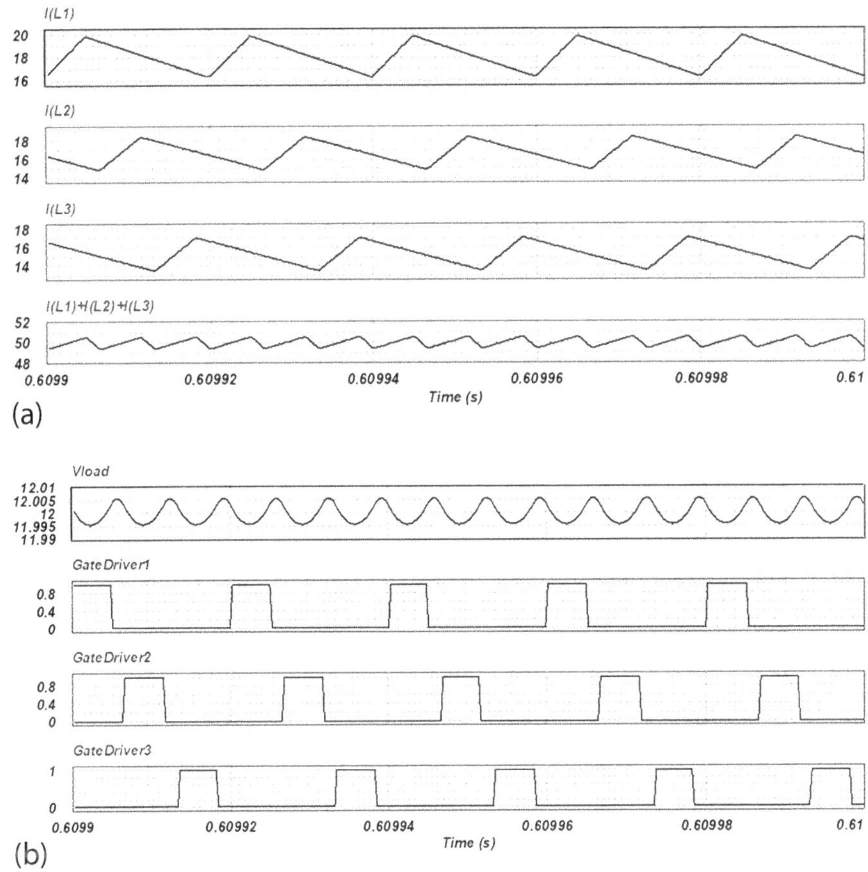

FIGURE 8.11 Control pulses and inductances' currents.

8.3.1 Flyback Converter

The flyback converter is illustrated in Figure 8.12.

When the transistor is turned on, the supply voltage is applied to the primary winding of the transformer. Since the voltage reflected on the secondary winding is negative, the diode maintains its off state. The transformer has a special construction leading to a finite magnetizing inductance. The collector/drain current is increasing linearly with a slope provided by the ratio between the supply voltage and the magnetizing inductance. Energy is thus stored within the magnetic field of the transformer.

After the transistor is turned off, the primary winding of the transformer tries to maintain the current circulation and generates a voltage in this respect. Since the current was flowing toward the transistor, this voltage would appear with a positive terminal near the transistor. This polarity of the voltage does not encourage any current flow on the primary side, but gets reflected into the secondary winding and

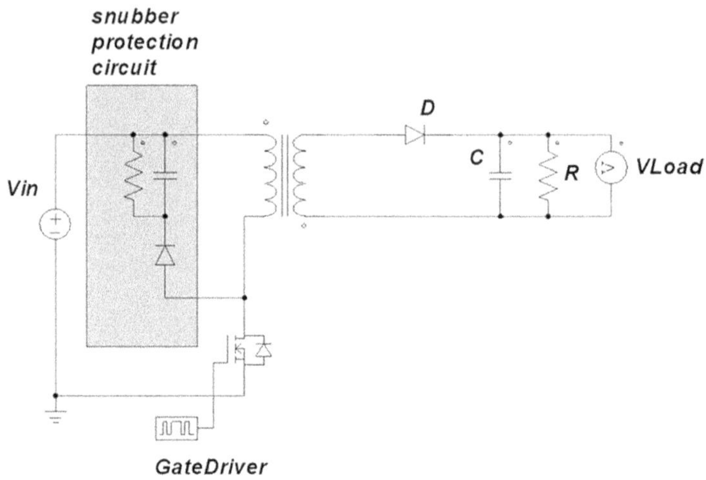

FIGURE 8.12 A flyback converter.

turns on the secondary side diode. The transformer ratio N_1:N_2 determines the output voltage, not unlike a duty cycle. The current circulation through the secondary side diode is now based on the energy stored previously in the magnetic field of the transformer and this looks similar to boost converter operation.

The circuit shown in Figure 8.12 also features a protection circuit called a *snubber.* Right after the transistor turns off, the leakage inductance on the primary side produces some overvoltage with negative effects for the power MOSFET. This snubber circuit takes over such overvoltage and limits the drain voltage.

The operation during the two conduction states of the transistor can be followed in Figure 8.13.

During the turn-on interval, equations similar to the boost converter can be written for the magnetic flux within the transformer, as shown in Figure 8.14.

$$\Phi(t) = \Phi(0) + \frac{V_{in}}{N_1} \cdot t \tag{8.14}$$

$$\Phi_{peak} = \Phi(0) + \frac{V_{in}}{N_1} \cdot t_{on} \tag{8.15}$$

The maximum flux (Φ_{peak}) occurs at the end of the conduction interval t_{on}. If the flux is not zero at the end of the period, after the interval with the transistor is in the off state, it yields:

$$\Phi(t) = \Phi_{peak} - \frac{V_o}{N_2} \cdot [t - t_{on}] \tag{8.16}$$

$$\Phi(T) = \Phi_{peak} - \frac{V_o}{N_2} \cdot [T - t_{on}] = \Phi(0) + \frac{V_{in}}{N_1} \cdot t_{on} - \frac{V_o}{N_2} \cdot [T - t_{on}] \tag{8.17}$$

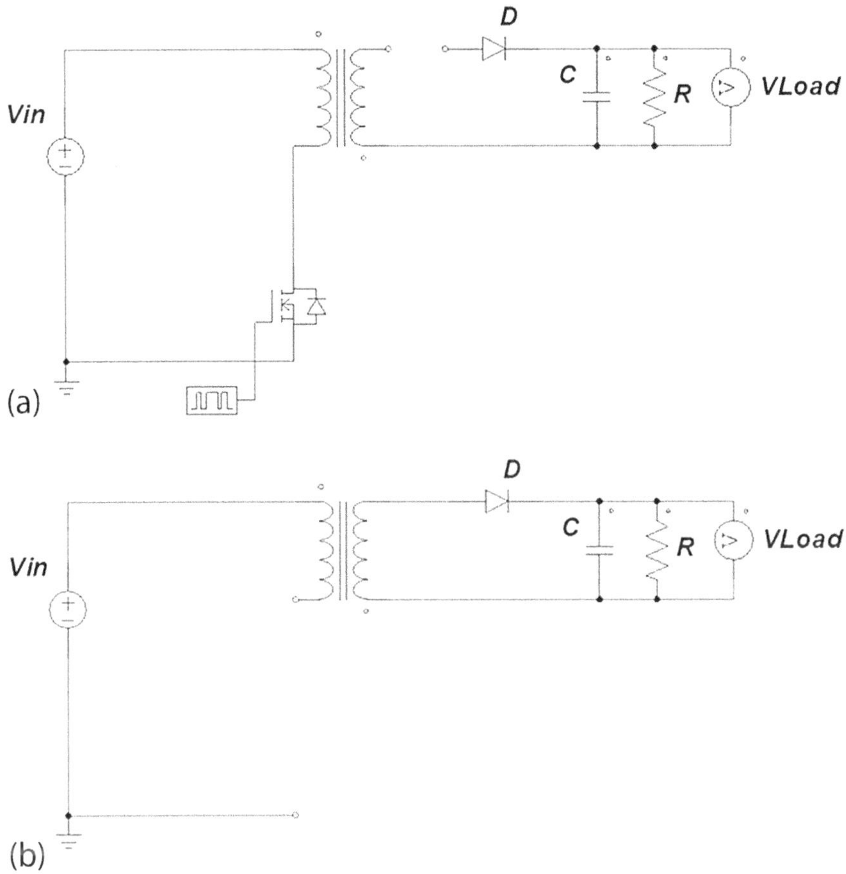

FIGURE 8.13 The two states for operation of a flyback converter in continuous conduction mode.

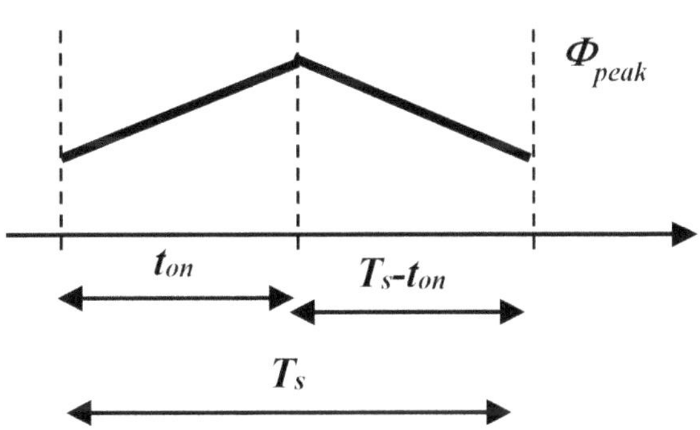

FIGURE 8.14 Magnetic flux of the transformer.

In the steady-state regime of operation, the cyclical waveform leads to Φ(T) = Φ(0). Hence

$$\Phi(0)+\frac{V_{in}}{N_1}\cdot t_{on}-\frac{V_o}{N_2}\cdot\left[T-t_{on}\right]=\Phi(0)$$
$$\Rightarrow\frac{V_{in}}{N_1}\cdot t_{on}=\frac{V_o}{N_2}\cdot\left[T-t_{on}\right]\Rightarrow V_o=V_{in}\cdot\frac{N_2}{N_1}\cdot\frac{t_{on}}{T-t_{on}} \quad (8.18)$$

If we neglect the caloric loss and consider a continuous flux, the transfer characteristic yields:

$$\frac{V_o}{V_{in}}=\frac{N_2}{N_1}\cdot\frac{D}{1-D} \quad (8.19)$$

The continuous conduction mode is defined when the transformer's magnetic flux does not reach zero as it can be observed with the waveforms from Figure 8.15. This numerical example shows results achieved for an output voltage of 5 V dc, an input voltage of 12 V dc, a duty cycle of 0.295, a transformer ratio N_1:N_2 = 1:1, and a design with nearly ideal components, while using the relationship:

$$V_o=V_{in}\cdot\frac{N_2}{N_1}\cdot\frac{D}{1-D}=12\cdot\frac{1}{1}\cdot 0.4167=\ 5\text{ V} \quad (8.20)$$

Similar to the boost converter, the discontinuous conduction mode occurs when the transformer flux yields zero before the end of the period. The secondary side diode turns off and a third interval appears when both the primary-side transistor and the secondary side diode are turned off. For the discontinuous conduction mode, the relationship (8.20) is not valid anymore, and the output voltage yields larger than this. Even if nonlinear, the discontinuous conduction mode presents the advantage

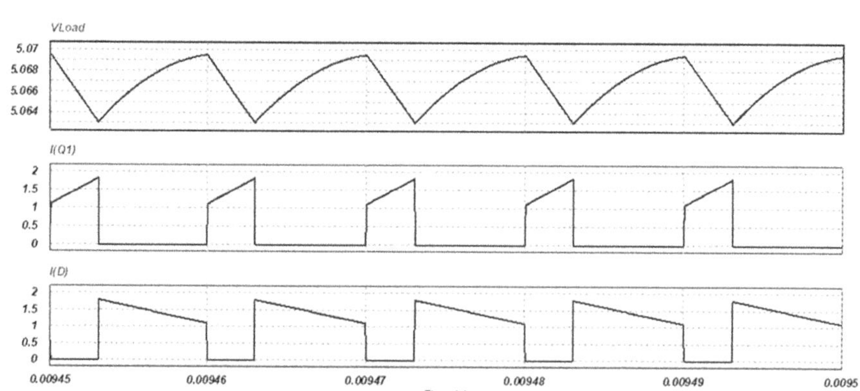

FIGURE 8.15 Waveforms for a flyback converter operated in continuous conduction mode.

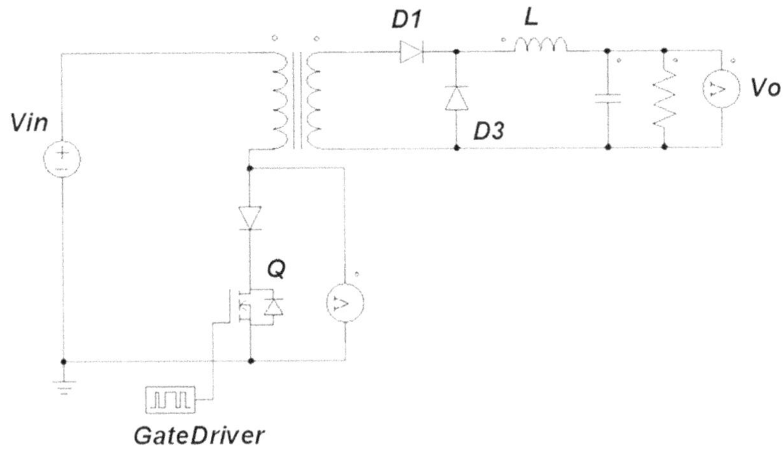

FIGURE 8.16 Forward converter.

of a reduction of the converter loss since the diodes are turned off at zero current without reverse recovery, while the transistor also enters into conduction at zero current.

8.3.2 Direct (Forward) Converter

The direct converter is illustrated in Figure 8.16. It looks somewhat similar to a *flyback converter*, yet it is different since the transformer is connected differently so that the secondary diode is turned on at the same time as the primary transistor. The transformer is as close to ideal as possible, without finite magnetization inductance, as was the case with a *flyback converter*. The diode D_3 works as a freewheeling diode in a buck converter.

When the transistor is turned on, the supply voltage is transferred to the secondary and the diode D_1 allows for charging the output capacitor. When transistor and diode D_1 turn off, the filter inductance L tries to maintain the current circulation through the generation of a voltage. Similar to the buck converter, this induced voltage turns on the freewheeling diode D_3. Since the energy is transferred directly, a simple transfer law yields as follows.

$$\frac{V_o}{V_{in}} = \frac{N_2}{N_1} \cdot D \tag{8.21}$$

The main waveforms are very intuitive, as shown in Figure 8.17 with a converter supplied with 12 V and able to deliver 5 V.

The limitation for the transfer of energy through the direct converter comes from the possibility of the transformer magnetizing due to the unipolar current pulse transferred through the forward converter. The solution consists of a discharge winding for the transformer energy called transformer reset winding, as is shown in Figure 8.18.

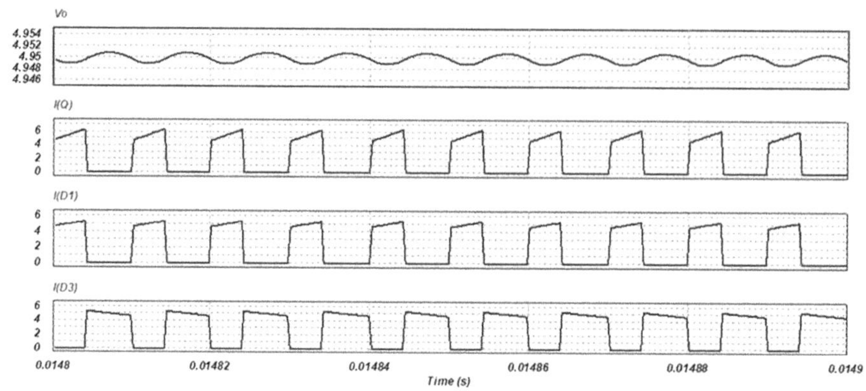

FIGURE 8.17 Main waveforms for a forward converter.

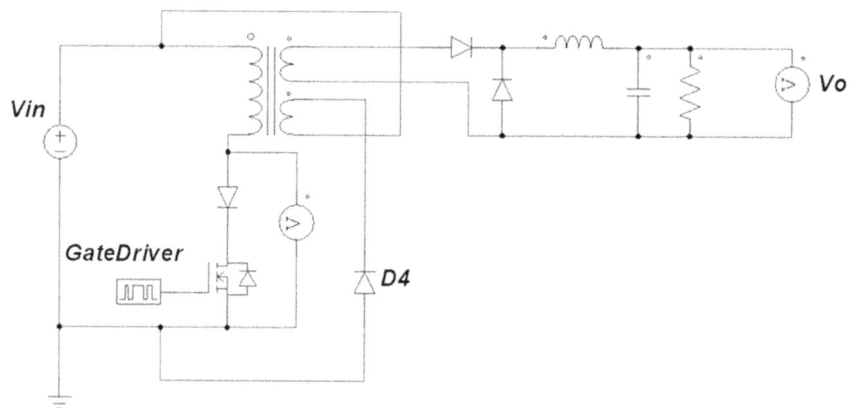

FIGURE 8.18 Forward converter with a discharge branch.

For this version of the converter, during conduction of the D_4 diode, the voltage reflected in the primary winding becomes as follows.

$$v_{IN} = -\frac{N_1}{N_3} \cdot V_d \tag{8.22}$$

The demagnetization current through diode D_4 yields as follows.

$$i_3 = \frac{N_1}{N_3} \cdot i_m \tag{8.23}$$

This constraint can be expressed as the condition of the current/flux through the transformer to get to zero before the next conduction interval of the transistor in order to complete a full discharge.

$$t_m < (1 - D) \cdot T \tag{8.24}$$

It yields for the conservation of the maximum (or peak) current:

$$\begin{cases} Sw = ON \Rightarrow V_d = L_m \cdot \dfrac{di_m}{dt} = L_m \cdot \dfrac{I_{m,\max}}{D \cdot T} \Rightarrow I_{m,\max} = \dfrac{V_d \cdot D \cdot T}{L_m} \\ Sw = OFF \Rightarrow -\dfrac{N_1}{N_3} \cdot V_d = L_m \cdot \dfrac{di_m}{dt} = L_m \cdot \dfrac{I_{m,\max}}{t_m} \Rightarrow I_{m,\max} = \dfrac{\left(\dfrac{N_1}{N_3} \cdot V_d\right) \cdot t_m}{L_m} \end{cases}$$

$$\Rightarrow D \cdot T = \left(\frac{N_1}{N_3}\right) \cdot t_m$$

$$\Rightarrow t_m = \left(\frac{N_3}{N_1}\right) \cdot D \cdot T < (1-D) \cdot T \Rightarrow D < \left(\frac{N_1}{N_3}\right) \cdot (1-D) \Rightarrow D < D_{\max} = \frac{1}{1+\dfrac{N_3}{N_1}} \tag{8.25}$$

$$D \frac{1}{1+\dfrac{N_3}{N_1}}_{\max} \tag{8.26}$$

For instance, when $N_3 = N_1$, then $D_{max} = 50\%$.

It is noteworthy that a very large isolation voltage is not needed between the primary winding and the demagnetization winding since both connect to the same dc bus. Furthermore, the current through the demagnetization winding is lower than the current through the primary or secondary winding.

8.3.3 Push–Pull Converter

The schematic of the push–pull converter is shown in Figure 8.19 and this converter can be seen as two forward converters that add their effects onto the same load. The operation is explained first for an ideal converter since the presence of passive components determines many operational aspects which can be revealed afterward.

When Q_1 is turned on, the high-side primary winding is supplied with the input voltage V_{in} and the energy is transferred to the secondary winding, determining the conduction of the diode D_{s1} and a direct transfer of energy to the load. After Q_1 is turned off by command, voltage across all transformer windings tend to be zero. Zero voltage on secondary side windings means these can be seen as short-circuit (a simple wire) connections. The inductor within the output filter tends to maintain the circulation of the current toward the load. This induced voltage brings or maintains in conduction both rectifier diodes, Ds_1 and Ds_2. Not unlike the buck converter, the conduction time for Ds_1 and Ds_2 can be until the next control of a transistor (which is called a continuous conduction mode) or shorter if their current yields zero (discontinuous conduction mode).

In the second half of the operation period, Q_2 is turned on. The low-side primary winding is supplied with input voltage V_{in}. The energy is transferred to the

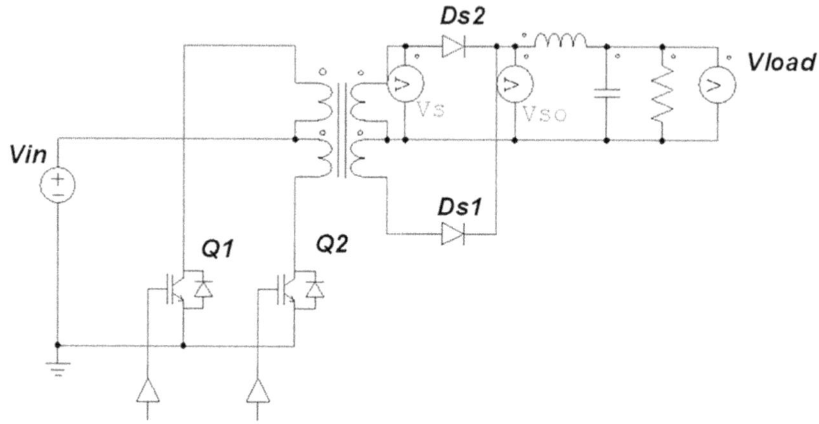

FIGURE 8.19 Push–pull converter.

secondary in a forward mode, determining the conduction of the diode D_{s2} and the transfer of energy to the load. During this interval, the transformer sees a flux in the opposite direction from the time interval when Q_1 was in conduction. The magnetic flux through the transformer therefore has opposite signs during the two transistor conduction intervals. This alternant operation avoids saturation of the magnetic core.

The duty cycle D is defined for the control of the two transistors Q_1 and Q_2, and it is measured for the entire period of operation. Thus, the maximum value of the duty cycle for a single transistor is $D_{max} = 0.5$. The transfer characteristic can be calculated as an average voltage for the train of pulses at V_{so}, as shown in Figure 8.19.

$$\frac{V_0}{V_d} = 2 \cdot \frac{N_2}{N_1} \cdot D \tag{8.27}$$

The ideal waveforms when considering an ideal transformer and continuous conduction mode are shown in Figure 8.20, for converter data: $V_{in} = 48$ V dc, N = 5:5:1:1, D = 0.26.

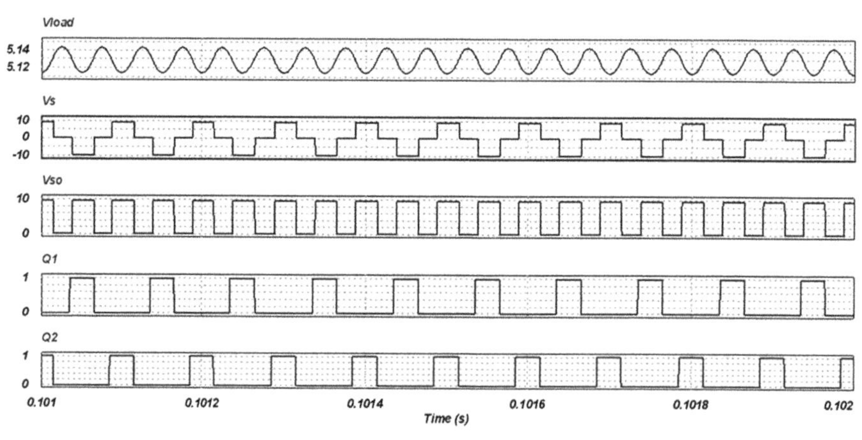

FIGURE 8.20 Main waveforms for operation of a push–pull converter.

The details of the waveforms depend in practice on the parasitic elements of the converter, such as the transformer parameters and load current. The main difference may consist of the time intervals of actual conduction through secondary side diodes. Whatever the peculiar details introduced by the parasitic elements, the principle of operation remains the same.

Finally, let us note that the anti-parallel diodes, near the transistors, work as discharge branches: after the conduction of Q_l, the diode near Q_2 becomes a discharge branch for the transformer, as does the diode near Q_1 after the conduction of the transistor Q_2.

8.3.4 Phase-Shift Converter

At higher power levels, the efficiency of the power converter becomes even more important and different solutions are proposed. Among them, the most used isolated converter in the kW range is the *phase-shift converter.* The operation of this converter is a little more complex and a simplified form is explained herein for a good understanding of the concept.

As shown in Figure 8.21, the phase-shift converter is built around a four-transistor bridge, which transforms the dc supply voltage into ac voltage with a high frequency (50–150 kHz). The ac pulsed voltage is sent through a transformer. A simple rectifier transforms these pulses into dc voltage within the secondary circuitry. Finally, an L–C filter improves the output dc voltage quality.

Each inverter leg is controlled with a duty cycle fixed at 50% and a variable phase shift is used between the control pulses for the left-side leg and the right-side leg. Adjusting this phase shift regulates the output voltage. To simplify the explanation, the right-side leg is considered to switch after (following) the left-side leg. In other words, the phase shift is measured under 90°. This is shown in Figure 8.22.

The immediate turn-on command for a transistor is avoided right after the other transistor is turned off. Due to finite time intervals, it would be possible for these two transistors to conduct simultaneously for a short time and produce a short-circuit in the supply voltage. A time interval is inserted to avoid simultaneous conduction, and this is called dead-time. This dead-time has a more extended role here.

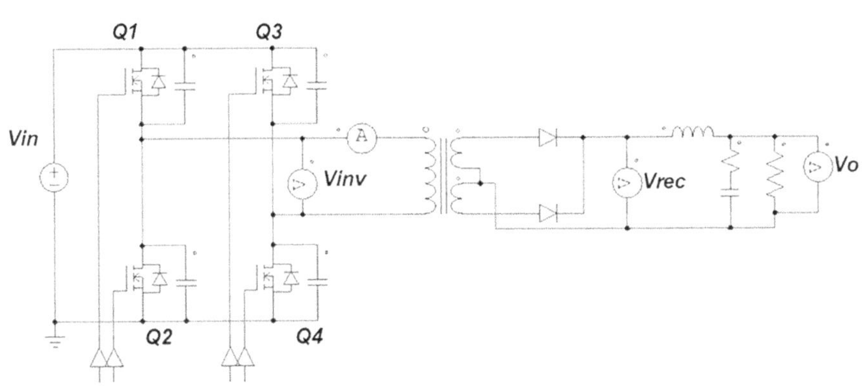

FIGURE 8.21 Phase-shift converter.

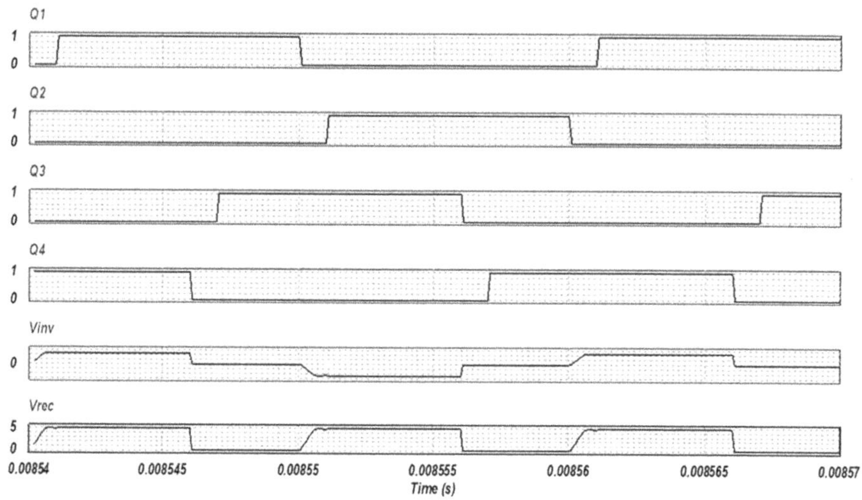

FIGURE 8.22 Main control waveforms.

Figure 8.21 also shows capacitors in parallel cu transistors. These can be considered as being the drain-source capacitance of the MOSFET transistors. These capacitors have an important role in energy saving since they slow down the transition between on and off states due to a current also passing through the inductance characteristic of the transformer. In this way, a transistor is controlled in conduction when the voltage across it is already near zero, and this saves a switching loss.

8.4 AUXILIARY POWER

8.4.1 Need for AC power

Along with the electronic control and actuation of various vehicle sub-systems, a series of loads are added for audio/video applications related to passenger comfort and convenience. While some of this equipment is directly supplied with electrical power from the low-voltage battery, others need a power converter from the conventional 12 Vdc battery to a local-grid voltage level of 120V@60Hz or 220V@50Hz. Until dedicated automotive devices come onto the market, auxiliary ac power is required through a dc/ac power converter for conventional household devices.

Appropriate converter products are present on the market and these can be found in specialty stores. Some of these products are dedicated to low power applications (< 100 W) and allow connection to the battery voltage through the cigar lighter connection, others require larger gauge wires connected directly onto the battery for higher power levels (100s of W to kW range).

The most important power electronics problem relates to boosting the voltage level from the battery voltage of 12 Vdc to more than 200 Vdc for a DC-powered inverter able to produce the local sinusoidal ac grid voltage. Such a high ratio of gain is difficult to be implemented within a single stage.

From the huge amount of possible converter solutions for this applications, two examples are included next.

8.4.2 Low Power Solutions

Since the installed power is low, the expectations for the cost of the power converter are also low. Hence, a solution with a reduced component count is sought after instead of a more conventional cascaded conversion through a high-frequency transformer able to raise the voltage level. Also, such low power applications may not impose important grounding restrictions, keeping in mind that the load may have an internal isolated power supply.

The most conventional solution would suggest the use of a boost converter. However, it is very difficult to produce a 12 V-to-200 V increase in a single step. This would mean a very low duty cycle D in boost operation (since $V_{out}/V_{in} = 1/(1–D)$), and a reduction of the control resolution. The alternative can be a cascaded structure with two boost converters sharing the step-up gains. Such a solution may unfortunately become too expensive for a low power application.

A possible cost reduction is suggested with the circuit in Figure 8.23, where a coupled-inductor allows the augmentation of the voltage along with the Z-source converter that is used to merge the inverter with the boost converter stage. The Z-source converter introduces additional zero vector states in the operation of a conventional inverter. During the zero states from the conventional inverter operation, the DC link is short-circuited like within the operation of a current source converter. Hence the "Z-source" name.

A low-pass passive filter is generally used to generate a smooth sinusoidal waveform on the load, which is not shown in Figure 8.23.

The operation of the Z-source converter implies that all the MOSFETs are turned on at the same time during a portion of the inverter's zero-state. Therein, the inductors accumulate energy during a boost converter stage since the two inductors have the capacitor voltage across, and their current increases linearly. There is no power transfer with the battery since the series diode D_1 is blocked.

For the rest of the inverter's zero-state, the load voltage is maintained at zero with the appropriate (high-side or low-side) pair of MOSFETs.

The boost inductor (secondary side of the transformer in our case) becomes connected in a series with a capacitor and current starts to flow from the battery into this LC circuit. A second circuit is identically formed in parallel, within the other inductor-capacitor group.

When the time is right from an inverter control perspective, the inverter operation requires the generation of an active state according to the PWM rule. The load becomes connected on the dc-side of the power converter and is supplied with the sum of voltages across capacitors minus the battery voltage. The capacitors should discharge a little under this load current.

Without entering in all the calculation details, it can be demonstrated that the theoretical relationship between the inverter dc-side voltage and the battery voltage yields:

$$V_{inv} = V_0 \cdot \frac{(1-x)\cdot\frac{N_1}{N_2} + x}{(1-x)\cdot\frac{N_1}{N_2} - x} = \frac{1+5\cdot x}{1-7\cdot x} \tag{8.28}$$

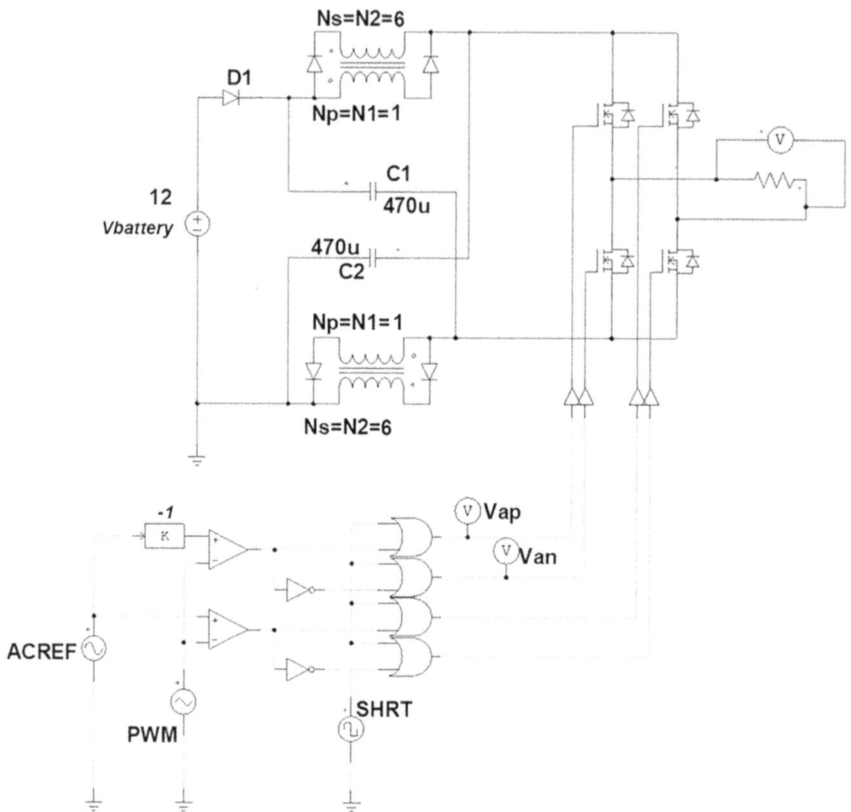

FIGURE 8.23 Principle of the non-isolated converter from dc battery to high-voltage ac load.

where x iss used to define the boost duty cycle (shoot-through interval):

$$x = \frac{T_o}{T} \tag{8.29}$$

T is half the PWM frequency of the inverter.

Figure 8.24 Converter waveforms: V_{ap} = control of the high-side inverter switch on the left branch; V_{an} = control of the low-side inverter switch on the left branch; I_{D1} = input current without filtering; $V_{inverter}$ = voltage at the inverter input; I_{D5}, I_{D6} = currents through primary and secondary windings of the boost inductor.

The dependency in Equation (8.28) is similar to the results for any Z-source inverter, with the multiplication introduced herein by the transformer(s). However, these results are impossible to achieve in practice. The transformer leakage inductance and the actual load current will influence waveforms and gain. The inverter dc side voltage is selected from the conventional operation of the PWM inverter with a modulation index that has to allow zero states larger than the shoot-through interval.

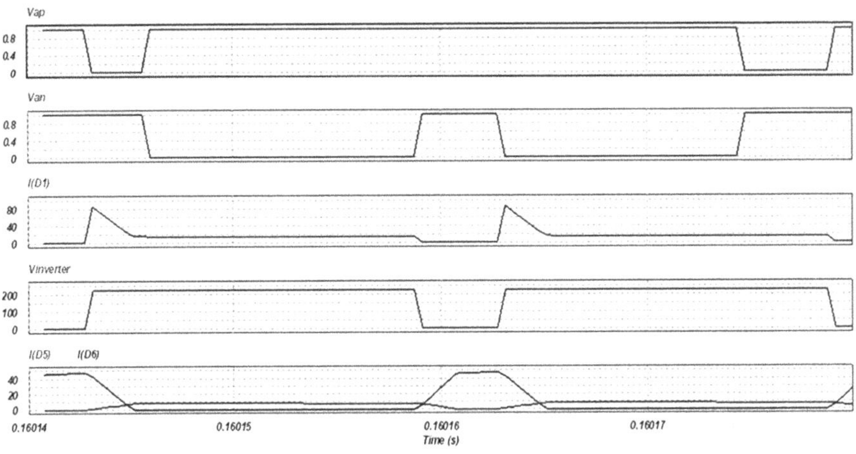

FIGURE 8.24 illustrates a small detail of the operation.

Furthermore, a drawback for this topology is the current peak on the battery and input diode. During the current commutation from one winding to the other, when both diodes are briefly conducting current, the equivalent inductance yields very low and the current spikes. For instance, a design for a 100 W load produces a repetitive spike of currents at 100 A. An increase in the leakage or other additional series inductance limits this spike, by design.

8.4.3 High Power Solutions

At higher power levels, an isolated converter may be required to separate the load grounding from the battery grounding. This solution can now benefit from a conventional ac plug and allow various household loads to be connected to this local grid.

The design requirements push for a high power density, which can be achieved with intensive integration. Integration of electronics is easier if the semiconductors are rated at a lower voltage, preferably under 60 Vdc.

Figure 8.25 illustrates an example, with a rather complex power converter structure, built after the flyback/forward converter principle, that allows energy transfer both as a flyback converter with the storage of energy and as a forward converter.

On the primary side of the isolation transformer, two switches are connected in a series with the primary winding and allow the supply according to the flyback converter principle. Furthermore, a second set of switches also allows the very simple winding to be supplied with a negative voltage. Overall, this setup allows for the production of the output voltage with both polarities as required by an ac waveform.

In order to simplify the control circuitry and to reduce loss, the primary side is operated under a DCM. The duty cycle is adjusted according to a rectified sinusoidal reference, and the average value of the current pulses in the secondary side follow this sinusoidal reference. Moreover, a four-switch bridge-like topology is considered for rectifying the secondary voltage onto the output capacitor(s). Their operation is similar to the synchronous rectification principles.

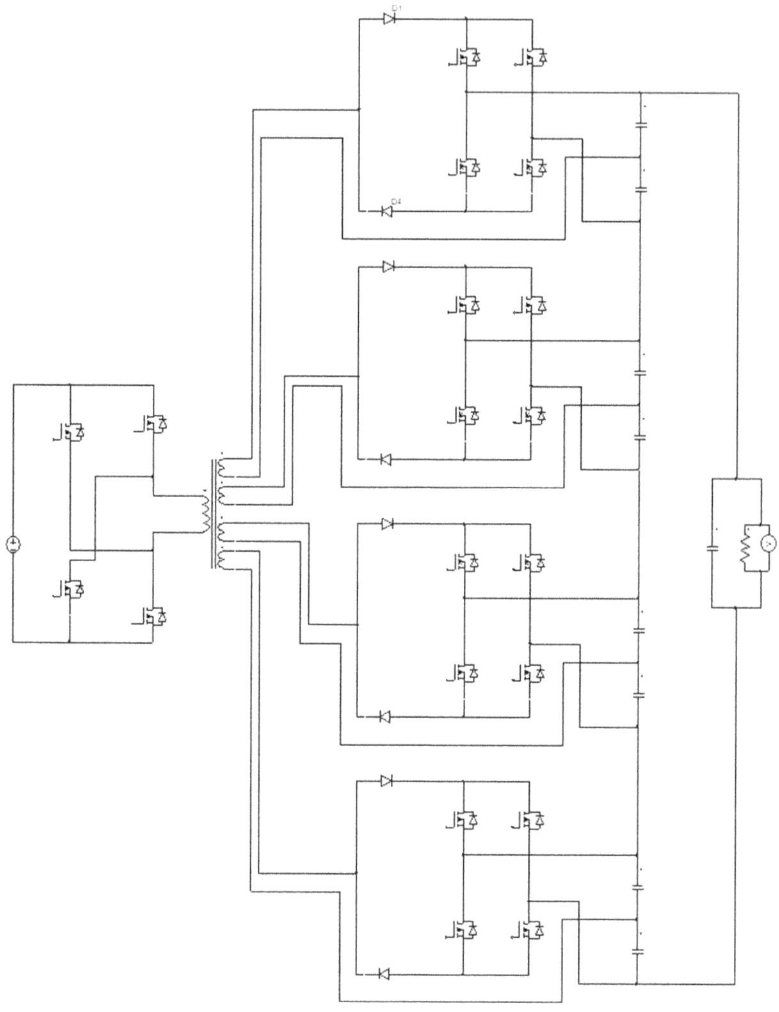

FIGURE 8.25 Isolated high-gain converter.

One topology that has proven its merits for high-gain conversion relates to the voltage multiplier rectification of a boost or flyback converter. Two secondary side stages are considered and different power flow is intended for each secondary side rectifier-capacitor stage.

During the "on" time of the main/primary-side switch, the magnetizing inductance of the transformer stores energy while a certain amount is transferred to the secondary winding through a forward conversion mode. When the main switch is turned off, the energy stored within the magnetizing inductance is transferred to the secondary through a conventional flyback mode.

There is a different voltage waveform for the two output capacitors V_{C1} and V_{C2}, at the right-side of each H-bridge. One capacitor benefits from the forward converter mode while the other from the flyback converter mode. During the second alternance of the sinusoidal waveforms, the roles reverse. Since the shapes of the voltages across the two output capacitors are not identical to each other, the control loop is closed from the sum of both voltages, which is also the voltage on the total output capacitor.

To increase the output voltage further, a multilevel converter structure is adopted on the output side. The major advantage of multilevel converters relies on the ability to generate high voltage with power semiconductors rated at lower voltage levels. This principle is used herein to add voltage from different circuits into a grid-level voltage.

Reasoning backward, the voltage across each multiplier voltage secondary should be up to 50 V, so that the output voltage can be generated up to 4 × 50 V = 200 V more than enough for the generation of the peak of grid voltage, with certain dynamics allowed. The voltage across each capacitor yields a maximum peak of around 25 V, obviously also considering the different waveform shapes for flyback and forward modes.

Based on these considerations, all the power semiconductors in the secondary circuit can be rated at 50 Vdc and 4 A, with minimal thermal requirements as only one device is on at a given moment. This means the it is possible to use conventional automotive MOSFET devices.

8.5 CONCLUSION

The dc/dc converters are used to convert electrical energy from a dc source to a dc load with a different dc level requirement. These converters can also offer isolation or not. Most non-isolated dc converters have a topology based on a single switch. The buck or boost converters can operate in a discontinuous or continuous conduction mode depending on the continuity of the current through the inductance. The analysis of these converters and the operation limits are discussed.

Other non-isolated topologies include the buck/boost converter, and the cuk converter for converters with a single switch, or the H-bridge converter with a bipolar operation. The synchronous converter operation is also involved in efficiency improvement.

In some cases, galvanic isolation is required in dc/dc conversion and this is achieved with either forward, flyback, push–pull, or phase-shift converters. These topologies have been briefly discussed in this chapter.

It is important to retain that the building of the converter strongly depends on the power level. Low-power-level converters can be built inside a three-pin integrated circuit, while higher power may require discrete implementation and advanced air or liquid cooling technologies. While the topology is the same, the actual converter may be very different as shape. This aspect differentiates power converters from most of the other electronic circuits.

Finally, this chapter discussed the use of power converters in auxiliary power sources able to supply household equipment from an automotive battery.

REFERENCES

Alexanderson, E.F.W., Phillipi, E.L.. 1944. History and development of the electronic power converter. *Electrical Engineering*, 63(9):654–657.

Erickson, R., Maksimovic, D. 2001. *Fundamentals of Power Electronics*. New York: Springer.

Mößlacher, C., Guillemant, O. 2012. Optimum MOSFET selection for synchronous rectification. *Infineon Application Note AN 2012–05*, version 2.4, May 2012.

Neacşu, D.O. 2010. *Towards an All-Semiconductor Power Converter Solution for the Appliance Market*. Presented in IEEE IECON Conference 2010, pp. 1677–1682.

Neacşu, D. 2018. *Telecom Power Systems*. Boca Raton, FL: CRC Press.

Nowakowski, R., Tan, N. 2009. Efficiency of synchronous versus non-synchronous buck converters. *Texas Instruments Analog Applications Journal*, 2009(4Q):15–20.

Panacek, J. 2017. *TI Designs: TIDA-01168 Bidirectional DC-DC Converter Reference Design for 12-V/48-V Automotive Systems*. Texas Instruments Design Note 79TIDUCS2B, June 2017, Revised March 2018, Internet reading at http://www.ti.com/lit/ug/tiducs2b/tiducs2b.pdf.

Ridley, R. 2000. Second-stage LC filter design. *Switching Power Magazine*, July 2000, pp.8–10.

Shan, Z.Y., Tan, S.C., Tse, C.K. 2013. Transient mitigation of DC–DC converters for high output current slew rate applications. *IEEE Transactions on Power Electronics*, 28(5):2377–2388.

9 Feedback Control Systems

HISTORICAL MILESTONES

1782–1806: Pierre-Simon Laplace used a similar transformation in his work on probability theory.

1892–1910: French mathematician Henri Poincaré is considered the founder of dynamical systems with two monographs, New Methods of Celestial Mechanics (1892–1899) and Lectures on Celestial Mechanics (1905–1910).

1922: A control law similar to the proportional-integrator (PI)/D controller was first developed using theoretical analysis by Russian American engineer Nicolas Minorsky.

1937: German mathematician Gustav Doetsch explored the modern formation and permanent structure of the Laplace transform and its application to dynamical systems in his work Theorie und Anwendung der Laplace-Transformation.

1952: Important control developments made by Ragazzini and Zadeh in sampled-data control group at Columbia University with the use of Z-transform.

1967–1971: The first microprocessors, the Four-Phase Systems AL1, Garrett AiResearch MP944, and Intel 4004 started the process of transitioning toward embedded power applications.

1975: The PWM control integrated circuit was invented by Bob Mammano and introduced to the market in 1976 by Silicon General Company as SG1524.

1983: D.H. Venable proposed a systematic design procedure for lead-lag compensation to power supplies using a K-factor.

9.1 FEEDBACK CONTROL OF DYNAMIC SYSTEMS

A *dynamic system* is a system where change exists and it is described as a function with time dependency. The origins of dynamic systems can be located in *Newtonian mechanics*. While the most intuitive definition refers to mechanical systems involving movement, time variation of physical measures could also define a dynamic system. Examples of the former include the swinging of a clock pendulum, the flow of water in a pipe, and the rolling of a vehicle on a highway. Examples for the latter include the variation of temperature in a room, current or voltage within an electrical circuit, or pressure in a vehicle's manifold.

Definitions related to dynamic systems and feedback control systems are introduced in Figure 9.1.

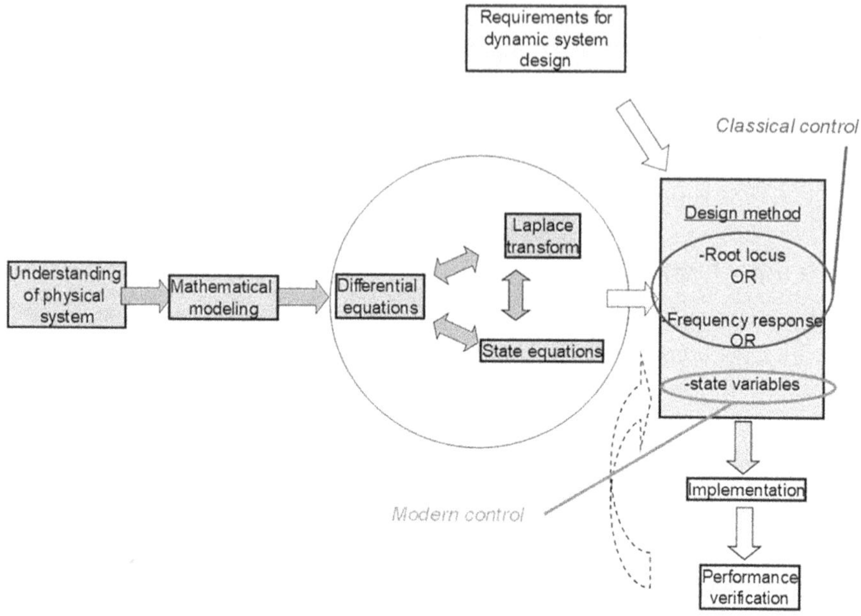

FIGURE 9.1 Definitions associated with feedback control systems.

Dynamic systems are characterized by differential equations with time as a variable. Hence, *mathematical modeling* is used to define these differential equations based on physical laws that characterize the operation of the system. Once the set of differential equations is defined, this is found to be equivalent to either a set of state equations or a Laplace model in the frequency domain.

Examples of the mathematical modeling of mechanical systems within a vehicle have been presented in Sections 4.9 "Cruise Control," 6.4. "Automotive Suspension," and can be easily extended in Section 6.3. "Electronic Control of Power Steering." These models can be further used for designing appropriate *regulators*.

The advantage of having a good mathematical model resides in the possibility to control various parameters of the original system with a *feedback control system*. In brief, a certain variable is measured and a proper action is quickly derived to keep that variable at a certain value or within certain dynamic evolution. The mathematical model (in either of its forms: differential equation, state equations, or Laplace frequency domain model) can be used to calculate this action of the control system. Since the operation follows the measurement of one or more of the system variables, the control system is called a *feedback control system*.

For a similar definition in automatic control, a *regulator* is a device that has the function of maintaining a designated characteristic. Chapter 14 "Propulsion Systems" presents numerous motor drive systems where regulators are used to control speed, torque, or currents. Those are presented therein without a comprehensive design.

Various design methods are involved in defining the right action of the control system against the controlled system. The best-known design methods are the *root locus method*, *frequency response design method*, or *state-space based design method*. These are used individually or together.

This chapter provides the basic information for understanding feedback control methods, as applied to power electronic converters. In most cases, the goal is to regulate the voltage supplied to a certain electronic circuit (load) in their nearby location. These examples of feedback control systems or regulators applied to dc/dc converters can be extended to more complex applications like motor drives. This chapter assumes a previous basic knowledge of the Laplace transform.

Design requirements for voltage or current regulators are usually defined for steady-state operation around a biased operation point. The output voltage variation with important operation parameters can also be characterized by the following performance indices. Parameters with subscript 1 refer to input, while subscript 2 refers to output.

- Regulation coefficient ($\Delta V_{out}/\Delta V_{in}$)

$$\frac{1}{S_0} = \left(\frac{\partial V_2}{\partial V_1}\right)\Bigg|_{\substack{I_2 = const \\ \theta = const}} \cong \left(\frac{\Delta V_2}{\Delta V_1}\right)\Bigg|_{\substack{I_2 = const \\ \theta = const}} \tag{9.1}$$

- Internal (output) resistance ($\Delta V_{out}/\Delta I_{out}$)

$$R_i = -\left(\frac{\partial V_2}{\partial I_2}\right)\Bigg|_{\substack{V_1 = const \\ \theta = const}} \cong -\left(\frac{\Delta V_2}{\Delta I_2}\right)\Bigg|_{\substack{I_2 = const \\ \theta = const}} \tag{9.2}$$

- Temperature coefficient ($\Delta V_{out}/\Delta T$)

$$K_\theta = \left(\frac{\partial V_2}{\partial \theta}\right)\Bigg|_{\substack{V_1 = const \\ I_2 = const}} \cong \left(\frac{\Delta V_2}{\Delta \theta}\right)\Bigg|_{\substack{V_1 = const \\ I_2 = const}} \tag{9.3}$$

Additionally, an important steady-state characteristic is the *steady-state error*.

In many applications, it is important to assess how fast a system responds to a reference change or to a change in the operational environment. The requirements may be expanded with necessities related to how the response evolves in time, with overshoot or stabilization time definitions. These are called *dynamic characteristics*.

A voltage regulator hardware is composed of either analog control circuits, or a combination of analog and digital circuits. Their design is made based on requirements expressed by previous performance indices.

9.2 IMPLEMENTATION WITHIN ANALOG-MODE POWER SUPPLY CIRCUITS

Before the advent of switched-mode power supplies, power semiconductor devices were used to regulate a voltage in analog mode. The principle is illustrated in Figure 9.2, where a variable resistor is connected between the supply voltage and the load. When the load voltage tends to decrease, the resistance is decreased so that the output voltage rises back to its desired value. The drawback of this solution consists of the large power loss across the series resistor.

For the implementation of this principle within a power supply, the series resistor is carried out with a power semiconductor transistor connected in a series, between the input and output. The collector-emitter voltage (for bipolar and IGBT transistors) or drain-source voltage (for MOSFET transistors) takes over the voltage difference between input and output (Figure 9.3). The feedback control system illustrated in Figure 9.2 is implemented within a circuit architecture, as shown in Figure 9.4.

The resistive divider R_1–R_2 senses the output voltage. A reference voltage is created with a Zener diode for a discrete circuit or with a band gap reference circuit (usually ~1.22V) when the controller is implemented inside an integrated circuit. The two voltages are compared with an *error amplifier* and the operation of the series transistor (also called *series regulator element* or *pass device*) is modified accordingly.

The *pass device* is the most important component herein since the entire load current passes through this transistor and the efficiency of the power supply and power loss depends on the operation of this transistor. It is therefore advantageous to work with a small voltage drop on the series device, which means there is a small difference between the input and output voltages. Such operation is referred to as *low dropout voltage* (LDO) circuits.

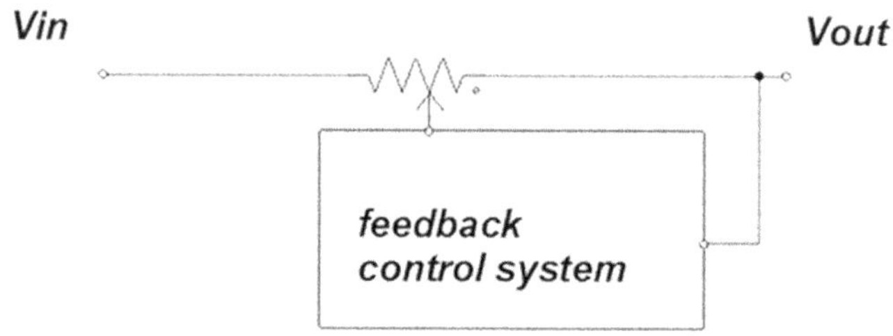

FIGURE 9.2 Principle of a power supply with analog implementation.

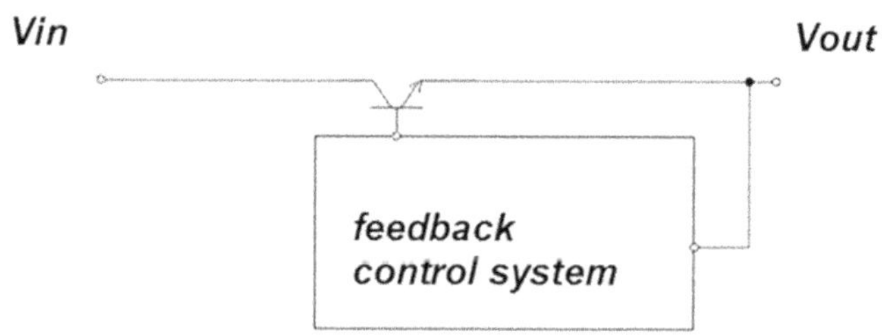

FIGURE 9.3 Implementation with a power transistor.

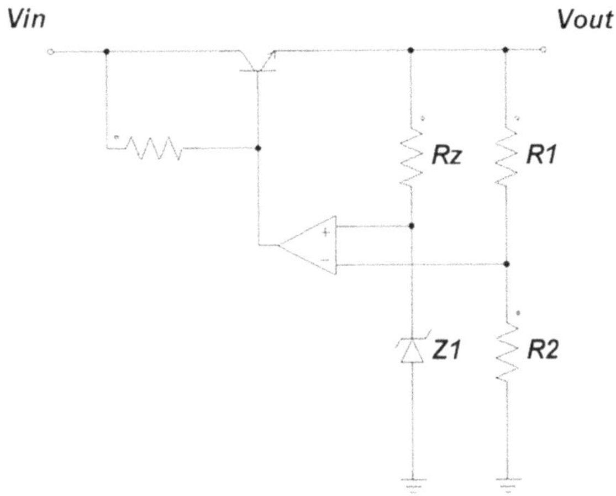

FIGURE 9.4 Analog-mode power supply.

TABLE 9.1
Performance Comparison

	(a) NPN	(b) Darlington NPN	(c) PNP	(d) PNP/ NPN	(e) PMOS (external to circ)
Minimum dropout voltage	1 V	2 V	0.1 V	1.5 V	$R_{dson} * I_L$
Current	< 1 A	> 1 A	< 1 A	> 1 A	> 1 A
Topology	*follower*, CC	*follower*	Inverting (CE)	Inverting (CE)	Inverting (CE)
Output impedance	small	small	large	large	large
Bandwidth	large	large	small	small	small
Do we need output capacitance for stability?	no	no	yes	yes	yes

All possible power semiconductor transistor solutions for the pass device are investigated in Table 9.1 with respect to controllability and power loss. The lowest voltage drop is obtained when using inverting topologies, which in turn carry stability problems.

9.3 DESIGN OF FEEDBACK CONTROL SYSTEMS

9.3.1 Definitions

Figure 9.5 illustrates the block diagram of a feedback control system. The output variable is detected by a sensor that is subject to electrical noise. The outcome is further compared with a reference, and a control law is applied to correct or compensate

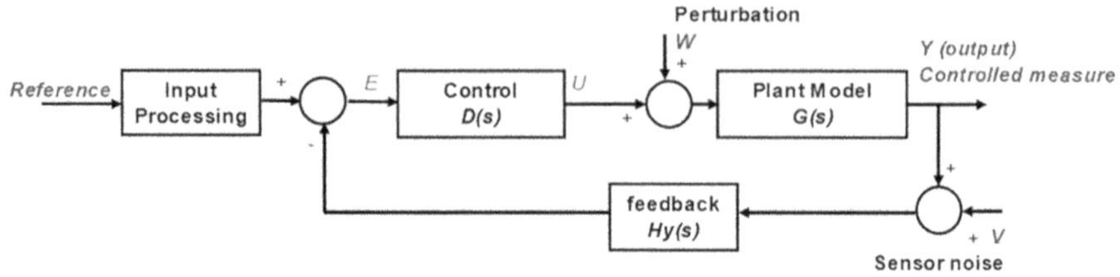

FIGURE 9.5 Feedback Control system.

for what is different from the reference. Since this is able to respond to fast variations, the entire diagram is considered with *frequency models*, also called *Laplace models*. In order to understand and design the control law, the mathematical model, in the frequency domain, for the plant (installation, equipment, process) is first required.

The electronic circuit which implements the control law *D(s)* is called the controller. The implementation with linear electronic circuits (amplifiers) defines a *linear control*.

Small-signal modeling is a very common technique used to approximate the dynamic behavior of a system around a known *bias* (fixed, nominal) *operation point*. This is possible if the dynamic component of the signals has a magnitude considerably smaller than the nominal component of the same signal. This small-signal model yields linear systems, and it can even approximate nonlinear systems for small variations around a nominal operation point. When applied to power converters, the small-signal model allows for the description of *system dynamics* around an operation point defined with the expected quasi-constant input voltage, output voltage, duty cycle, and load current.

In most cases, dynamic design requirements are provided with respect to a response of the system to a step (sudden) change in the input. Figure 9.6 defines the most important design parameters derived from the response to a step excitation.

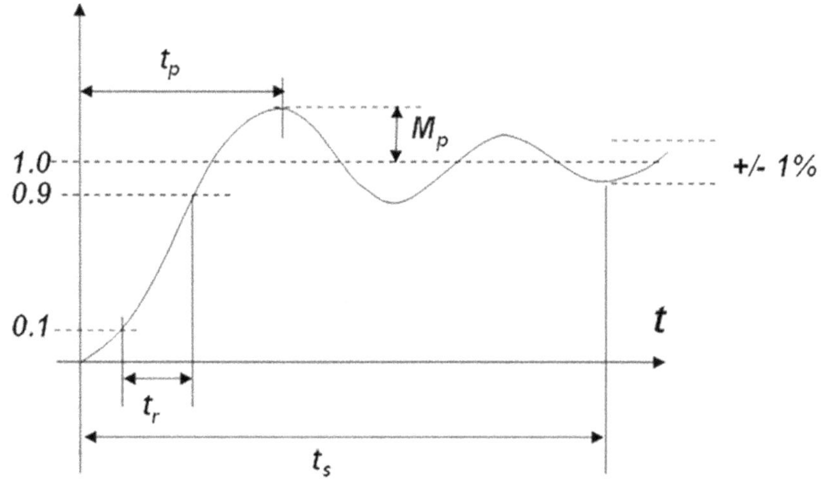

FIGURE 9.6 Dynamic parameters derived from a step response.

These are herein shown for a second-order system, which is the most common case. The parameters are: *overshoot* (M_p), *rise time* (t_r), and *stabilization time* (t_s).

The most important concern is *stability of the system*. By definition, a system is stable if any initial condition converges to zero, which is equivalent to saying that any transient response, in small-signal variables, decreases toward zero. Conversely, the system is unstable if the output becomes divergent in time.

As a criterium for stability, a linear and time-invariant system is stable if the Laplace transfer function has all the denominator's roots (also called *poles*) with a negative Real component. In other words, poles are placed on the left side of the complex plane.

The dynamics of the power converter can be described with a Laplace transfer function *G(s)* and the compensation law can be described with a Laplace transfer function *D(s)*. Using the negative feedback configuration shown in Figure 9.5 and neglecting noise and perturbation produces a closed-loop transfer function:

$$H_{cl}(s) = \frac{D(s)\cdot G(s)}{1 + D(s)\cdot G(s)} \tag{9.4}$$

This becomes unstable if the denominator equals zero. Hence, the stability is secured when:

$$1 + D(s)\cdot G(s) \neq 0 \Leftrightarrow D(s)\cdot G(s) \neq -1 \tag{9.5}$$

Within the complex plane, Equation (9.2) is false when the magnitude equals unity and the phase equals –180°, with the latter being the analytical form of the negative sign.

Since the frequency domain small-signal model for the power supply is defined approximately, and bearing in mind a possible variation of the parameters, the measurement of the system's distance from an instability condition becomes very important.

The analysis is made with *Bode diagrams*, which separate magnitude and phase, and present results with respect to frequency. A *gain margin* and a *phase margin* can be defined as the distance to the instability constraint in a polar (magnitude, phase) representation. The *gain margin* is the distance from unity gain to the gain measured at the frequency where the phase is –180°, while the *phase margin* is measured as the phase distance to 180^0 when the gain is unity, which means 0 dB. The gain margin shows how much the gain can be increased until it reaches instability, while the phase margin shows how much the phase can be decreased until instability occurs.

More importantly for the design, the *phase margin* also corresponds to details of the system's transient response. In most cases, a system can be approximated with a second-order dominant system and design rules are easily established for this second-order system. A direct relationship between phase margin, attenuation, and overshoot from a hypothetical step response exists and it is illustrated in Figure 9.7.

As a design tool, Figure 9.8 illustrates the relationship between design requirements and the desired phase margin. The equations used to derive this result are explained next. For a second-order system, a direct relationship between phase

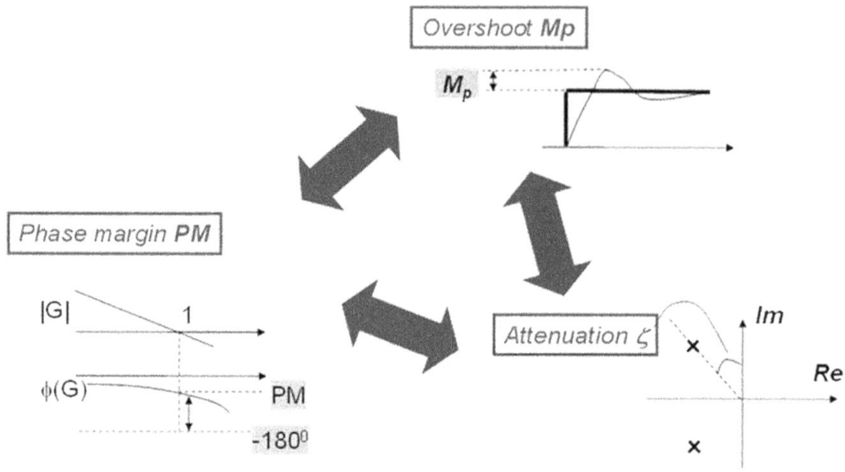

FIGURE 9.7 Relationship between phase margin, attenuation, and overshoot for a second-order system.

margin, attenuation, and overshoot exists, as it is explained in Figure 9.7. A system with unity feedback yields:

$$\begin{cases} open\_loop & G(s)\cdot D(s)=\dfrac{\omega_n^2}{s\cdot(s+2\cdot\xi\cdot\omega_n)} \\ closed\_loop & T(s)=\dfrac{\omega_n^2}{s^2+2\cdot\xi\cdot\omega_n\cdot s+\omega_n^2} \end{cases} \tag{9.6}$$

where ω_n represents the system natural frequency and ξ represents the attenuation.

This determines a phase margin of (given herein without demonstration)

$$PM=\tan^{-1}\left(\frac{2\cdot\xi}{\sqrt{\sqrt{1+4\cdot\xi^4}-2\cdot\xi^2}}\right) \tag{9.7}$$

Additionally, the overshoot recognized within the response to an input step variation applied to a second-order system is dependent upon the attenuation of that second-order system with the following relationship.

$$M_p=e^{-\frac{\pi\cdot\zeta}{\sqrt{1-\zeta^2}}} \tag{9.8}$$

9.3.2 Requirements for Feedback Control of a Power Supply

Power supply design experience acquired over the last 50 years has converged into the creation of general rules for the definition of stability margins and design constraints for power supplies (Figure 9.9).

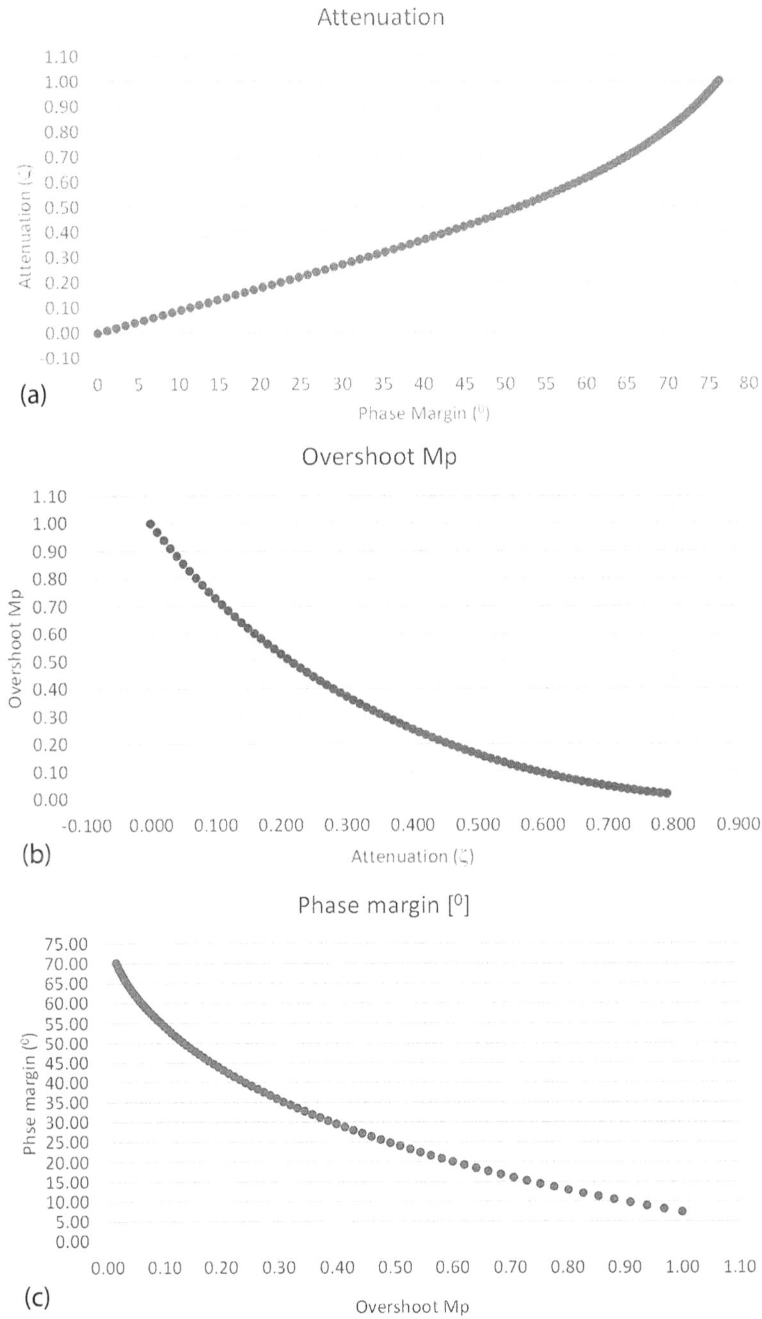

FIGURE 9.8 Dependency within a second-order system.

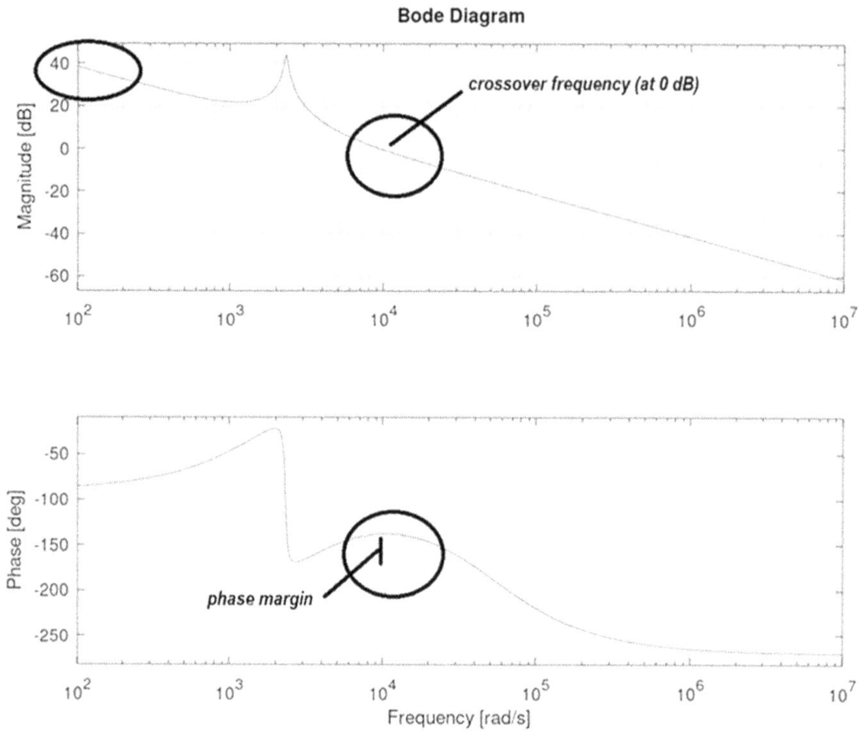

FIGURE 9.9 Graphical illustration of design rules, typical for a power supply.

1. A large dc gain at zero frequency is desired in order to reduce the steady-state error. This is usually accomplished with an integrative element (similar to a capacitor) placed within the compensation law.
2. A large crossover frequency (the frequency where gain = 1 or 0 dB) for the frequency response of the closed-loop system is required for a fast response. The crossover frequency is suggested at under 20% of the switching frequency of the switch-mode power supply and several times above the resonance frequency of the passive components.
3. A phase margin larger than at least 45° is desired, and preferable at around 60°.

These rules are visually introduced in Figure 9.9.

9.4 CASE STUDIES: FEEDBACK CONTROL FOR VARIOUS POWER SUPPLIES

Figure 9.10 shows the most typical power supply configurations. Since most power supplies fall within one of these four categories, each case is considered herein for the design of the control law. This starts with a small-signal model for the power circuitry, which is also called *plant* in control systems terms.

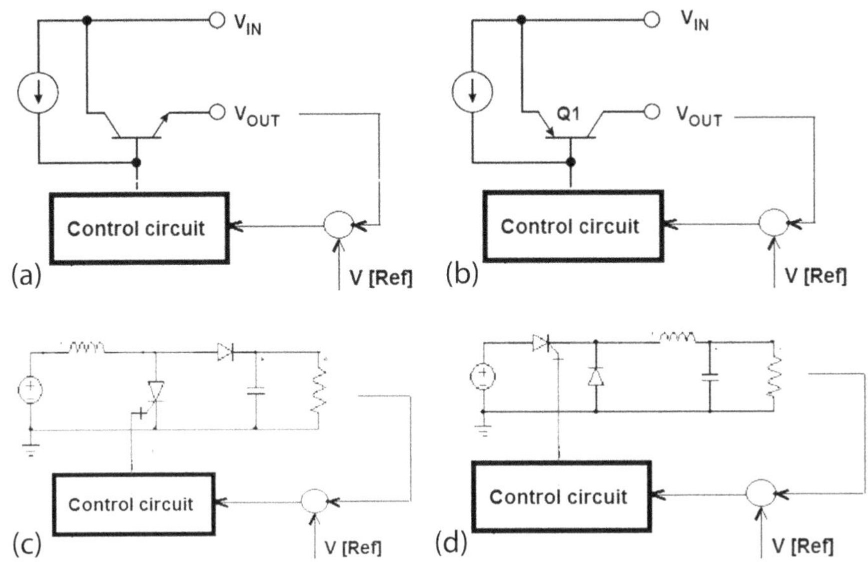

FIGURE 9.10 Typical power supply circuits considered as design examples.

(a) Analog-mode power supply in follower topology
(b) Analog-mode power supply with inverting topology
(c) Boost converter
(d) Buck converter

9.4.1 Analog Circuit in Follower Connection

The complete electrical circuit is illustrated in Figure 9.11 and the equivalent small-signal circuit is further represented in Figure 9.12. The latter can be described with the equation for the transfer function.

$$\begin{cases} A_{vi}(s) = \dfrac{V_o}{V_i} = \dfrac{(\beta+1)\cdot(CR_{load})}{r_\pi + r_x + (\beta+1)\cdot(CR_{load})} = \dfrac{(\beta+1)\cdot R_{load}}{(r_\pi + r_x)\cdot(1+s\cdot R_{load}\cdot C)+(\beta+1)\cdot R_{load}} \\ (CR_{load}) = \dfrac{R_{load}}{1+s\cdot R_{load}\cdot C} \end{cases}$$

$$A_{vi}(s) = \frac{(\beta+1)\cdot R_{load}}{s\cdot(r_\pi + r_x)\cdot(R_{load}\cdot C)+\left[r_\pi + r_x + (\beta+1)\cdot R_{load}\right]} \tag{9.9}$$

The pole in the power circuit model is moved at a higher frequency, namely from $[1/R_{load}\cdot C]$ to

$$\left[1+\frac{(\beta+1)\cdot R_{load}}{(r_\pi + r_x)}\right]\cdot\frac{1}{R_{load}\cdot C} \tag{9.10}$$

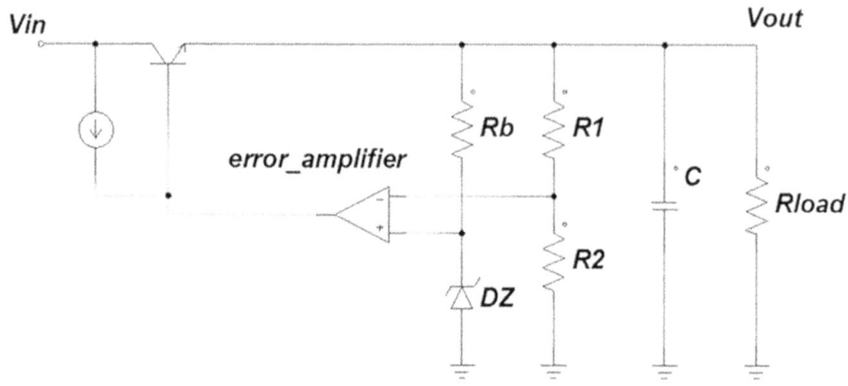

FIGURE 9.11 Circuit for the linear circuitry, follower (npn power transistor with common collector).

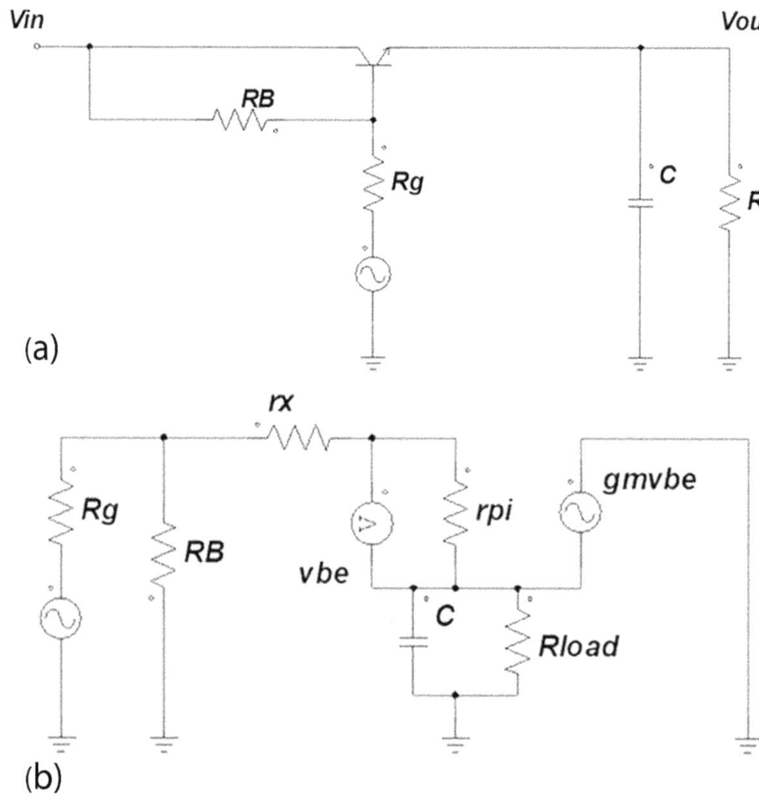

FIGURE 9.12 Equivalent circuit for the small signal analysis.

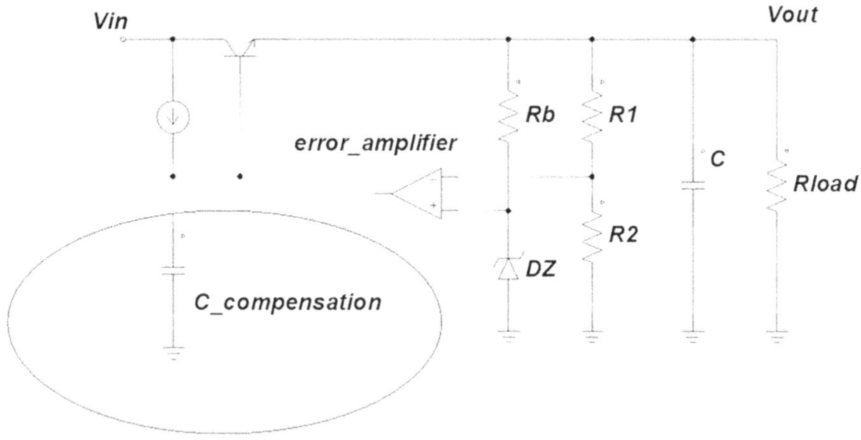

FIGURE 9.13 Compensation circuitry with an internal capacitor.

Since the main task relates to the reduction of the steady-state error, a capacitor within the feedback path of the error amplifier is used. This can sometimes be implemented inside the integrated circuit supporting the control of the power supply. An example is shown in Figure 9.13.

A low frequency pole (P1) around 100 Hz is therefore added to the transfer function which was previously derived for the output voltage in Equation (9.9). The closed-loop system starts with a characteristic decrease of 20 dB/dec and a phase of around –90°. The effect of the second pole (P_2, the same as in Equation (9.10)) on the power circuitry is at larger frequencies than the crossover frequency. This solution is implemented within many integrated circuits. For a practical example, consider the integrated circuit LM309, a voltage stabilizer, with three terminals, for output at 5V/1A. This is not a high performance circuit as it takes a large operating current I_{ground} of around 5 mA. Other similar circuits belong to the series 780x (x = output voltage, 7803, 7805, 7812, ...).

9.4.2 Analog Circuit with an Inverting Topology

A low frequency small-signal model is derived in Figure 9.15, for the circuit in Figure 9.14. The capacitor shown on the load side is important to system stability since the capacitor's ESR determines a pole of variable frequency, as is next demonstrated.

After neglecting r_x and R_E from the transistor model shown in Figure 9.15(b), the transfer function yields

$$\begin{cases} \left|A_{vi}(s)\right| = \dfrac{V_o}{V_i} = g_m \cdot \left(CR_{load}\right) = \dfrac{g_m \cdot R_{load}}{1 + s \cdot R_{load} \cdot C} \\ \left(CR_{load}\right) = \dfrac{R_{load}}{1 + s \cdot R_{load} \cdot C} \end{cases} \tag{9.11}$$

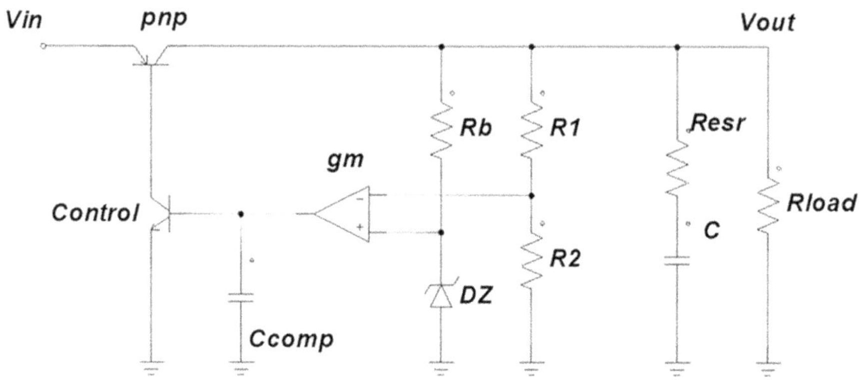

FIGURE 9.14 Power supply circuit with an inverting topology.

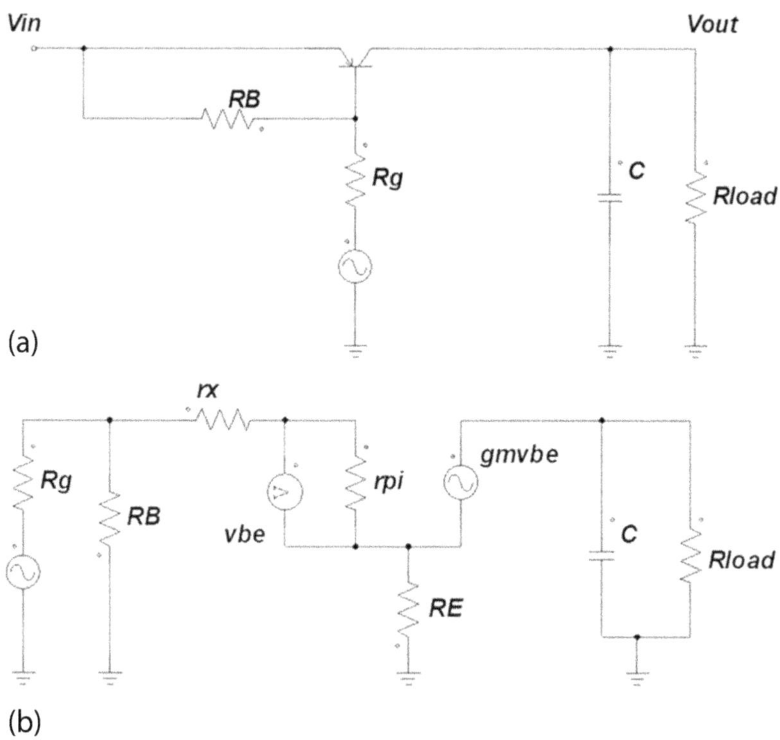

FIGURE 9.15 Small-signal model of power supply with an inverting topology.

Inverting transistor connection determines a pole with variable frequency in the power circuit at relatively low frequencies. Fortunately, the capacitors are not ideal components and we have a zero produced with the series connection of capacitor's ESR resistance.

$$\left|A_{vi}(s)\right| = \frac{V_o}{V_i} = g_m \cdot \left(X_s R_{laod}\right) = g_m \cdot \frac{\left(R_{esr} + \frac{1}{s \cdot C}\right) \cdot R_{load}}{R_{esr} + \frac{1}{s \cdot C} + R_{load}} \tag{9.12}$$

$$= \frac{g_m \cdot R_{load} \cdot \left(1 + s \cdot R_{esr} \cdot C\right)}{1 + s \cdot C \cdot \left(R_{esr} + R_{load}\right)}$$

When the steady-state error is addressed with a simple integrative compensation law at low frequency, the integrator adds the effect of the existing pole (shown in Equation (9.5)), the phase component yields at 180°, and the phase margin becomes zero. This produces instability.

An inherent improvement occurs when the small-signal model includes the loss resistance of the output capacitor bank. The effect of this *zero* can be seen in Figure 9.16 and demonstrates that the selection of the output capacitor is made for stability of the closed-loop system and a finite ESR is welcome since it adds a phase to the overall characteristics. Figure 9.17 shows that a suitable ESR can secure stability for any load current and defines the range of suitable values.

9.4.3 Boost/Buck Converters

The small-signal model for a boost or buck converter is calculated for a variation of the control duty cycle (D) to a variation of the output voltage (V_{out}). Such small-signal models can be derived for any of the topologies previously introduced in Chapter 8 as non-isolated or isolated converters. It can be demonstrated that these models can ultimately be reduced to one of the two forms: buck or boost converters.

A generic transfer function for the power converter is provided herein without demonstration:

$$G_{d-v}(s) = G_{d0} \cdot \frac{\left(1 - \frac{s}{\omega_0}\right)}{1 + \frac{1}{Q} \cdot \left(\frac{s}{\omega_0}\right) + \left(\frac{s}{\omega_0}\right)^2} \tag{9.13}$$

This generic transfer function is used along with the particular details presented in Tables 9.2, 9.3, and 9.4. It can be seen that the major difference from the buck converter to the boost converter is the presence of a Real-number positive zero, located on the right-hand side (RHP) of the complex plane. The meaning of this positive zero is illustrated in Figure 9.18 where the step response starts toward negative values, then recovers toward the desired steady-state value. A RHP zero may produce a phase reversal at higher frequencies, which is a threat to system stability. The compensation herein is achieved by adding a phase at the frequency of interest.

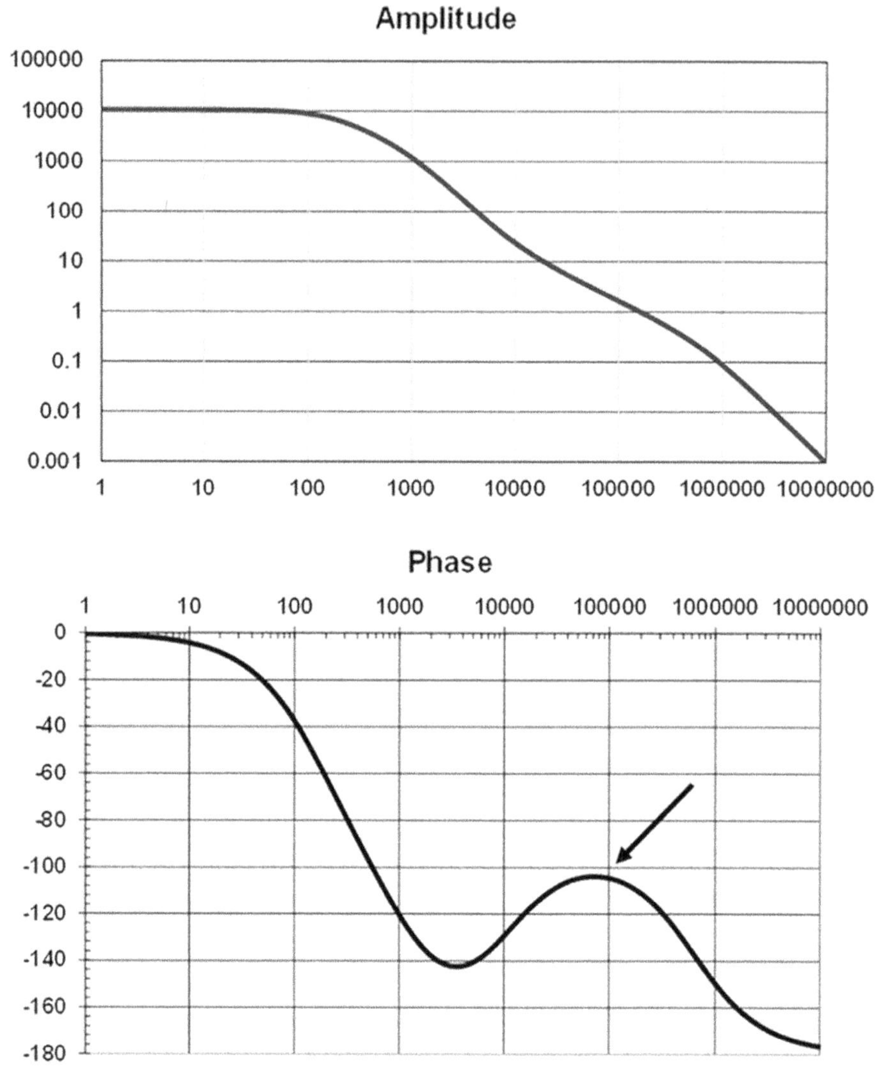

FIGURE 9.16 Stability achieved with a proper output capacitor (phase margin around 70°).

9.5 ANALOG-MODE FEEDBACK CONTROL SOLUTIONS

Since the control of most power supplies is implemented within mixed-mode integrated circuits, the actual compensation law is carried out with operational amplifiers and a passive compensation network. Moreover, a standardization is achieved when the number of possible compensation laws is limited. This means the same circuitry used for the compensation network can satisfy requirements for various power supplies, just by changing the passive components within the network.

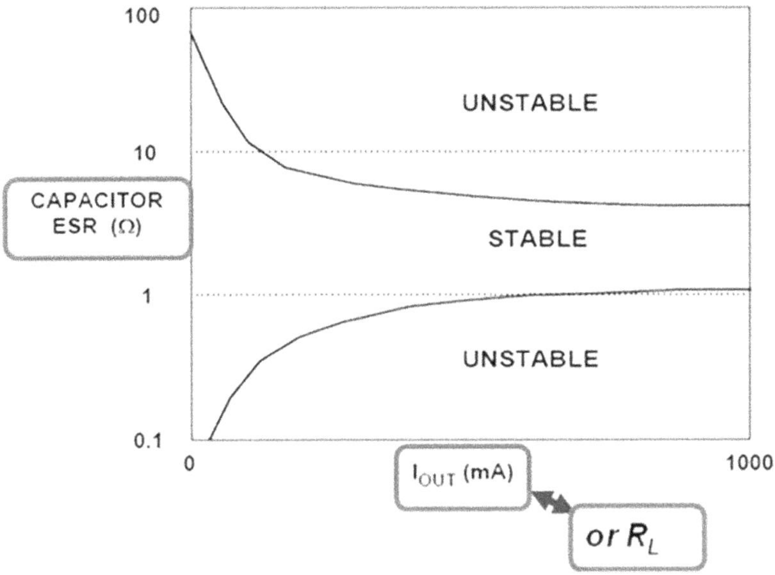

FIGURE 9.17 Constraints in the selection of the output capacitor.

TABLE 9.2
Details of Transfer Functions for Various Converters, with Ideal Output Capacitor

Converter	G_{do}	ω_0	Q	ω_z
Buck	$\frac{V_{in}}{1-D}$	$\frac{1}{\sqrt{L \cdot C}}$	$\frac{R}{\sqrt{L/C}}$	∞
Boost	$\frac{V_{in}}{1-D}$	$\frac{1-D}{\sqrt{L \cdot C}}$	$\frac{(1-D) \cdot R}{\sqrt{L/C}}$	$\frac{(1-D)^2 \cdot R}{L}$
Buck/boost	$\frac{V_{in}}{D \cdot (1-D)^2}$	$\frac{1-D}{\sqrt{L \cdot C}}$	$\frac{(1-D) \cdot R}{\sqrt{L/C}}$	$\frac{(1-D)^2 \cdot R}{D \cdot L}$

9.5.1 Type I Compensation

This is the simplest compensation law, which is reduced to just one integrator term. The circuit is shown in Figure 9.19 and its effect in Figure 9.20. The actual Laplace compensation law yields as follows.

$$\frac{V_{out}}{V_{in}} = -\frac{1}{R_1 \cdot C_1 \cdot s} \tag{9.14}$$

TABLE 9.3

Details of Transfer Functions for Various Converters, with Ideal Output Capacitor: *R*-Load Resistance, (*L*, r_l) = Inductance, (*C*, r_{ESR}) = Capacitance

Converter	G_{do}	ω_0	Q	ω_z
Buck	$\frac{V_{in}}{D}\cdot\frac{R}{R+r_L}$	$\frac{1}{\sqrt{L\cdot C\cdot\frac{R+r_{ESR}}{R+r_L}}}$	$\frac{1}{\frac{\sqrt{L/C}}{r_L+R}+\frac{r_C+\left(r_L\parallel R\right)}{\sqrt{L/C}}}$	$\frac{1}{r_{ESR}\cdot C}$
Boost	$\frac{V_{in}}{1-D}$	$\frac{1}{\sqrt{L\cdot C}}\cdot(1-D)$	$\frac{1}{\frac{r_L}{L}+\frac{1}{\left(r_{ESR}+R\right)\cdot C}}\cdot\frac{1}{\sqrt{L\cdot C}}\cdot(1-D)$	$\frac{1}{r_{ESR}\cdot C}$, ω_p
Buck-boost	$\frac{V_o}{1-D}\cdot\frac{1}{R+r_{ESR}}$	$\sqrt{\frac{r_L+(1-D)^2\cdot R}{L\cdot C\cdot\left(R+r_{ESR}\right)}}$	$\frac{\sqrt{L\cdot C\cdot\left(R+r_{ESR}\right)\cdot\left(r_L+(1-D)^2\cdot R\right)}}{C\cdot\left(r_L\cdot\left(R+r_{ESR}\right)+(1-D)^2\cdot R\cdot r_{ESR}\right)+L}$	$\frac{1}{r_{ESR}\cdot C}$, ω_p

TABLE 9.4
Actual Converter Transfer Functions for Buck and Boost Converters

Boost Converter	Buck Converter
$G_{d-v} = G_{d0} \cdot \dfrac{\left(1+\dfrac{s}{\omega_z}\right)\cdot\left(1-\dfrac{s}{\omega_p}\right)}{1+\dfrac{s}{Q\cdot\omega_0}+\left(\dfrac{s}{\omega_0}\right)^2}$	$G_{d-v} = G_{d0} \cdot \dfrac{1+\dfrac{s}{\omega_z}}{1+\dfrac{s}{Q\cdot\omega_0}+\left(\dfrac{s}{\omega_0}\right)^2}$

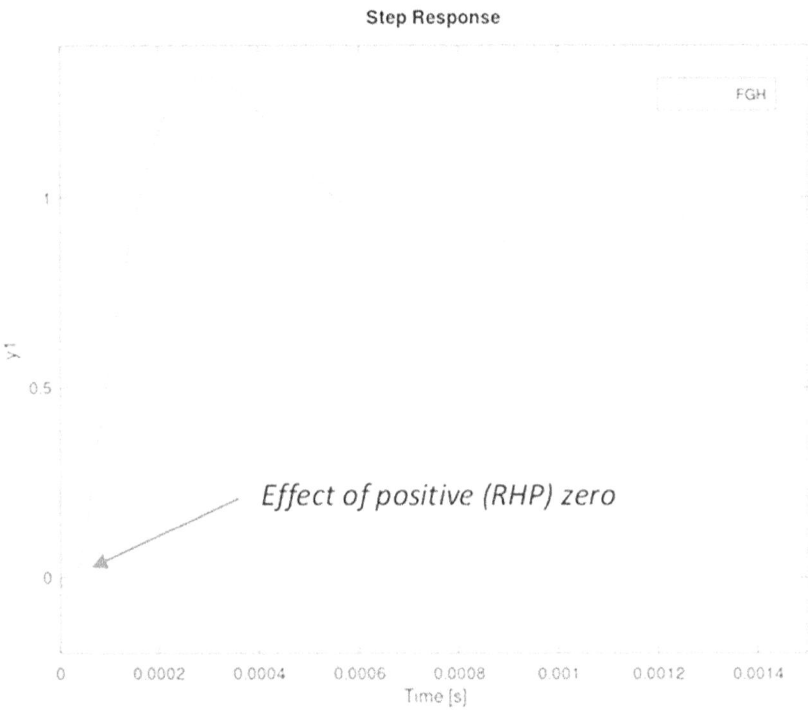

FIGURE 9.18 Step response applied to a transfer function with a positive (RHP) zero.

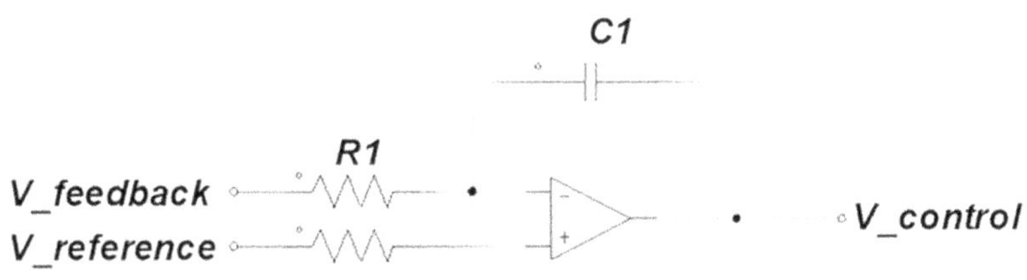

FIGURE 9.19 Type I compensation.

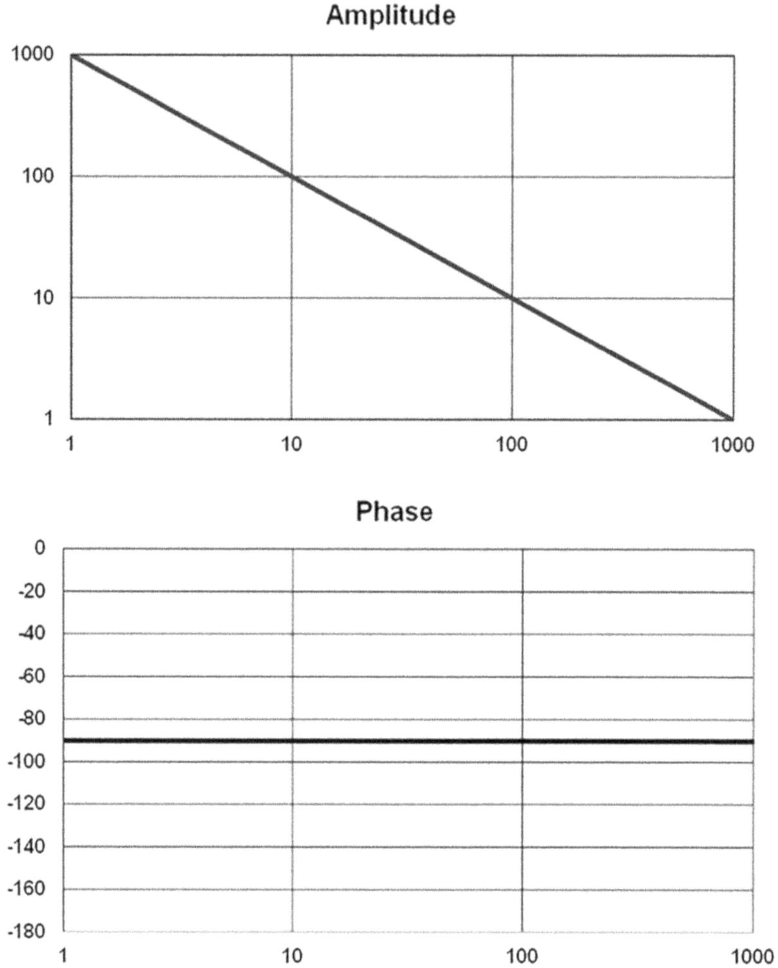

FIGURE 9.20 Characteristic of the Type I compensation.

It can be seen that the magnitude falls by 20 dB/decade while the added phase equals 90°. The magnitude characteristics crosses unity gain at a frequency where the absolute value of the C_1 capacitor's reactance ($1/\omega C_1$) equals the resistance (R_1).

9.5.2 Type II Compensation

The circuit for implementation of the compensation law is shown in Figure 9.21 and the Bode characteristics of the compensation law is shown in Figure 9.22. Analytically, the compensation law yields:

$$\frac{V_o}{V_{in}} = -\frac{R_2 \cdot C_2 \cdot s + 1}{R_1 \cdot (C_1 + C_2) \cdot s \cdot \left(R_2 \cdot \frac{C_1 \cdot C_2}{C_1 + C_2} \cdot s + 1 \right)} \tag{9.15}$$

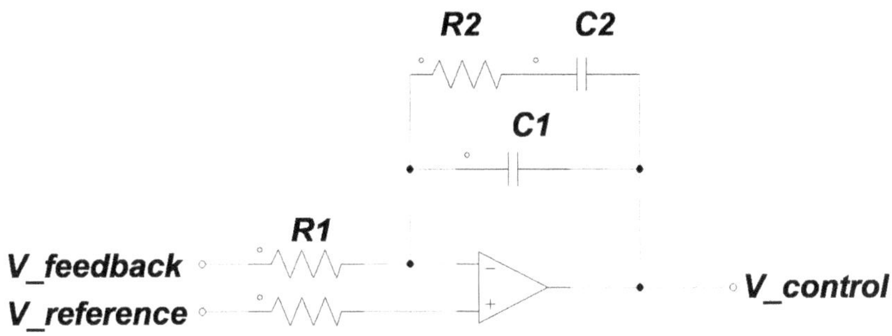

FIGURE 9.21 Type II compensation circuitry.

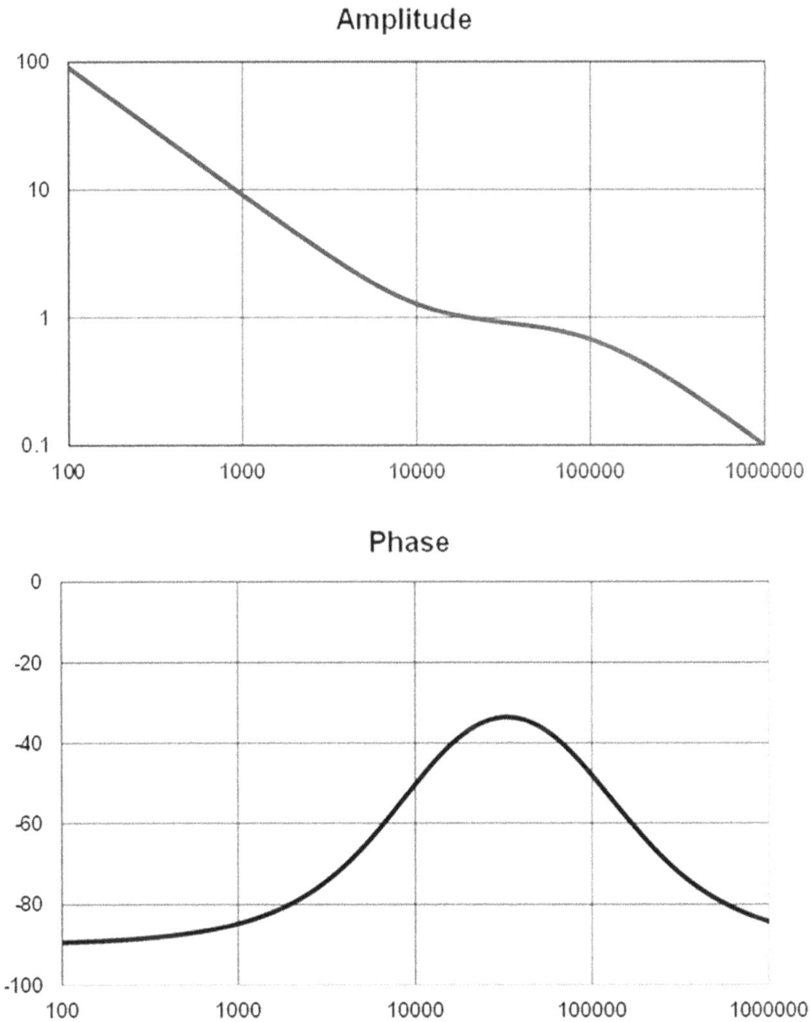

FIGURE 9.22 Bode characteristics for a Type II compensation.

This compensation law features a pole in origin with effect of integrator for steady-state error reduction, and a zero-pole pair at a higher frequency. The zero occurs at a frequency where the absolute value of the C_2 capacitor reactance equals the resistance R_2. The pole occurs at the frequency where the absolute value of the C_1 capacitor reactance equals the resistance R_1. The zero-pole pair creates a region with zero gain (a flat characteristic) which also features a phase increase. A phase up to 90^0 can be added at a desired frequency in order to increase the phase margin and improve stability. If 90^0 is not enough, the next compensation network is used.

9.5.3 Type III Compensation

A circuit which is able to implement a Type III compensation is shown in Figure 9.23, and its effect in Figure 9.24. A typical transfer function for the compensation network yields as follows.

$$\frac{V_o}{V_{in}} = -\frac{(R_2 \cdot C_2 \cdot s + 1) \cdot \left[(R_1 + R_3) \cdot C_3 \cdot s + 1\right]}{R_1 \cdot (C_1 + C_2) \cdot s \cdot \left(R_2 \cdot \dfrac{C_1 \cdot C_2}{C_1 + C_2} \cdot s + 1\right) \cdot (R_3 \cdot C_3 \cdot s + 1)} \tag{9.16}$$

This transfer function features a pole in origin, a pair of zeros which can be equal to each other, and another pair of poles which can also be equal to each other. The pole in origin helps reduction of the steady-state error with an integrator effect. If the two zeros are identical, and the two poles are identical, the slope of the magnitude plot easily transitions from –20 dB/decade to +20 dB/decade and back to –20 dB/decade (like in Figure 9.24), but this condition is not mandatory. A phase up to 180^0 is added at a desired frequency in order to increase the phase margin and to improve stability. The amount of added phase depends on the distance between the zeros and poles.

9.6 DESIGN PROCESS FROM CONSTRAINTS TO COMPONENT SELECTION

A complete design example is herein included for a boost converter able to transfer power between 12 V and 48 V distribution buses within a modern automotive application. Similar converters are manufactured by various companies and target a

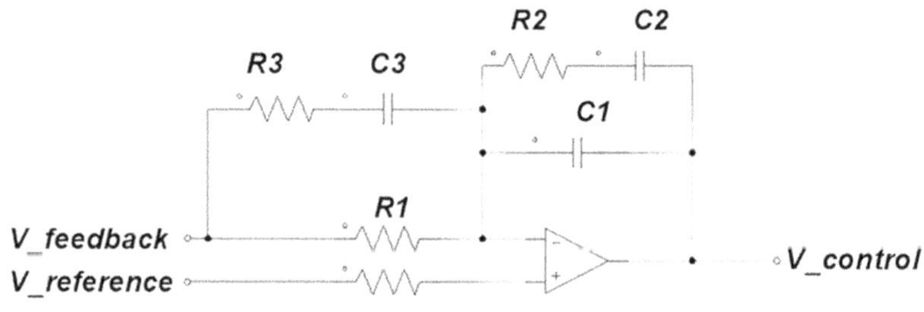

FIGURE 9.23 Type III compensation.

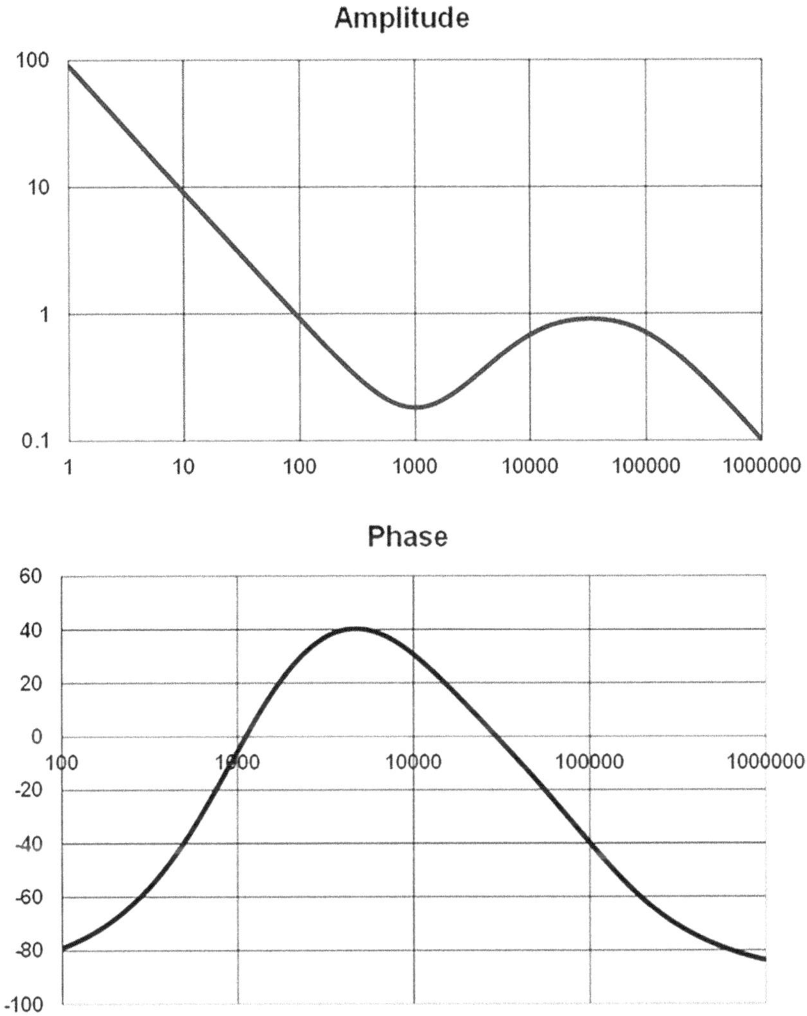

FIGURE 9.24 Bode characteristics for a Type III Compensation network able to add 130° of phase (–90° to 40°).

power level of several kW. At such power level, these are air-cooled. Various design constraints usually converge toward a multi-phase boost converter topology, able to reduce current within each power MOSFET and to smooth the output voltage ripple.

To simplify the design of the lead-lag compensation to the feedback control loop in this example, a model is considered with an equivalent single-phase converter, observing that the six-phase boost converter is equivalent to a simple boost converter. The circuitry for this model is shown in Figure 9.25.

The simple boost converter considered as example has the following data:

- $V_{in} = 12$ Vdc, $V_{out} = 48$ Vdc, $D = 0.25$, $R = 4.8$ Ohm (10 A load on each phase)
- $L = 36\mu$H /6, $R_L = 30$ mOhm /6

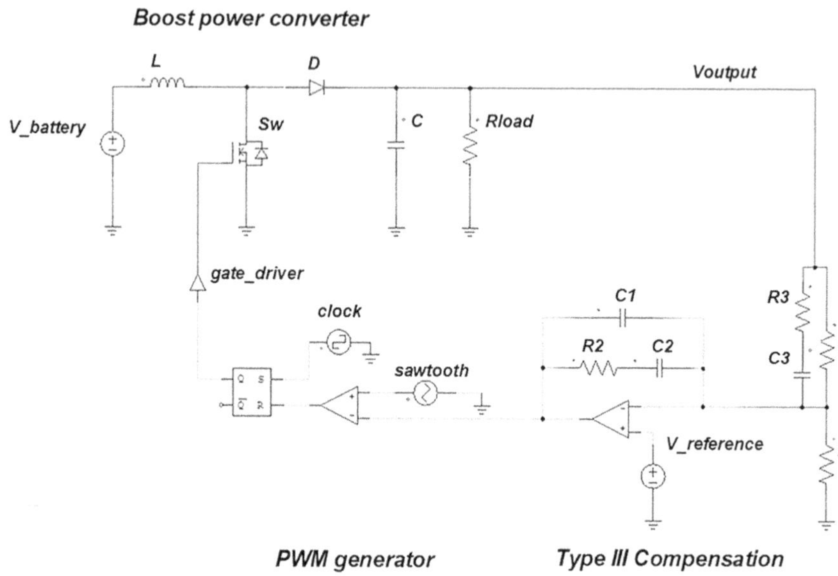

FIGURE 9.25 Example of a boost converter controlled with Type III compensation.

- $C = 6^* \, 330\mu F$, ESR = 18 mOhm /6
- Desired crossover frequency $f_c = 1{,}500$ Hz

Analytical design methods are known and applicable, especially when design requirements are provided as target numbers. When precise requirements are missing, the design of the compensation law can be achieved from empirical remarks able to explain the entire process. This is the process adopted herein. The compensation law is designed to provide a crossover frequency of 1.5 kHz and a phase margin of around 45^0, while the other core remarks are commented on next.

- A pole in origin is considered for reduction of the steady-state error with an integrator.
- A double zero at a frequency equal to the resonance frequency within the converter model (resonance from boost inductance and output capacitor).
- A second pole equals the zero produced by output capacitor's ESR.
- A third pole equals the RHP zero as absolute value, for the boost converters.

If any of these poles or zeros are calculated above the switching frequency, its value is limited to the switching frequency. This set of rules achieves direct compensation of the poles in small-signal model of the power converter with zeros introduced in the compensation law. Finally, two poles are added at higher frequencies to create multiple lead compensation terms with the previous zeros.

First, the design determines the gain of the converter model (plant) at the required crossover frequency of 1.5 kHz. The inverted value for this gain is used as the gain of the compensation law so that their product provides unity gain at that frequency. Next, the stability of the power supply is ensured by adding phase near the crossover frequency. A Type III compensation is directly considered herein. Bode characteristics for the compensation and plant are shown in Figure 9.26 while the entire system's transfer function is illustrated in Figure 9.27. In this design, the phase margin yields at 42°.

9.7 ON THE USE OF CONVENTIONAL PI/D CONTROLLERS

PI/D controllers are very used in industry due to their simplicity and generality. In order to understand their meaning, a proportional gain is first considered to act against the variations of the output measure from a reference. Unfortunately, this simple proportional gain controller can lead to steady-state error. In order to get rid of the steady-state error, an integrative term is added. Thus, the PI controller is formed. While this is excellent in annihilating the steady-state error in many cases, the integral effect may produce a slower response to a reference change and even damage the dynamic performance with small oscillations. Therefore, in cases where fast dynamics or other special requirements for the transient response are desired, the transient response is improved with an additional derivative term. The effect of this derivative term is more obvious in higher-order plant systems.

In the practical case of a power supply, where the main goal is reduction of the steady-state error and the small-signal transfer function of the plant is merely a second-order system, a PI controller is usually enough.

The compensation law for the complete PI/D controller yields

$$u(t) = k_p \cdot e(t) + k_I \cdot \int_{t_o}^{t} e(\tau)\,d\tau + k_D \cdot \frac{de(t)}{dt} \tag{9.17}$$

where the first term corresponds to the proportional law, the second term represents an integrator, and the last term shows the derivative term. This form of the PI/D control is inherently in the time domain. For completeness of information, the Laplace transform looks like

$$\frac{U(s)}{E(s)} = k_p + \frac{k_I}{s} + k_D \cdot s \tag{9.18}$$

Analogously, a parallel form is more attractive for industrial control since it expresses each term with time constants (T_I, T_D) which are easier to understand and use in practical equipment.

$$\frac{U(s)}{E(s)} = k_P \cdot \left(1 + \frac{1}{T_I \cdot s} + T_D \cdot s\right) \tag{9.19}$$

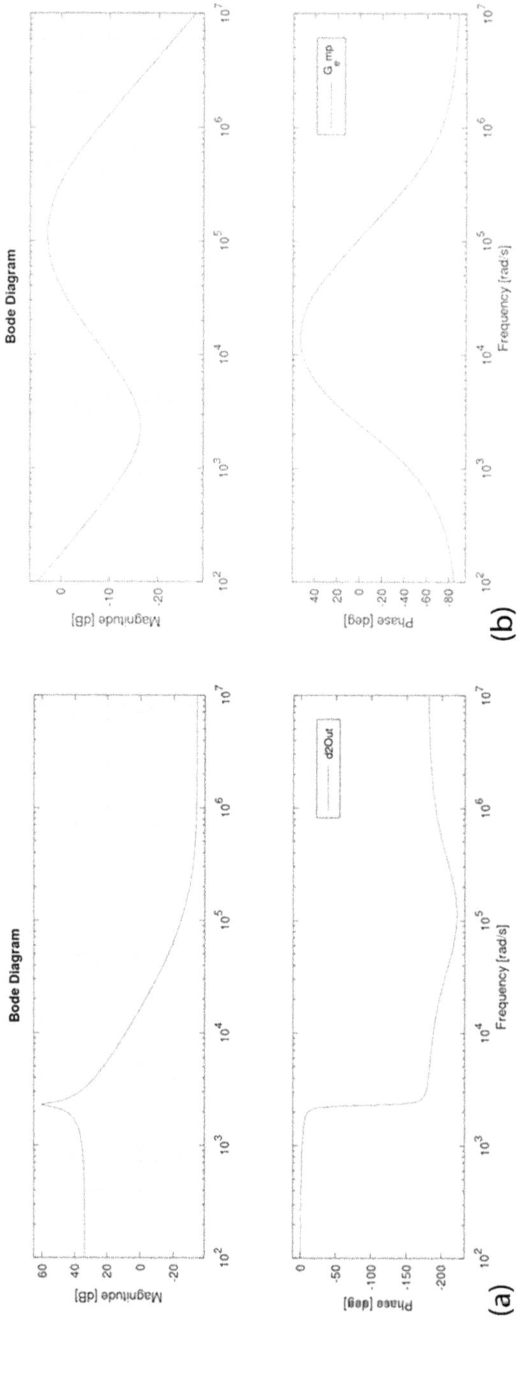

FIGURE 9.26 Design with Bode characteristics for (a) plant model and (b) compensation law.

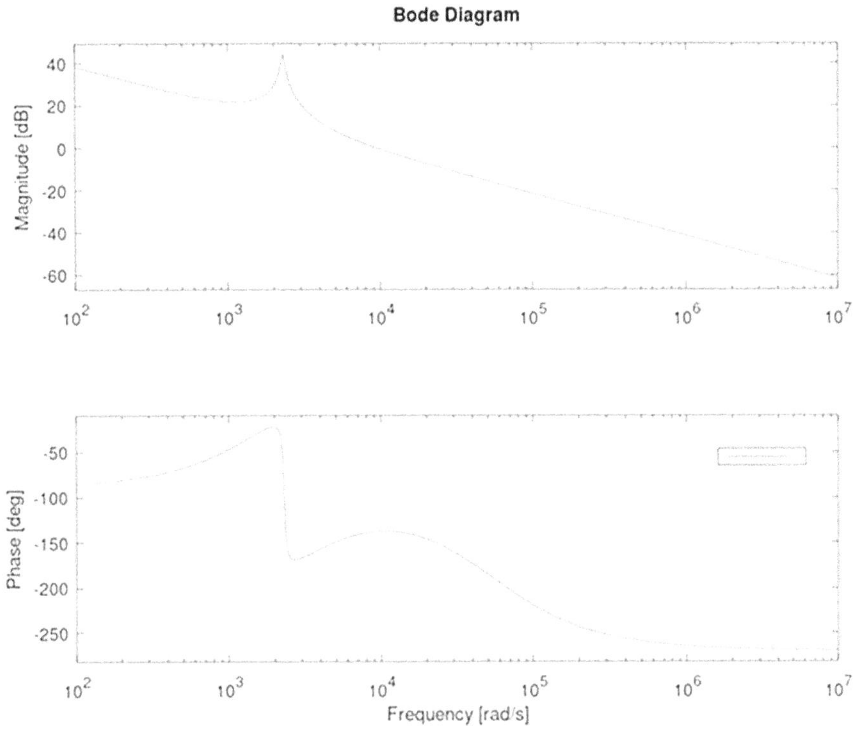

FIGURE 9.27 System after application of the Type III compensation to the boost converter.

9.8 CONVERSION OF ANALOG CONTROL LAW TO DIGITAL SOLUTIONS

The advent of embedded power and the requirements for software control brought up the need for a conversion of the analog control law into a digital form. The design process first defines an analog controller and next converts the compensation law into a digital equivalent. The analog and digital control laws are not identical since the embedded system introduces the need for sampling. The input and reference measures are sampled at a fixed rate, the software is run between the sampling moments, and corrective action is taken. Figure 9.28 illustrates this process.

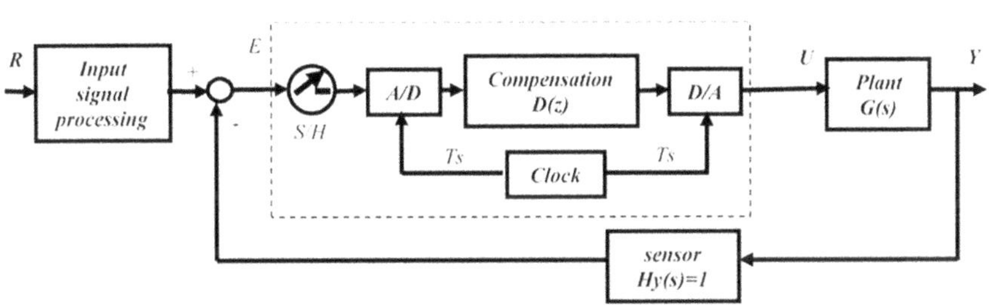

FIGURE 9.28 Implementation of a digital control law in practice.

The input signals are acquired at a constant sampling rate within the digital controller through an analog–digital converter. These samples are called *discrete signals*, while the data acquired in digital form is called a *digitized signal.* The digitized signal is quantized on a number of bits which define the *digital resolution.*

The selection of the sampling rate is an important design decision. On the one hand, a very large sampling frequency is excellent since it makes the system behavior closed to its analog equivalent, but it is difficult to implement due to physical limitations of the system. On the other hand, a very low sampling frequency tends to slow down the system's response. In order to address this compromise, an empirical design rule suggests that the sampling rate is chosen so that there are 5–10 samples of the inputs signal during the slope of the step response. Advanced digital control methods can compensate for sampling effect and reduce further the sampling frequency to just 2–3 samples during the rising time.

The actual compensation law is implemented in software or hardware with a relationship using *finite differences*, like small changes from a sampling instant to the next. The output of the compensation law is converted back to analog with a digital–analog converter, and sent to the real world for control of the power converter (in control systems terms, the plant).

While design methods specific to digital systems exist and these are mostly used for higher-order systems, design is herein performed in analog, following the Laplace transform method. The Laplace form of the compensation law is converted into *Z-transform* to account for the effect of sampling. As a last step, the Z-transform is further converted into a relationship using finite differences, where succeeding samples of the input or output measures are used to compose a compensation law.

It is very important to understand that this conversion to digital is not unique. There are numerous methods for the definition of a digital equivalent for the analog compensation law. Examples include the matched pole-zero method, zero-order hold method, linear interpolation of inputs method, and other similar methods. Whatever the method, the digital form and all the coefficients from the digital form also depend on the actual value of the sampling time T_s. Some examples are herein included.

The conversion to digital of the PI/D controller is next developed.

$$U(s)=\left(k_P+\frac{k_I}{s}+k_D\cdot s\right)\cdot E(s) \tag{9.20}$$

$$u(t)=k_P\cdot e(t)+k_I\cdot\int_0^t e(\tau)d\tau+k_D\cdot\frac{de}{dt}=u_P+u_I+u_D \tag{9.21}$$

Each term is re-written after sampling.

- Proportional term

$$u_P(k\cdot T_s+T_s)=k_P\cdot e(k\cdot T_s+T_s) \tag{9.22}$$

- Integrator term

$$u_I\left(k\cdot T_s+T_s\right)=k_I\cdot\int_0^{k\cdot T_s+T_s}e(\tau)d\tau=k_I\cdot\left(\int_0^{kT_s}e(\tau)d\tau+\int_{kT_s}^{kT_s+T_s}e(\tau)d\tau\right)$$

$$=u_I\left(k\cdot T_s\right)+\left[area\_under\_e(t)\right] \tag{9.23}$$

$$\approx u_I\left(k\cdot T_s\right)+k_I\cdot\frac{T_s}{2}\cdot\left\{e\left(kT_s+T_s\right)+e\left(kT_s\right)\right\}$$

- Derivative term

$$\frac{T_s}{2}\cdot\left\{u_D\left(k\cdot T_s+T_s\right)+u_D\left(k\cdot T_s\right)\right\}=k_D\cdot\left\{e\left(k\cdot T_s+T_s\right)-e\left(k\cdot T_s\right)\right\} \tag{9.24}$$

If the Laplace form has multiple poles and zeros, an operator $z=e^{s\cdot T_s}$ is defined to facilitate the usage of the Z-transform. This has the properties

$$\begin{aligned}u\left(k\cdot T_s\right)&\leftrightarrow U(z)\\ u\left(k\cdot T_s+T_s\right)&\leftrightarrow z\cdot U(z)\end{aligned} \tag{9.25}$$

Each term of the previous PI/D compensation law yields differently than Equations (9.22–9.24).

- for the integrator term

$$z\cdot U_I(z)=U_I(z)+k_I\cdot\frac{T_s}{2}\cdot\left[z\cdot E(z)+E(z)\right]\Rightarrow U_I(z)=k_I\cdot\frac{T_s}{2}\cdot\frac{z+1}{z-1}\cdot E(z) \tag{9.26}$$

- for the derivative term

$$U_D(z)=k_D\cdot\frac{2}{T_s}\cdot\frac{z-1}{z+1}\cdot E(z) \tag{9.27}$$

The compensation law in the Z-transform yields

$$U(z)=\left(k_P+k_I\cdot\frac{T_s}{2}\cdot\frac{z+1}{z-1}+k_D\cdot\frac{2}{T_s}\cdot\frac{z-1}{z+1}\cdot\right)E(z) \tag{9.28}$$

Similar results can be achieved when replacing the operator "s" with

$$s\leftrightarrow\frac{2}{T_s}\cdot\frac{z-1}{z+1} \tag{9.29}$$

and this is called the *Tustin method*.

The actual implementation requires an important effort in scaling the input and output measures.

Furthermore, many software systems within microcontrollers use fractional data format. This means all data has sub-unitary values and each variable has to previously be scaled within a sub-unity range.

9.9 CONTROL SYSTEM INFLUENCE ON POWER ELECTRONICS HARDWARE

A good understanding of the control system behind the power electronics hardware is necessary not only for horizon development, but also because the control system influences the design decisions in hardware. The system response is influenced by *ranges in operation* and by the *control effort.*

Advanced control methods can be employed to account for the range of variation for all parameters shown in Figure 9.5 (see "Parameters"). Minimizing the range for the input voltage or temperature in the operation of a power supply simplifies the design of the control system.

Figure 9.29 reiterates Figure 9.5 and points out a new measure, called the *control effort.*

This is the variable applied to the plant following the calculation of the control law. It is an excellent tool for responding to the question of whether the physical system is able to generate the action desired by the designer of the control system. The actual physical systems are of limited power, and the actual action may be limited and far from the one desired by the control system designer. For instance, the cruise control for a vehicle may see its action limited by the actual power of the engine, or a heating system based on natural gas may see a limitation due to the pressure of the incoming gas.

It is also easy to note that at large errors, the control effort yields are large as the control system tries to use everything to correct the error. A control system for the voltage at the output of a power supply would respond a large decrease in voltage with application of the largest current available from the power supply onto the output capacitor bank. If such a current is limited to 1 A, it may take longer to correct the voltage error than in the case of a current limit at 10 A, even if the design of the control system is identical. Analogously, a 600 HP engine would respond faster than a 55 HP engine.

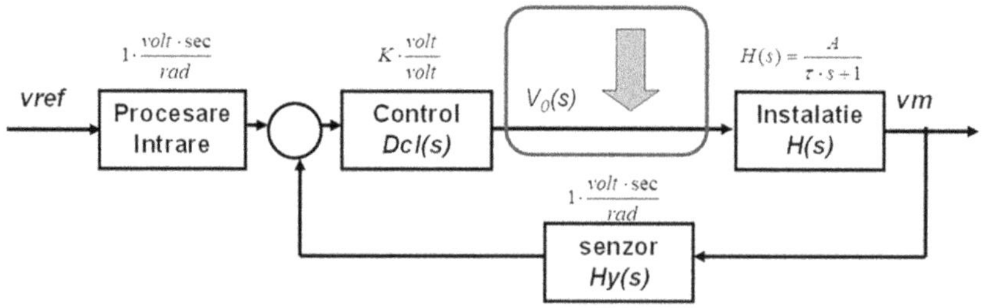

FIGURE 9.29 Definition of the *Control effort* for a power supply.

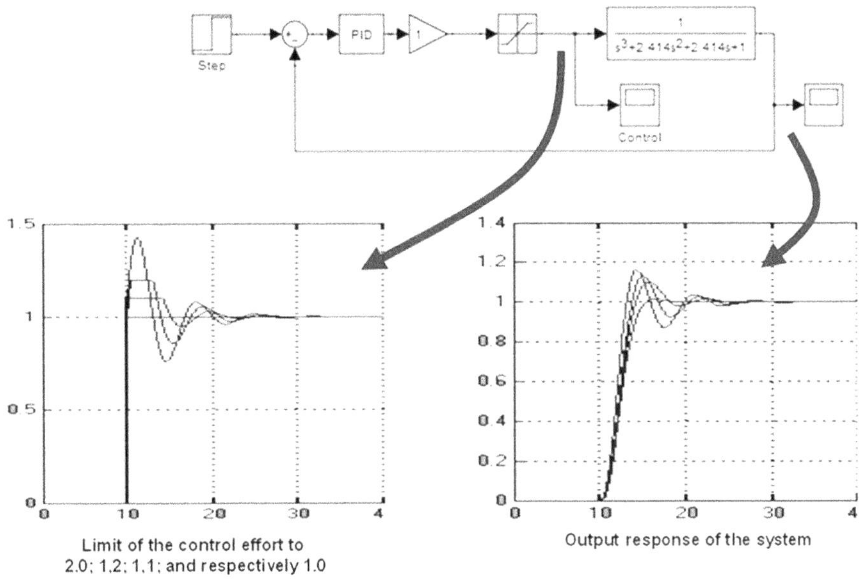

FIGURE 9.30 Example of the effect of actuator limitation.

At small errors, the control effort yields are smaller and the response is not very influenced by the actuator limitations.

Figure 9.30 illustrates an example for a generic system. The slowness of the response can be seen for the case when the actuator limits the control effort.

Without entering the actual theory, it is noteworthy that the control system can be optimized in respect to the control effort rather than the output dynamics. More complex control systems, like the linear quadratic controller, allow for optimization of both the output and the control effort through a weighing relationship.

9.10 CONCLUSION

This chapter attempted an introduction to the control of power converters that are used in automotive application. While the main focus resided in definitions and general theory, this knowledge can be extended toward control of other automotive systems like those modeled in Sections 4.9 "Cruise Control," 6.4. "Automotive Suspension," and 6.3. "Electronic Control of Power Steering."

Linear analog circuits with amplifiers are used for control of many power supplies. The output voltage conditioning is made in a closed loop, with feedback. Electronic control circuits with continuous operation necessitate a series-connected pass device, which would dissipate a large amount of energy.

Both the power supplies with analog operation and the switched-mode power supplies may use analog circuits with amplifiers and passive components for implementation of the control law, which is also called *frequency compensation*. In order to simplify the design, these analog control circuits have predefined structures—Type I, Type II, or Type III—which allow for the addition of passive components along the existing integrated circuits. Design examples were briefly presented in this chapter.

Digital form for the compensation law is next derived for software implementation. This allows for a correlation with digital power management.

REFERENCES

Franklin, G.F., Powell, J.D., Emami-Naemi, A. 2014. *Feedback Control of Dynamic Systems.* 7th edition, Upper Saddle River, NJ: Prentice-Hall.

Ghosh, A., Prakash, M., Pradhan, S., Banerjee, S. 2014. *A Comparison among PID, Sliding Mode, and Internal Model Control for a Buck Converter.* Presented at IEEE IECON Conference, Dallas, TX, USA, pp. 1001–1006.

Jones, M. 2011. *Digital Compensators Using Frequency Techniques.* EEWeb Documentation, Posted November 2011.

Kelly, A., Rinne, K. 2005. *Control of DC-DC Converters by Direct Pole Placement and Adaptive Feed-Forward Gain Adjustment.* Presented at IEEE APEC Conference, Austin, TX, USA, pp. 1970–1975.

Maksimovic, D. 2013. *Digital Control of High-Frequency Switching Power Converters.* Tutorial Presented at The 14th IEEE Workshop on Control and Modeling for Power Electronics (COMPEL), Salt Lake City, UT, USA.

Neacşu, D.O., Bonnice, W., Holmansky, E. 2010. *On the Small-Signal Modeling of Parallel/Interleaved Buck/Boost Converters.* Presented at IEEE International Symposium in Industrial Electronics, Bari, Italy, pp. 2708–2713.

Ogata, K. 1995. *Discrete Time Control Systems.* Upper Saddle River, NJ: Prentice-Hall.

Venable, D.H. 1983. *The K Factor: A New Mathematical Tool for Stability Analysis and Synthesis.* Presented in Proceedings of Powercon Conference, San Diego, CA, USA, pp. 1–12.

Zaitsu, R. 2009. *Voltage Mode Boost Converter Small Signal Control Loop Analysis Using the TPS61030.* Texas Instruments Application Report SLVA274A, May 2007, Rev. January 2009.

10 Power MOSFET

HISTORICAL MILESTONES

1959–1971: The MOSFET (MOS field-effect transistor, or MOS transistor), invented by Mohamed M. Atalla and Dawon Kahng at Bell Labs in 1959, led to the development of the power MOSFET by Hitachi in 1969.

1977: Alex Lidow and Tom Herman co-invented the HexFET, a hexagonal type of Power MOSFET, at Stanford University.

1977–1979: The insulated-gate bipolar transistor (IGBT), which combines elements of both the power MOSFET and the bipolar junction transistor (BJT), was developed by B. Jayant Baliga at General Electric.

10.1 POWER MOSFET IN AUTOMOTIVE APPLICATIONS

A metal–oxide–semiconductor field-effect transistor (MOSFET) is a field-effect transistor (FET) where the voltage applied to gate determines the conductivity of the device. The ability to change conductivity with the amount of applied voltage can be used for amplifying or switching electronic signals. MOSFETs are now even more common than BJTs (bipolar junction transistors) in digital and analog circuits, and by far the most used device in automotive power electronics.

The power MOSFET is used as a switch in a power distribution circuit, either as a stand-alone device or within a solid-state relay. A MOSFET based switch is controlled to connect a load to the battery. A solid-state relay consists on an optocoupler and a power MOSFET with gate driver and protection, all packaged within the same case, and it is intended to replace an electromagnetic relay. While more details about the circuit use of an electronic relay are provided within Chapter 11, this chapter focuses on the power semiconductor device called MOSFET.

This can be used within an electronic relay, a power supply, or in a in chopper with unidirectional control for an actuator. The latter application benefits from a low on-state voltage drop, and low drive power requirements. An *actuator* is a component of a machine that is responsible for moving and controlling a mechanism or system, for example by opening a valve.

A more elaborated use considers the power MOSFETs in H-bridge converters for a reversible control of an actuator. In such application, two devices are connected in series with the motor instead of a single device for a unidirectional drive. In order to maintain the same drop voltage through the same drop resistance $R_{ds(on)}$, each MOSFET must have a double area and this leads to eight times more semiconductor material than the chopper case. As an operational detail, the simplest control for the bidirectional drive is accomplished when high-side switches used to control direction and low-side switches constantly switched at 20–50 kHz.

Due to space and reliability concerns within the automotive environment, the devices used for electronic control of solenoids or motors are highly integrated. This grants the possibility to include the power MOSFET inside the integrated circuit. The technological question is *How much integration is possible?*

The answer relates to a trade-off that has to be solved since the internal MOSFET will be limited in maximum current due to reduced semiconductor area available. This has consequences in higher R_{dson}, and a higher voltage drop.

Another limitation of the maximum current possibly passing through an integrated power MOSFET comes from the limited thermal capability of the integrated circuit package. For example, 2.7 W determine a die temperature of 107 °C for the improved PowerPAK (and for DPAK) and 148 °C for the standard SO-8. Even a heatsink based on the printed circuit board cannot help since 1 cmc represents about 1,000 °C/W and a board with 10 cm × 5 cm barely gives 20°C/W.

Finally, there is a limit of the maximum current passable through a pin of the integrated circuit. Most power integrated circuits allocate more pins for the drain and source connection to the internal power MOSFET.

Considering all of the above, the use of an internal power MOSFET is usually limited at 3–5 A, also depending on the bus voltage. Typical setup for a printed circuit board soldering solution is sketched in Figure 10.1.

To illustrate this trade-off between using an external MOSFET or a fully integrated solution, Figure 10.2 discusses the power-management integrated circuit

FIGURE 10.1 Surface-mount power MOSFET devices for printed circuit board use.

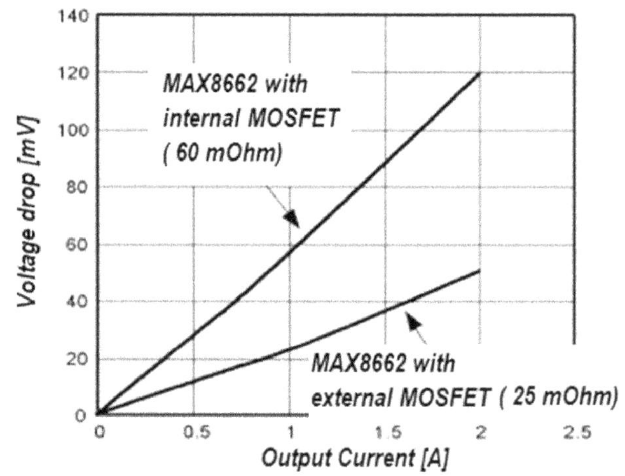

FIGURE 10.2 Using MAX8662 with internal or external MOSFET.

MAX8662 (*Power-Management ICs for Single-Cell, Li+ Battery-Operated Devices*). When using the internal MOSFET, a series drop resistance of 60 mΩ is seen due to exploited internal available semiconductor area. The alternative consists of using an external power semiconductor device for the switch function. In such case, the series drop resistance can be lower, at 25 mΩ. Depending on the processed current, the voltage drop differences can be important. This also influenced the efficiency of the power converter since the voltage drop means more loss.

10.2 THE IDEAL SWITCH

Before the description of the power MOSFET specifics, the requirements for an ideal power switch worth a look. A simple application circuit is shown in Figure 10.3.

The power switch used in the circuit from Figure 10.3 needs to offer:

- The capacity to block any voltage applied directly or in reverse.
- The possibility to conduct relatively large currents, whenever the circuit is turned on.
- The transition from conduction into blocking state or the reverse transition from blocking into conduction state should be done instantaneous, without energy loss.
- The transition from conduction into blocking state or the reverse transition from blocking state into conduction state should be done with zero gate control loss.

Changes within the conduction states of the power semiconductor device are leading to waveforms from Figure 10.4, when all secondary aspects such as the recovery currents are neglected. Figure 10.4 also helps understanding the power loss which is defined as the instantaneous voltage-current product. Since power semiconductor circuits process high-level currents, the power loss is important. Furthermore, the integration (summation) of the power components provides the energy loss within the circuit.

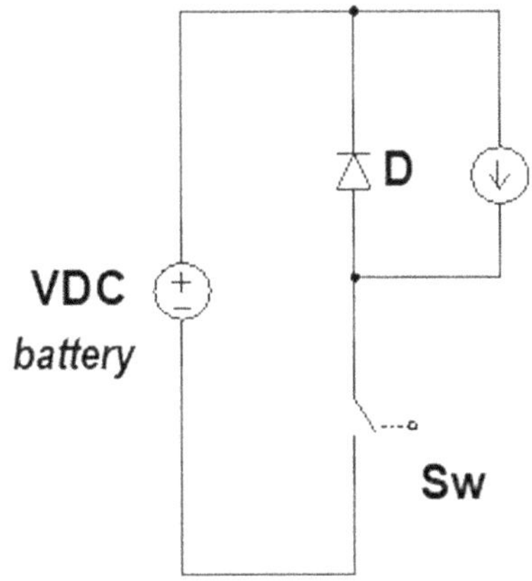

FIGURE 10.3 Simplified power circuit for a switch.

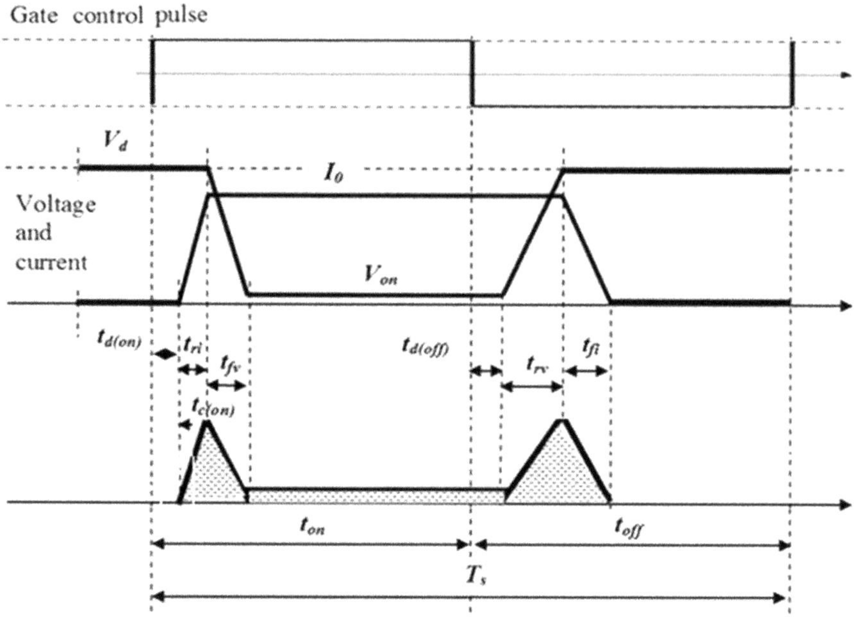

FIGURE 10.4 Definition of loss power and energy.

The power loss in a switching circuit has two components. During each change of state, there is a peak in instantaneous loss power. This component is defined as *switching loss*. The definition can go further by identifying the loss associated with the turn-on process as being *turn-on switching loss*, and the component seen for turn off identified as *turn-off switching loss*. During the continuous conduction of the power semiconductor device, a quasi-constant loss component occurs when the current is also constant. This component is called *conduction loss*.

A time-integral for any of this loss components provides the corresponding energy loss, as these are also identified with the hatched areas in Figure 10.4. For the simplified circuit, with linear variation of currents and voltages, simple equations relate loss components with supply voltage and load current.

$$W_{c,on} = \frac{1}{2} \cdot V_d \cdot I_o \cdot t_{c,on} \tag{10.1}$$

$$W_{c,off} = \frac{1}{2} \cdot V_d \cdot I_o \cdot t_{c,off} \tag{10.2}$$

$$W_{c,on} = V_{cond} \cdot I_o \cdot t_{on} \tag{10.3}$$

where V_d is the supply voltage, I_o is the load current, V_{cond} represents the voltage drop across the power semiconductor device during the on-state, $t_{c,on}$ represents the duration of the turn-on process, $t_{c,off}$ represents the duration of the turn-off process, and t_{on} the duration of the conduction state.

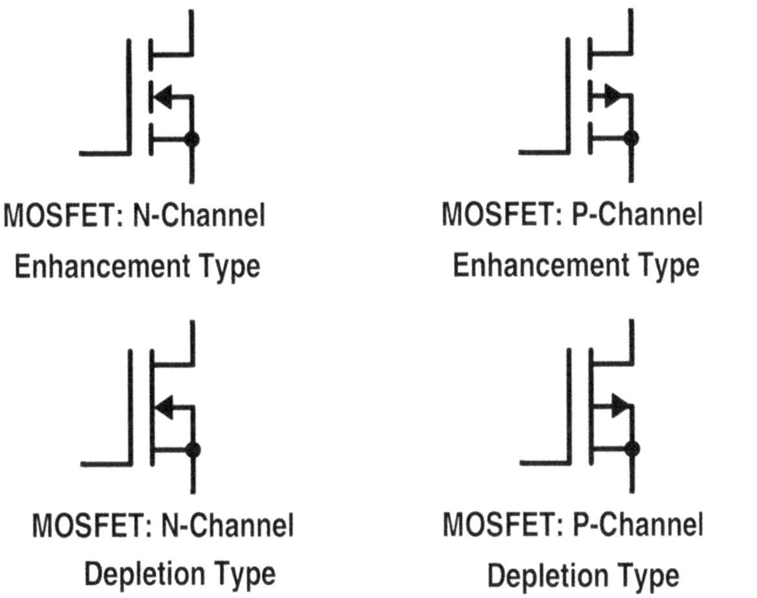

FIGURE 10.5 Circuit symbols for MOSFET transistors.

10.3 ENHANCEMENT-MODE AND DEPLETION-MODE MOSFETS

The power MOSFETs have been optimized for their circuit use, and two categories have been designed in: the *enhancement-mode* and the *depletion-mode* MOSFETs. Another possible classification comes from the technology used for fabrication, defining PMOS and NMOS transistors. Circuit symbols for these classes of MOSFET transistors are shown in Figure 10.5.

The NMOS and PMOS transistors differ from construction perspective. NMOS transistor is built with n-type source and drain and a p-type substrate, while PMOS transistor is built with p-type source and drain and a n-type substrate. In a NMOS transistor, carriers are electrons, while in a PMOS transistor, carriers are holes. When a high positive voltage is applied to the gate, NMOS transistor will conduct, while PMOS transistor will not. Conversely, a negative gate voltage would turn on the PMOS transistor.

The *enhancement-mode MOSFETs* are the common switching elements in most applications because they act as open circuits (are in the off state) when there is no voltage applied to gate ($V_{GS} = 0$), and act as closed-circuit (are in the on-state) when gate is controlled. The control voltage for the gate circuit depends on the type of the MOSFET device. The NMOS can be turned on by pulling the gate voltage higher than the source voltage, while PMOS can be turned on by pulling the gate voltage lower than the source voltage. Either way, it means that pulling gate voltage toward its drain voltage turns it on.

Enhancement power MOSFETs are either *N-channel* enhancement-mode power MOSFET or *P-channel* enhancement-mode power MOSFET. The former is the most used MOSFET in switch mode circuits, and the circuit usage is shown in Figure 10.6.

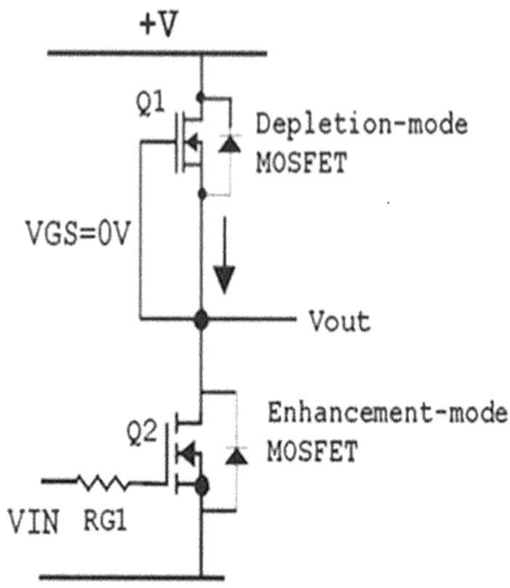

FIGURE 10.6 Depletion and enhancement-mode MOSFETs.

The *depletion-mode MOSFETs* are mostly used as load "resistors" in logic circuits, Figure 10.6. They are normally in conduction (on-state) at zero gate-source voltage. For NMOS type, it can be turned off by pulling the gate below the negative threshold voltage. This means that the drain, by comparison, is more positive than the source in NMOS. In PMOS, the polarities are reversed.

As a variety, the *junction field-effect transistors* (JFETs) are depletion-mode transistors, since the gate junction becomes bias forward if the gate is more than a little from source toward drain voltage. The simple called depletion power MOSFETs are actually *N*-channel depletion-mode power MOSFETs.

Getting past classifications allows a view into the operation of the power MOSFET. This device is the most used transistor as a switch in automotive applications. This implies a relatively low voltage, compatible with the dc distribution bus and battery, with a rating under 200 V. It is worthwhile that MOSFET transistors can be manufactured up to 1,000 V and currents typically under 100 Amperes.

The *enhancement-mode MOSFET—channel N* transistors are the most used as a switch. The main advantages consist in reduced energy loss and low supply currents. A voltage needs to be applied onto the gate circuit to maintain the device in conduction while a small current is demanded from the control circuit. The main shortcoming consists in the relatively large voltage drop in the conduction state. This is even higher when the transistors are dedicated to operation at higher voltages. The static characteristics of the enhancement-mode NMOS is shown in Figure 10.7.

The switch mode operation is achieved when the transistor is operated in its linear region, also called triode. This means the voltage drop in conduction state is proportional with current through the $R_{ds(on)}$ resistance. This is an important datasheet parameter with minimal dependency on temperature and current.

$$v_{ds(on)} = R_{ds(on)} \cdot I \tag{10.4}$$

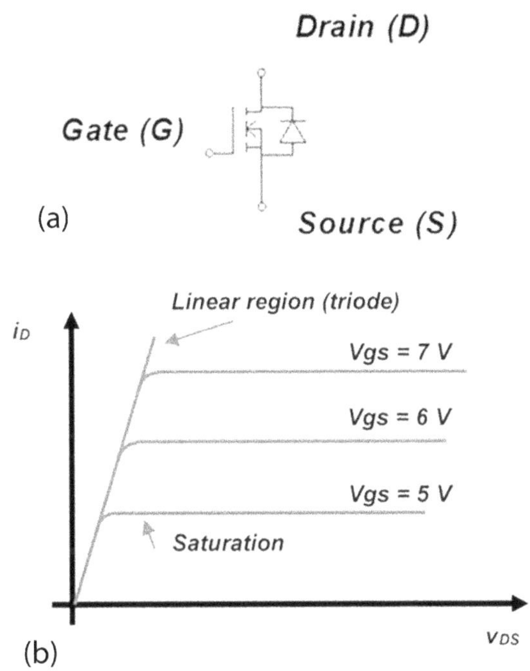

FIGURE 10.7 Enhancement-mode NMOS: (a) symbol and (b) I–V characteristics.

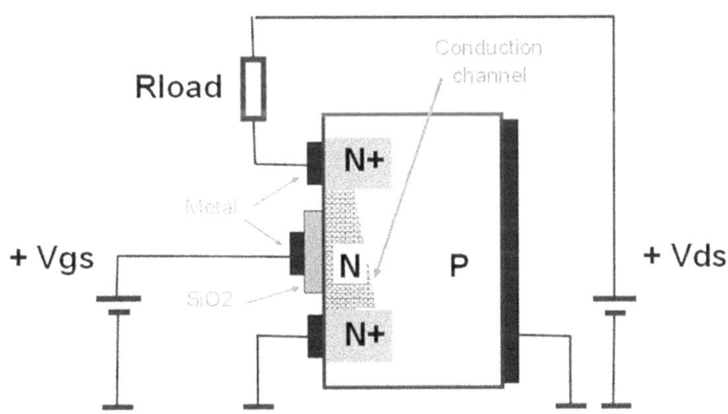

FIGURE 10.8 Simple circuit for switch mode operation of a NMOS device.

10.4 OPERATION PRINCIPLE

A simple circuit for use of an enhancement-mode NMOS transistor in a switch mode application is shown within Figure 10.8. Therein, the control voltage is applied between the gate and source terminals to produce an electric field able to set the channel up and allow current circulation. The current flows from drain to source inside the transistor, and from positive, through the load resistor, to the negative

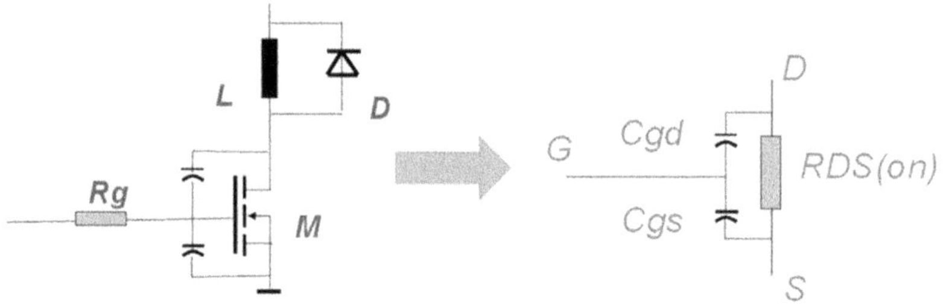

FIGURE 10.9 Dynamic model of a MOSFET.

battery terminal in the circuit. Because the change in the state of conduction is fast, the role is very similar to a switch, that is to connect and disconnect a load.

When the voltage applied to gate satisfies $v_{DS} < v_{GS} - V_{GSth}$, the MOSFET enters the ohmic region.

For a fast transition, the gate voltage used in switch mode circuit is selected to greater than the threshold, that is $v_{GS} >> V_{GSth}$. Hence, the boundary for the ohmic region becomes $v_{DS} < v_{GS}$.

Changes in the conduction state are better understood with a dynamic model of the transistor, uttered with an equivalent circuit model for the ohmic region. Figure 10.9 illustrates the circuit implications of the dynamic model, where $R_{DS(on)}$ represents the ohmic losses, mostly arising from drain drift region. Furthermore, capacitances C_{gd} and C_{gs} are varying with the voltage across them (Figure 10.10), with a substantial change in C_{gd} that can be approximated with a two-step variation. Gate charge characteristics are given in respect to the drain-source voltage (also called gate-emitter voltage from ancillary IGBT device).

Both gate-source and gate-drain charges are used within the gate driver circuit design. The gate-drain capacitance—though smaller in static value than the gate-source capacitance—goes through a voltage excursion that is often more than 20 times that of the gate-to-source capacity.

A global gate charge can be also considered for design. The gate charge necessary for switching is very important for establishing the MOSFET's switching performance. The lower the charge, the lower is the gate drive current needed for a given switching time. Its meaning is demonstrated with Figure 10.11.

The previous dynamic model can be used to explain details of the turn-on and turn-off processes.

Figure 10.12 presents all waveforms involved during the turn-on process. When applying voltage on the gate circuit, the gate circuit current can rise suddenly while the gate voltage rises with a certain slope due to the input capacitance. After this gate voltage reaches a certain level of voltage (*threshold*, V_{GSth}), the MOSFET transistor starts to conduct current from drain to source. The rise of the drain current is made with a slope *di/dt* defined by components in the external circuitry. For an inductive load, this rise of the current is approximately linear. When the drain current reaches a nominal value dictated and clamped by the external circuitry, the drain-source voltage drop starts to decrease.

Sometimes the diode connected in Figure 10.1 in parallel with the load is considered in conduction before the transistor is controlled to turn-on. The load

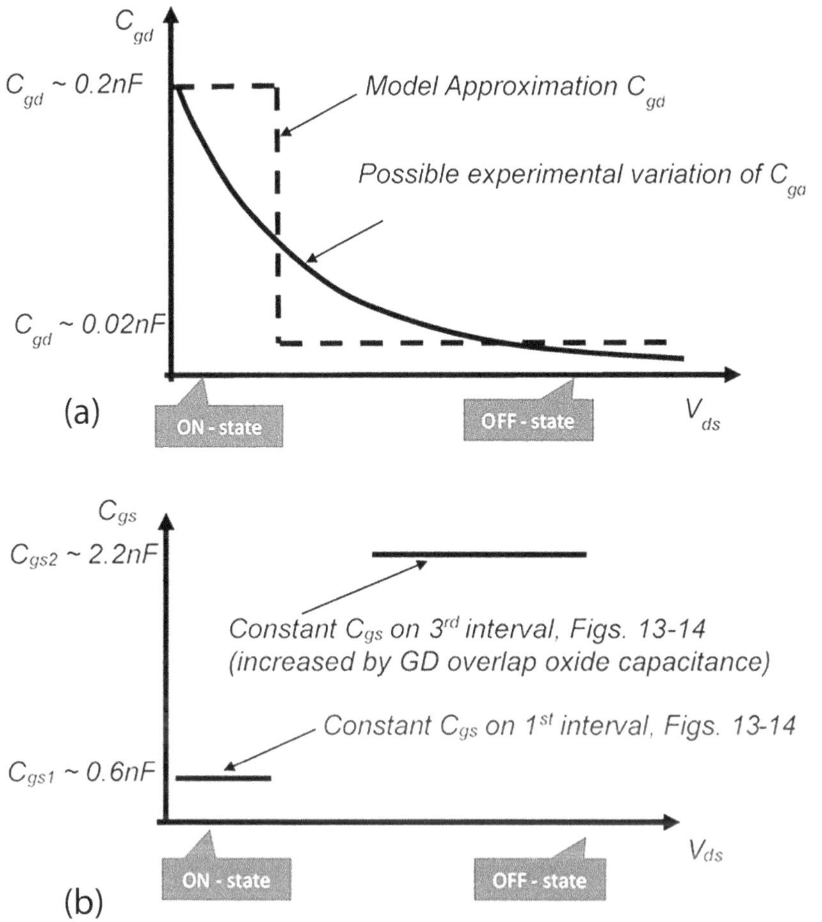

FIGURE 10.10 Variation of capacitances with voltage: (a) C_{ds}; (b) C_{gs}. Numerical values are just an example.

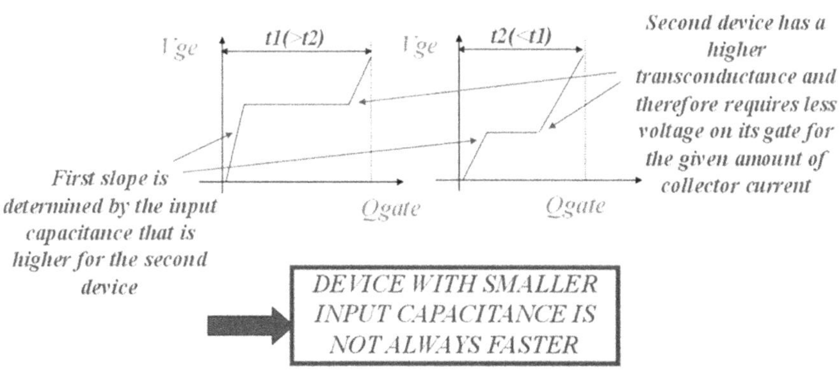

FIGURE 10.11 Factors determining the speed of transition.

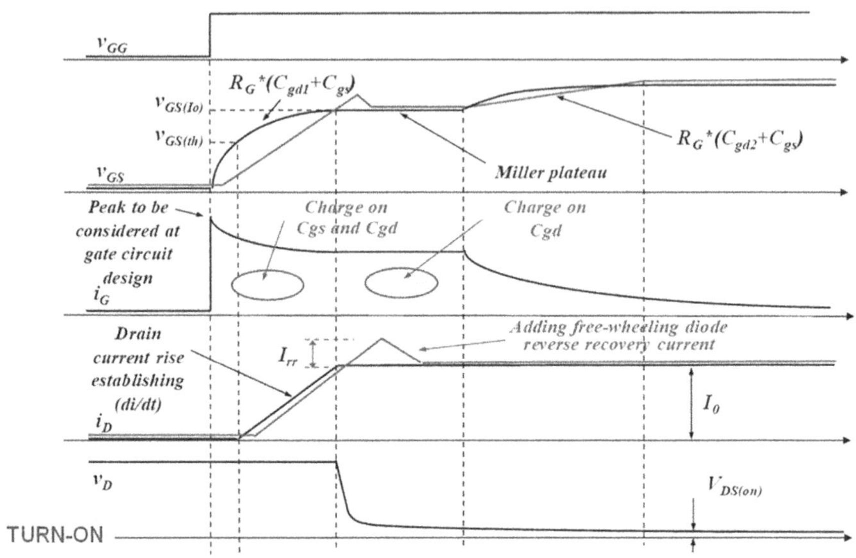

FIGURE 10.12 Waveforms at turn-on of the MOSFET.

current was previously directed through diode toward the load and now switches over into the MOSFET. Depending on the diode selection, the diode's recovery current can produce an additional current to the drain current and this is shown with a dotted line.

During the interval featuring a decrease of the drain-source voltage drop, the gate voltage does not change, and it is clamped to a level called Miller plateau. The final voltage drop across the MOSFET yields from the equivalent resistance and the load current. It is worthwhile to note that the entire switching process ends long after the current reaches the maximum value I_0 within the circuit.

Conversely, Figure 10.13 illustrates waveforms during the turn-off of the MOSFET. After a sudden decrease in control voltage, the voltage in the device's gate decreases slowly and the gate current discharges the capacitive charge. When the gate voltage reaches the threshold called Miller plateau, the collector voltage starts to increase. Meanwhile the voltage keeps constant on the gate. When the drain-source voltage reaches the maximum voltage in the circuit, that is usually the supply voltage, the collector current starts to decrease. The gate voltage decreases in the same time toward the control voltage.

10.5 SAFE OPERATION AREA

MOSFET devices have an ideal rectangular *Safe Operating Area (SOA)* as sketched in Figure 10.14. Within a real circuit, the operation points are along the trajectory shown in Figure 10.14 due to a switching inductance (series inductance for the controlled circuit). A proper design necessitates that these points are inside the rectangular SOA shown within the datasheet.

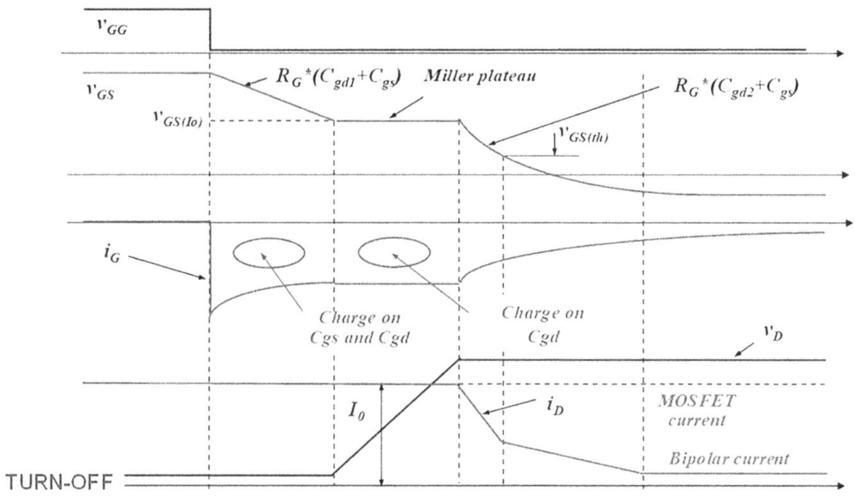

FIGURE 10.13 Waveforms at turn off.

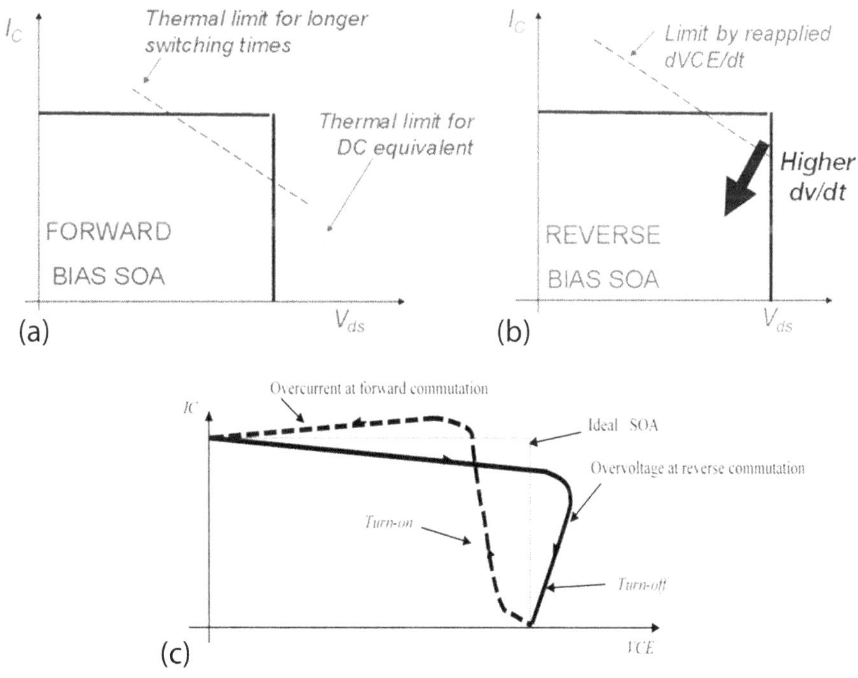

FIGURE 10.14 Safe operating area: (a) ideal forward bias, (b) ideal reverse bias, (c) in actual circuit.

10.6 GATE DRIVER REQUIREMENTS

For the application engineer, the requirements for the control circuit are very important. The *enhancement-mode MOSFETs* need a gate voltage of same polarity as the drain voltage. The gate control voltage V_{GS} should be limited to protect the silicon oxide layer between the gate and the source regions. A low impedance gate circuitry needs to be designed to avoid overvoltages. If such voltages occur, a protection can be provided with Zener diodes after inserting a small resistor between Zener and gate to prevent oscillations. The voltage supply of the gate driver is usually at 10–30 V, or 4.5 V for "logic level MOSFETs." Although the peak current on the gate circuit is large, but the power dissipation may be very low since the average current is very low.

For instance, a gate driver circuit—supplied with a voltage of 15 V and switching at a frequency of 100 kHz—controls the gate of a MOSFET with a gate charge of $Q = 74$ nC and a desired transient time of $t = 40$ nsec. The maximum gate current is calculated as $Q/t = 74$ nC/ 40 nsec = 1.85 A, which seems a lot. However, the low gate power dissipation yields as follows.

$$P = V_{GS} \cdot Q \cdot f_{sw} = 15 \cdot 74n \cdot 100k = 111\,\text{mW} \tag{10.5}$$

The integrated gate drivers are usually made in CMOS technology. This offers symmetrical source/sink current capability that can be approximated with a driving resistance of 500 Ohm above 8 V and 1,000 Ohm below 8 V within the standard buffered CMOS drivers. This possible drawback of a large gate resistor can be eliminated by paralleling gate drivers.

Since the CMOS technology implies operation with gate voltage above 5 V, the turn-on of the power MOSFET is guaranteed. To avoid cross-conduction, the turn-on may be slowed-down with the turn-off as fast as possible. This is possible with different gate resistors acting against almost the same gate capacitance. Since most gate drivers keep the circuitry at a minimum with a single control channel for both turn on and turn off, different gate resistors can be secured with a diode connected backward from the gate to the gate driver, Figure 10.15.

A special design issue relates to the control of a high-side MOSFET. A high-side MOSFET is a power transistor connected within a circuit and not referenced to ground. The gate control signal must be sent to the gate with a change in ground

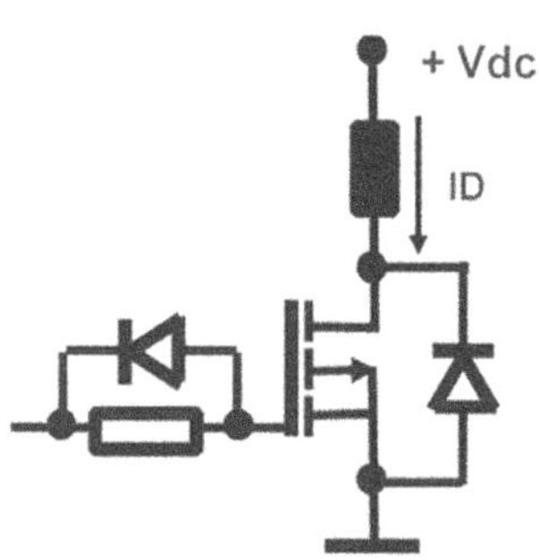

FIGURE 10.15 Separate gate resistance for turn on and turn off.

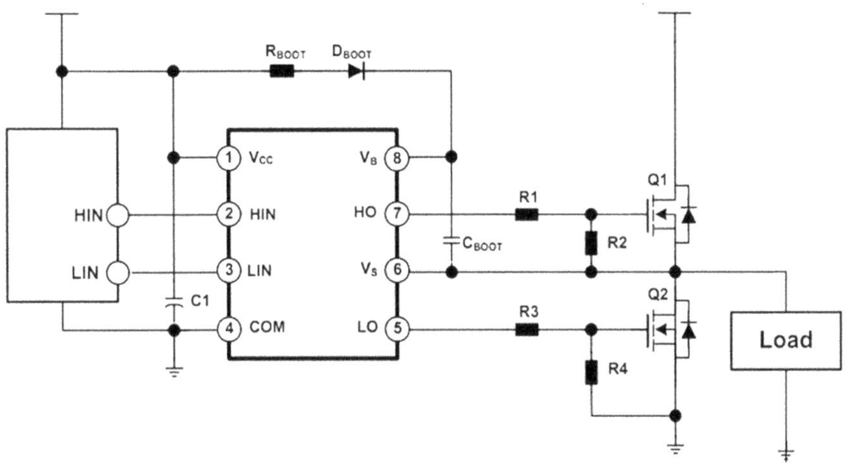

FIGURE 10.16 Example for a bootstrap circuit.

plane. The most immediate solution would be by means of optically coupled isolators, but this is expensive, especially for low voltage automotive applications. Alternative solutions include pulse transformers, or by means of DC to DC chopper circuits, or by level shifting techniques.

Additional to the control information, the control of the high-side power MOSFET needs to secure the power level with a local power supply. A common isolated dc/dc converter is again a very costly solution for this class of applications. A more affordable solution makes use of a bootstrap circuit or a charge pump.

Figure 10.16 provides an example for the bootstrap circuit. This is suitable to converters with a complete leg composed of a high-side transistor and a low-side transistor. The two transistors must be controlled alternatively, at certain duty ratio. When the low-side transistor is brought to conduction, it also provides a path for a current from the power supply used to supply the gate circuit of the low-side transistor, through a diode and D_{boot} and a capacitor C_{boot}. If the in rush current for this setup is too large, a limiting series resistance can be added as well. When the low-side transistor is turned off, this current is interrupted and the series diode D_{boot} is also turning off. Next, either the power diode near the high-side switch or the high-side switch turns-on and moves the mid-point closer to the positive rail. This definitely applies a negative voltage to the D_{boot} and keeps it in an off state. The voltage is preserved on C_{boot} which can now be used to supply the high-side gate driver. The energy stored and used from this capacitor can be calculated to be enough to turn on and to keep on the high-side switch. Obviously, the circuit accounts for the low consumption of the gate circuit of a MOSFET device and short conduction time. Furthermore, this supply voltage will see a certain ripple, but this is acceptable for the gate circuitry.

Overall, this is a very used and very affordable solution for a wide class of applications.

A second solution uses a charge pump. An example is provided with Figure 10.17.

Closing switches marked with S_1 allows to charge the external capacitors C_1 and C_2 from the external power supply V_{cc} to ground. After these switches are turned off,

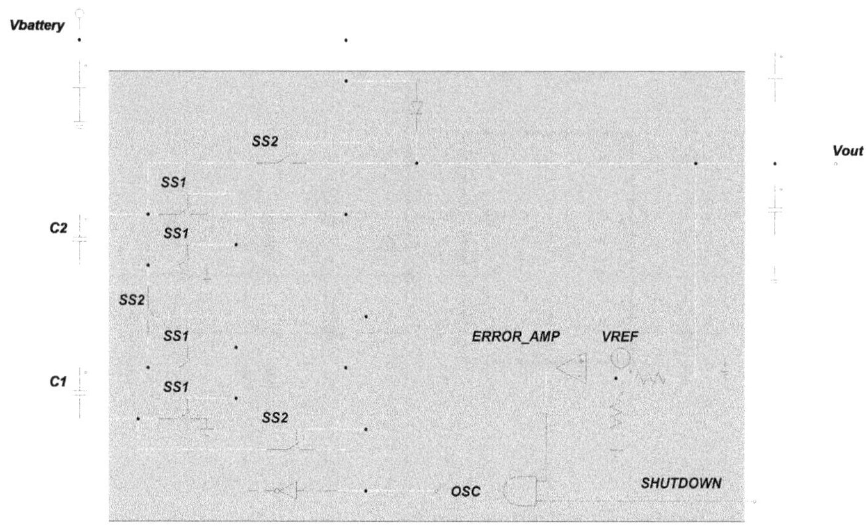

FIGURE 10.17 Charge pump supply of the high-side MOSFET (example based on Maxim MAX662A).

the voltage is preserved on the capacitors. Next, switches S_2 are closed and the sum of the capacitor voltages is applied to V_{out}, in top of V_{cc}. Any other similar solution can be imagined.

In rare cases, the power electronic converter uses logic level MOSFETs. While standard MOSFETs are controlled from 10 or 20 V and are not suitable for direct interface with logic [TTL] controllers, the "logic level MOSFETs" are designed for operation at 5 V or lower to 2.5 V and have guaranteed low on-resistance at these gate voltages. Since this implies devices with thinner gate oxide and different doping concentrations, the input characteristics yield different from conventional power MOSFET. Thus, the transconductance is higher, the input capacitance is higher, the gate threshold voltage is lower, and the gate-source breakdown voltage is lower. On the other hand, the output characteristics are maintained the same with the conventional MOSFET, that concerns the reverse transfer capacitance, the on-state resistance, the drain-source breakdown voltage, the avalanche energy rating, the output capacitance. Most circuit application requires a pullup resistor since the TTL does not provide directly 5 V levels.

10.7 USING P-CHANNEL MOSFET DEVICES

The *P*-channel MOSFET devices are enhancement-mode P-channel MOSFET devices and a switch substitute for a *N*-channel device. They require several extra gate circuit components which allow a circuit simplification from the *N*-channel MOSFET. On the downside, they have significantly higher power loss [*Rdson* ~2–3 times higher than *N*-channel], and they have larger capacitances, larger threshold, snappier recovery of internal diode.

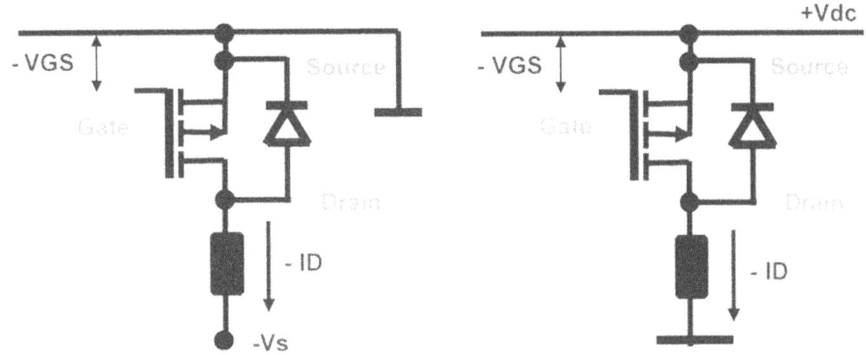

FIGURE 10.18 Using p-channel MOSFET.

Current flows through device after a voltage is applied between the gate and the source terminals, while no drain current flows when the gate is shorted to the source. The drain current flow is in negative direction for a negative G-S voltage, and this is mode used in practice.

The application circuits are shown in Figures 10.18. It can be seen that the simple switch function is more suitable herein for circuits with negative voltage (the positive end of the battery is connected to ground).

The foremost advantage of the *P*-channel MOSFET is the usability with grounded loads and a voltage supply under ~15 V, that was the case with most previous automotive circuits. In such cases, the load is tied directly to the drain of the MOSFET in order to use a single power supply. An *N*-channel MOSFET would otherwise require additional power supply to secure gate voltage above the drain connection to the positive dc bus. Using p-channel MOSFET requires no level shifting circuit or isolation of the high-side control in most automotive applications (Figure 10.19). However, above 15 V supply, level shifting is still required and the *P*-channel MOSFET is not very attractive.

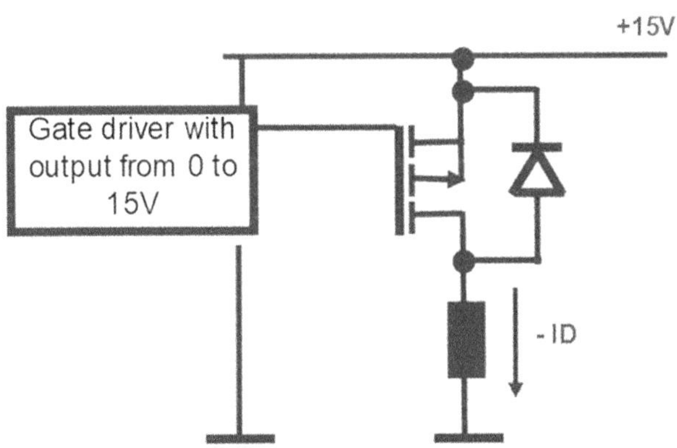

FIGURE 10.19 Gate driver requirements for the *P*-channel MOSFET.

10.8 PARAMETERS USED IN MOSFET SELECTION

Power MOSFETs are manufactured with a threshold voltage having values of 2.7 V, 4.5 V, or 10.0 V. This should be considered when selecting the gate driver voltage.

Automotive electronics is always rated for a wide temperature range, from –40°C to 125°C, but all power MOSFETs can work up to 175–200°C junction temperature.

The drain-source voltage is provided as rated voltage conditions for both normal operation and abnormal operation. Power MOSFET devices are marketed in voltage series of 55 V, 60 V, 75 V, 80 V, or 100 V.

Since most automotive applications work with a large inrush current, the power MOSFETs need to be overrated in current. Marketed series include ratings for the drain current, between 1 A and 170 A.

As it has been shown, the dynamic performance depends heavily on the gate charge since this determines the switching speed of the MOSFET. Datasheet information identifies automotive MOSFETs with gate charge in the range 2.6 to 100 nC.

While switching loss would depend on the gate charge, the conduction loss depends on the drain-source resistance in the linear region of the MOSFET. The $R_{ds(on)}$ determines the voltage drop and ultimately the efficiency of the power stage. Many relay-type of automotive applications do not care about the value of efficiency or the voltage drop. Conversely, switch mode power supplies, motor drives or even relays with continuous action are qualified based on efficiency.

The drain-source resistance $R_{ds(on)}$ represents the sum of different components within the semiconductor: the source region resistance, the channel resistance, the accumulation layer resistance, the drain region resistance, and the drift region resistance. The latter is the most important in high voltage MOSFETs. During operation, the $R_{ds(on)}$ increases somewhat with current for the same device.

From fabrication, the $R_{ds(on)}$ can be decreased by heaviest doping in each region, by dimensional control to minimize the length of current path, or by using as high as possible gate voltage (gate voltage threshold). Unfortunately, $R_{ds(on)}$ increases with the rated voltage since there is a need for more semiconductor material to withstand the larger voltage drop.

As a figure of merit, usually, $R_{ds(on)}$ are around 4–200 mOhm.

The *Repetitive Avalanche Energy* (EAR) is the maximum permissible reverse-voltage breakdown energy in continuous operation while observing the maximum permissible chip temperature. Using this EAR allows a lower voltage device to be used in the same application (for example 40 V instead of 55 V). A lower voltage rated device will have lower on-resistance, and therefore lower conduction losses. This means a smaller die may be used for substantial cost savings. The designer targets the use of a MOSFETs with EAR as high as possible.

10.9 SYNCHRONOUS RECTIFICATION

Switching converters with low output voltage encounter large loss due to the diode voltage drop during its conduction state. Schottky or ultrafast diodes can be a solution since these do not have reverse recovery but they have a relatively large

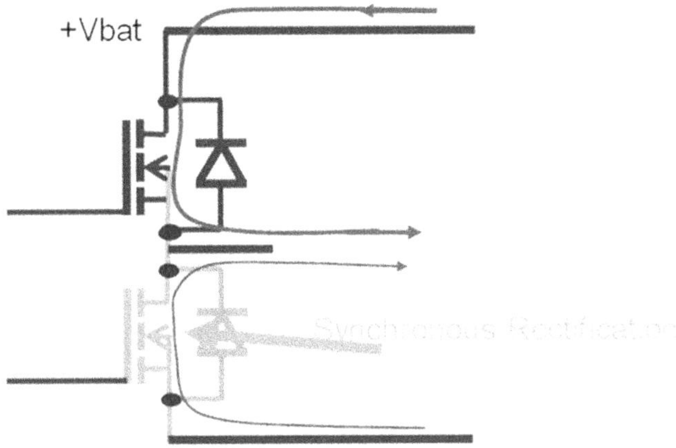

FIGURE 10.20 Synchronous rectification.

voltage drop during conduction (~ 0.4 V). Low voltage MOSFETs with low $R_{ds(on)}$ are used to replace diodes. In this respect, these need to be controlled exactly when the diodes are supposed to conduct current. This principle is illustrated with Figure 10.20.

The main caveats of this technology relate to taking care to avoid shoot-through, and observing a trade-off between the power for the gate driver and the diode loss.

A simple control of the replacing MOSFET transistor may encounter problems at or near zero current crossing. This is shown with Figure 10.21. The problem can be solved with one of the following three solutions.

- Hold the synchronous switch on ~pulls current from output;
- Disable the synchronous rectification at light loads;
- Sense current and shutdown the gate at zero current ~complex and costly solution.

To improve further the converter efficiency, diodes are kept in circuit and still used in anti-parallel with the power MOSFETs. This is also a safety reason to avoid

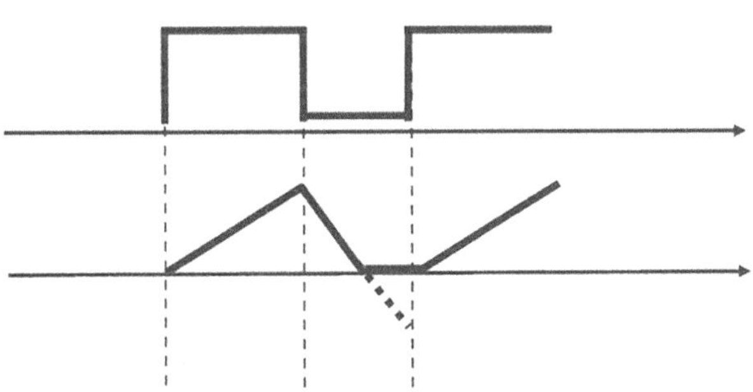

FIGURE 10.21 Zero current crossing.

TABLE 10.1
Amount of Semiconductor Material Required to Isolated 10,000 V

Si	GaAs	GaN	SiC	Diamond
1,000 μm	1,000 μm	100 μm	90 μm	20 μm

shoot-through. During transients of turn-on and -off, diode is shortly conducting current providing reduced power loss.

Another improvement concerns the use of programmable dead-time in order to eliminate MOSFET body diode conduction. The negative effects of dead-time can be counter-balanced with dead-time compensation based on the load current sense.

It is finally noteworthy that synchronous rectification is used in both DC/DC and DC/AC applications.

10.10 ADVANCED FET DEVICES

Analysis of Si-based MOSFET technologies revealed an unfortunate material-related performance limitation. An order-of-magnitude performance improvement can be achieved only with new substrate materials like silicon-carbide (SiC) and gallium-nitride (GaN), both being operated with larger *critical electric field*. The critical electric field strength shows when the electron velocity saturates.

With higher electron mobility, a GaN and SiC device can have a smaller size for a given on-resistance and breakdown voltage than a silicon semiconductor. This is shown in Table 10.1.

Analytically, the drain-source resistance can be expressed as:

$$R_{ds(on)} = \frac{4 \cdot BV^2}{\epsilon \cdot \mu_n \cdot E_c^3} \tag{10.6}$$

where BV represents the breakdown voltage, and E_c represents the critical electric field. Being able to design a device with a higher critical electric field at the same breakdown voltage, inherently allows a reduction in the drain-source on-state resistance. This helps with the reduction in conduction loss, and improve efficiency. The graphical conclusion is shown in Figure 10.22.

Using a new material technology for the fabrication of a MOSFET device introduces some differences in control and operation. For instance, the GaN devices require a negative gate voltage. Since this may pose an important practical barrier for technology acceptance, a new cascode depletion-mode device has been manufactured to ensure control compatibility at the price of slightly increased conduction voltage drop. The new device is codenamed *eGaN*.

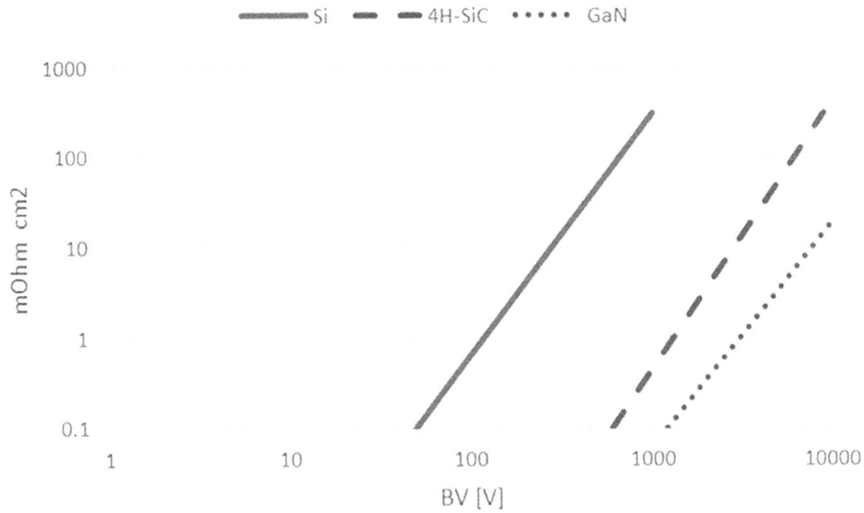

FIGURE 10.22 Achievable R_{dson} illustrated with a graphical dependency of specific on-resistance ($mOhm \cdot cm^2$) on breakdown voltage.

10.11 CONCLUSION

The automotive power electronics is dominated by the usage of power MOSFET devices as power switches. The power MOSFETs used in automotive need to be designed for rugged operation, extended temperature ranges, and increased reliability. Main properties of this power semiconductor device are described in this chapter.

REFERENCES

Anon. 2008. *Power Semiconductors Market: Latest Trends 2008*. Published as Yano Research Report.

Anon. 2012. *About the World Power Semiconductor Discretes and Modules –2012 Report Edition*. Published as IMS Research Report.

Anon. 2015. *Automotive Power Distribution*. TE Connectivity Corporation Documentation.

Anon. 2017. *Automotive Relay Replacement Reliability Meets Space Savings*. Nexperia Corporation Documentation.

Anon. 2018. *MLA Automotive Varistor Series*. Littelfuse Corporation Documentation.

Anon. 2018. *Power MOSFET Selecting MOSFFETs and Consideration for Circuit Design Application Note*. Toshiba Corporation Documentation.

Brown, J., Moxey, G. 2003. *Power MOSFET Basics: Understanding MOSFET Characteristics Associated With The Figure of Merit*. Vishay-Siliconix Application Note AN-605.

Denton, T. 2017. *Automobile Electrical and Electronic Systems*. 5th edition, Abingdon-on-Thames: Routledge.

Fujihira, T., Kaneda, H., Kuneta, S. 2006. Fuji' electric semiconductor: Current status and future outlook. *Fuji Electric Review*, 52(2):42–47.

Gerster, C.H. Hofer-Noser, P. 1996. Gate controlled dv/dt and di/dt – limitation in high power IGBT converters. *EPE Journal*, 5(3/4):11–16.

Lidow, A. 2015. *The GaN Effect - How GaN is Changing the Way We Live.* Presented at Darnell Energy Summit, Los Angeles, CA, USA.

Lutz, J., Schlangenotto, H., Scheuermann, U., De Doncker, R. 2011. *Semiconductor Power Devices - Physics, Characteristics, Reliability.* New York: Springer.

Neacşu, D. 2004. *Power Semiconductor and Control for Automotive Applications.* Tutorial Presented at IEEE APEC, Anaheim, CA, USA.

Seki, Y., Hosen, T., Yamazoe, M. 2010 The current status and future outlook for power semiconductors. *Fuji Electric Review*, 56(2):47–50.

Zverev, I, Konrad, S Voelker H, Petzoldt J, Klotz, F. 1997. *Influence of the Gate Drive Technique on the Conducted EMI Behaviors of a Power Converter.* Presented at IEEE PESC Conference, St. Louis, MI, USA, vol. 22, pp. 1522–1528.

11 Fuses and Relay Circuits

HISTORICAL MILESTONES

1809: Samuel Thomas von Sömmerring designed an electrolytic relay (electrochemical telegraph) as part of his electrochemical telegraph.
1823: Solenoid was defined by André-Marie Ampère to designate a helical coil.
1831: Michael Faraday experimentally establishes the principle of magnetic induction.
1835: American Joseph Henry claims to have invented an electromagnetic relay.
1864: A variety of wire or foil fusible elements were used to protect telegraph cables and lighting installations.
1890: Thomas Edison patented a fuse as part of his electric distribution system.
1970: The first and standard type of blade fuse was patented by Littelfuse Corp.

11.1 INTELLIGENT SWITCH VERSUS SOLID-STATE RELAY

Many applications within a motor vehicle require the "*switch*" function in order to connect a load to the power distribution system. A multitude of products have been designed to meet this goal. Among these, the intelligent switch and the solid-state relay are the most significant. Even if these terms are often interchangeable, the two devices are different.

- ***An Intelligent (or Smart) Switch*** is a controlled MOSFET with certain added protection features and with the entire assembly considered under a given power supply.
- ***A Solid-State Relay*** is also based on a controlled MOSFET but it includes a control through a galvanic isolation so that the power circuitry has nothing to do with its control.

In other words, a relay can make ("close") or break ("open") any circuit, while the intelligent switch can work under the same power supply.

"Intelligent switch" or "smart switch" are used synonymously by different manufacturers. An example of an intelligent switch is the high-side switch BTS7008-1EPP from the PROFET family from Infineon which is illustrated in Figure 11.1, with the application circuit in Figure 11.2.

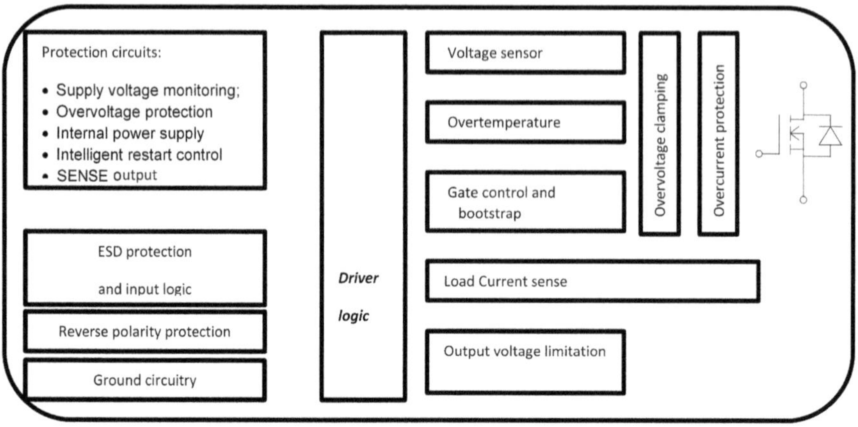

FIGURE 11.1 Features of the BTS7008-1EPP intelligent switch from Infineon Corporation.

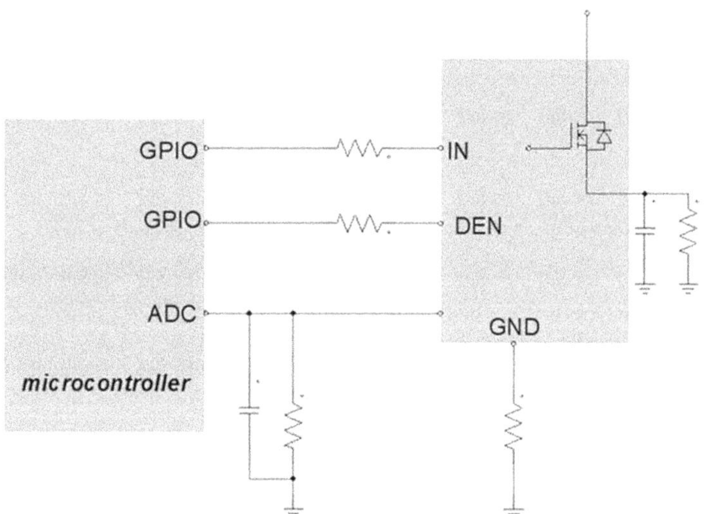

FIGURE 11.2 Application circuit for BTS7008-1EPP from the Infineon Corporation.

11.2 ELECTROMAGNETIC RELAYS

11.2.1 Using Electromagnetic Relays

Any transfer of control information to the actuator needs be done with a relay. The actuator in motor vehicle applications is usually a small electric motor or a solenoid. The use of relays provides a galvanic separation from the electronic control circuit to the power system hardware. In this way, any malfunction in the power system will not produce damage to the control circuit. Furthermore, the wires used for the control are small in diameter, while the cables used in the power system are ready to carry multi-A currents.

The automotive electrical system features a considerable number of relays. For instance, a complete electrically controlled seat uses four positioning motors and one smaller motor to operate a pump, which controls the lumber support bag. Each motor can be considered to operate using a simple rocker-type switch that controls the two relays. Overall, nine relays are required for the modern automotive seat, two for each motor, and one to control the main supply.

Electromagnetic (EM) relays are so far favored over semiconductor relays. The advantages offered by conventional electromagnetic relays consist of almost zero drop resistance, and lack of critical operation regions. Critical operating regions for a semiconductor relay come from linear load constraint (which is critical for capacitive loads), and avalanche operation (which is critical for inductive loads). Hence, electromagnetic relays are insensitive to load characteristics. This explains why there are so many electromagnetic relays in an average car, mostly related to body control applications, where we witness lots of small electric motors. An average modern car has around 30 relays. Their use and location can be listed herein.

1. Ventilator coolant
2. Gas pump
3. Wiper motor
4. Air Blower motor
5. Electric heat seating
6. Seat adjustment
7. Heated rear window
8. Brake light
9. Central locking system
10. Power windows
11. Power exterior mirrors
12. Starters
13. Horns
14. ABS
15. Power distribution
16. Blower fans
17. Car alarm
18. Hazard warning signal
19. Heated front screen
20. Lamps front/rear/fog light
21. Interior lights
22. Main switch/supply relay
23. Seatbelt pretensioner
24. Sunroof
25. Turn signal
26. Valves
27. Global positioning systems
28. In-vehicle entertainment systems
29. Security devices
30. Driver assistance systems

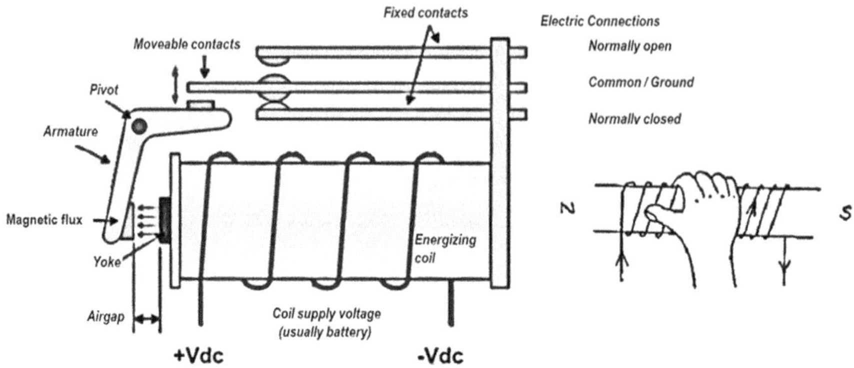

FIGURE 11.3 Principle of an electromagnetic relay.

11.2.2 Construction

Various circuits for lights, motors, heating resistors, and so on, need to be closed by electronic control through electromagnetic relays. The physics of operation are explained in Figure 11.3. The energizing coil is supplied with a dc voltage from the battery. This produces a dc current which generates a magnetic flux in a direction dictated by the "right hand rule." The magnetic flux produces movement of the mobile armature which in turn acts against the movable contacts. Therefore, supplying voltage to the coil determines a change in the electric connections.

Based on the same principle, various constructive versions are marketed and used within the motor vehicle. Their names and classification are shown in Figure 11.4. A relay switches one or more poles, each of whose contacts can be thrown by energizing the coil. Hence, the classification follows.

- SPST-NO (single-pole single-throw, normally open) relays have a single contact or make contact. These have two terminals which can be connected or disconnected. Including the two for the coil, such a relay has four terminals in total.
- SPST-NC (single-pole single-throw, normally closed) relays have a single break contact. As with an SPST-NO relay, such a relay has four terminals in total.

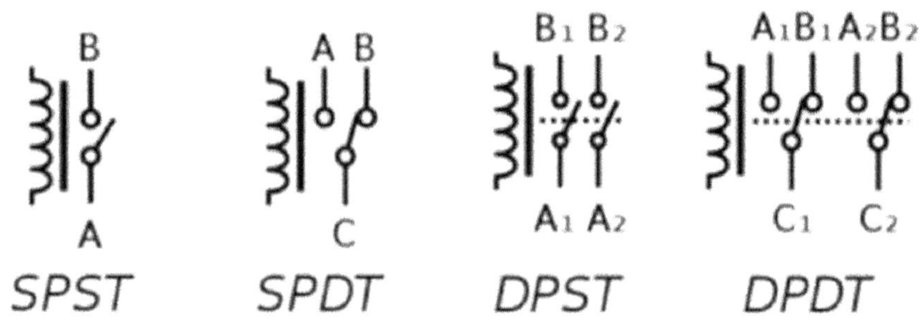

FIGURE 11.4 Various configurations for the electromagnetic relays.

- SPDT (single-pole double-throw) relays have a single break-before-make or transfer contacts. That is, a common terminal connects to either of two others, never connecting to both at the same time. Including two for the coil, such a relay has a total of five terminals.
- DPST—double-pole single-throw relays are equivalent to a pair of SPST switches or relays actuated by a single coil. Including the two for the coil, such a relay has a total of six terminals. Designations of either NO or NC may be used.
- DPDT—double-pole double-throw relays have two sets of contacts. These are equivalent to two SPDT switches or relays actuated by a single coil. Such a relay has eight terminals, including the coil.

Due to the wide use of electromagnetic relays, standardization is considered for packaging in order to allow inter-changeability. Packages for small-form or printed-circuit board locations are as follows.

- Micro relays, with through-hole mounting (TH) and for wave (THT) and reflow (THR/pin-in-paste) solderable versions, are typically under 30 A.
- SMT—surface mount relays, typically under 10 A:

For higher currents, relays may be grouped within a box (Figure 11.5).

These are either box and socket relays, or miniature automotive relay enclosures. Interestingly enough, both electromagnetic and opto-electronic relays use the same packaging, for further inter-changeability, Figure 11.6.

Higher current rated relays sometimes have a unique rating, confusedly written as 10/25 A. This means they can be operated at 10 Arms with natural convection cooling and up to 25 Arms with forced air cooling.

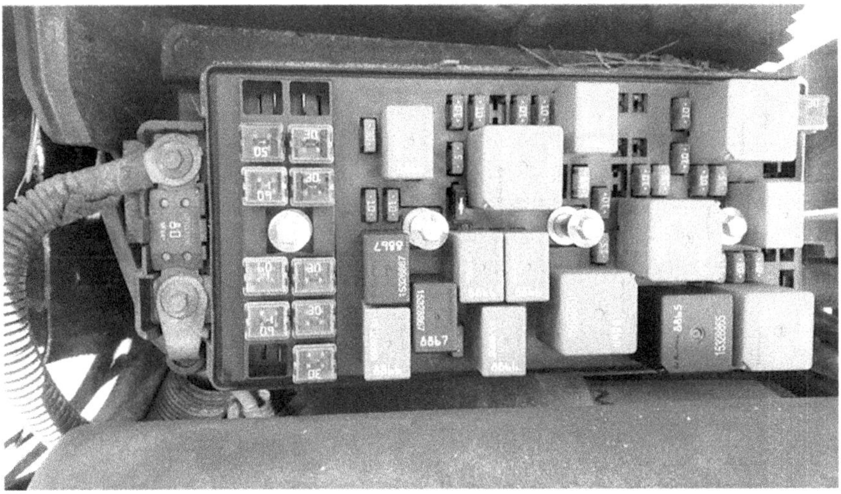

FIGURE 11.5 Fuse box with multiple relays (there are more relays on the fuse box placed near the engine).

FIGURE 11.6 Electromagnetic and optoelectronic relays sharing the same package.

TABLE 11.1
Product Codename for Solid-State Relays

Manufacturer	Product name
Toshiba	Photo relay
Matsushita Electric Works	Photo MOS relay
NEC	MOSFET relay
OKI Electric Industry	Photo MOS switch
Okita Works	Photo DMOS-FET relay
HP	Solid-state relay
OMRON	MOS FET relay
IXYS	OptoMOS solid-state relay

11.3 SOLID-STATE RELAYS

In most cases, the switch function in a dc circuit is achieved with a power MOSFET. *Solid-state relays* are available from multiple manufacturers under different names. Table 11.1 illustrates this several product examples.

Solid-state relays can be unidirectional or bidirectional. The unidirectional devices are ready to allow circulation of the current in a given direction. A simple MOSFET would do this function. Bidirectional relays allow circulation of current in both directions, and they can be seen as an ac device, thus using a triac or a combination of MOSFET devices. Figure 11.7 provides internal an structure for the two classes of device. It is noteworthy that both devices use an optocoupler to send the control information to the gate of the power semiconductor device.

In most automotive applications, the solid-state relay function is achieved with a hybrid integrated circuit, with a possible structure suggested in Figure 11.8. A driver circuit causes current to flow through the LED, turning it on. This circuit can be as simple as a pullup resistor and a switch. The LED light travels through the silicone

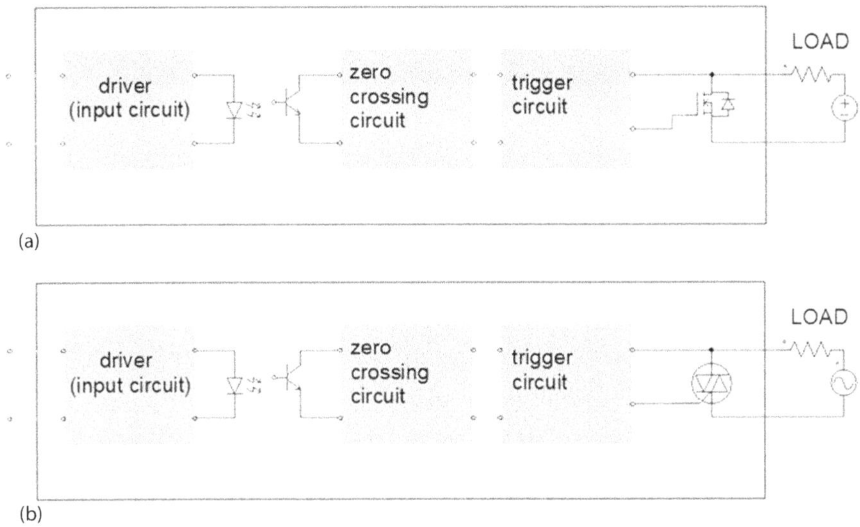

FIGURE 11.7 (a) Unidirectional and (b) bidirectional relays.

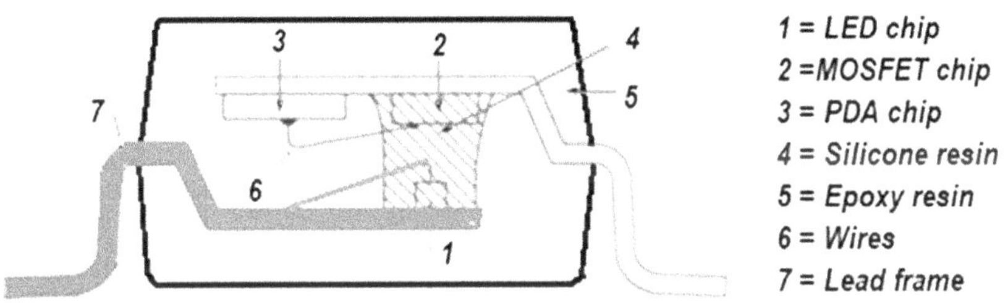

FIGURE 11.8 Possible internal structure of a solid-state relay offered as a hybrid integrated circuit.

resin and it is next converted to voltage. This voltage drives the power MOSFET via a gate driver circuit. For more general use, the output may contain a double MOSFET, allowing it to supply both ac and dc loads.

Solid-state relays offer numerous advantages and it seems to be the solution in the long term. These include:

- Lower maintenance costs.
- A smaller footprint with possible compatibility to other electronic printed-circuit boards and able to switch up to 10 A in a small printed-circuit-board package.
- A longer lifetime with greater than 500 million operations or 100,000 hours of continuous operation.
- A mechanical contact always produces audible noise during switching and also mechanical vibrations that are a source of electromagnetic compatibility.

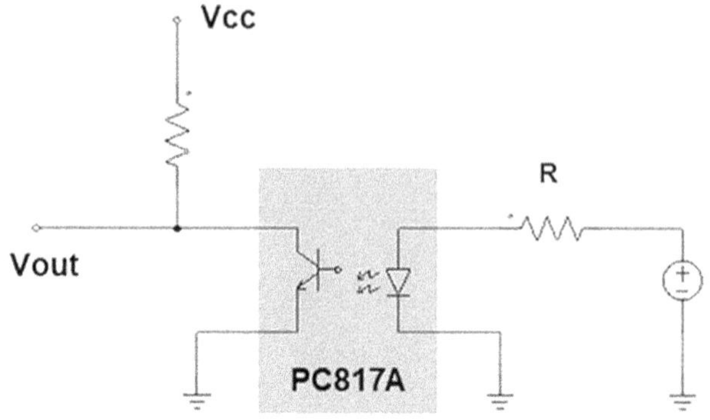

FIGURE 11.9 Circuit example for the optocoupler.

- Higher-speed switching.
- On-resistance as low as 5 mΩ.
- Off-state capacitance as low as 0.3 pF.
- Control currents through LEDs as low as 0.2 mA, which has a good compatibility with a direct connection to the microcontrollers.
- Multiple solutions, with single- and dual-contact models available with both normally open and normally closed contacts.
- Cost of semiconductor devices such as power MOSFETs and bipolar transistors is dropping rapidly and becoming more financially attractive than electromagnetic relays.

Since the optocoupler part of the solid-state relay is connected with the electronic circuit, a brief design example is herein included. The optocoupler is composed of two devices, an emitting LED and a receiving phototransistor. The LED driver needs to ensure a fast transmission of information, to protect against negative voltages, to limit the driving current under the LED limits—usually with a series resistances, to sustain operation for the entire temperature range, and to address lifetime and reliability concerns. An example circuit is shown in Figure 11.9.

The control/command voltage (herein denoted with V_{dd}) is derived from a microcontroller or DSP. The biasing resistance R_f is set such that the current rating of the digital circuit or device is not exceeded. For MCU and DSP, the *sink and source currents* usually range from *4 mA* to *9 mA* (check datasheet). Supposing the current rating is only *4 mA* maximum, set the actual forward current to *70 %*, or at most *80 %* of it. The biasing resistor yields as follows.

$$R_f > \frac{V_{dd} - V_f}{80\% \cdot I_{\text{rating}}} \tag{11.1}$$

The R_c resistor on the transistor side yields from the datasheet CTR_{device} current gain.

$$\frac{I_c}{I_f} < CTR_{\text{device}}$$
$$I_c = \frac{V_{cc} - V_{CEsat}}{R_c} \quad (11.2)$$
$$R_c > \frac{V_{cc} - V_{CEsat}}{CTR_{\text{device}} \cdot I_f}$$

11.4 INTRODUCTION TO FUSES

Given the complexity of the automotive power system, numerous loads are protected with *fuses* or *fusible links*. If an overload of current occurs, then the fuse will melt and disconnect the circuit before any serious damage is caused. Figure 11.10 illustrates the grouping of fuses in a fuse box or distribution box.

As a variant for fuses, the fusible links are different in design and identical in purpose and functionality. Instead of a location in a dedicated fuse box, the fusible links are placed on the wire. The fusible links are used in the main feeds from the battery as protection against major short circuits in the event of an accident or error in wiring connections. These links are simply heavy duty fuses and are rated in values such as 50, 100, or 150 A. Figure 11.11 illustrates the fusible links. It is noteworthy that certain fusible links have a small LED allowing quick access when the link is operational.

Concerning its construction, a fuse consists of a metal strip or wire fuse element, of a small cross-section compared to the circuit conductors, mounted between a pair of electrical terminals, and (usually) enclosed by a non-combustible housing. The fuse element is made of zinc, copper, silver, and aluminum.

FIGURE 11.10 Example of the fuse box in automotive applications.

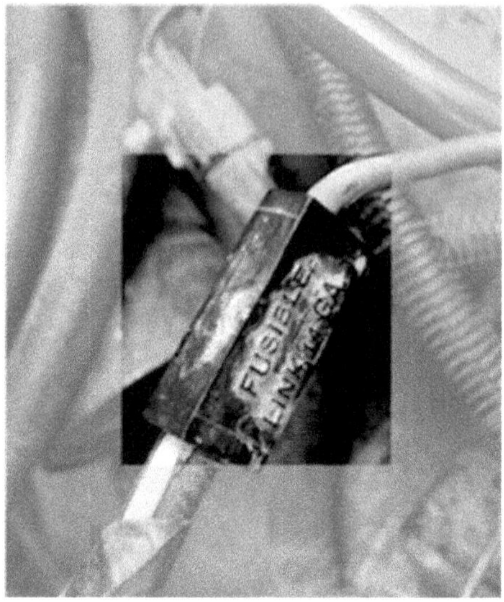

FIGURE 11.11 Fusible links.

The main parameters for a fuse are as follows.

- The *I2t rating* represents the amount of energy let through by the fuse element when it clears the electrical fault.
- The *breaking capacity* is the maximum current that can safely be interrupted by the fuse.
- The *voltage rating* of the fuse must be equal to, or greater than, the open-circuit voltage.
- The manufacturer may specify the *voltage drop* across the fuse at a rated current.

From a technological perspective, fuses can be classified into slow and fast fuses. A slow blow fuse can withstand transients or the surge current upon power-on/off, thus ensuring the equipment works normally. These are used mostly to protect against thermal rundown. Therefore, slow blow fuses are often called time-delay fuses. Conversely, a fast-(or quick) blow fuse is a fuse that bursts instantly when a high-power current is passed through it. The fast fuse is the fuse we find in most everyday electronic equipment where a quick disconnect is required.

There are many different types of fuses used within a motor vehicle. Most modern cars and trucks use one or more of the following types of bladed fuses, presented herein in descending order of size.

- Maxi (APX) heavy duty fuses
  - The largest type of blade fuse.
  - Used in heavy duty applications.
  - Available with higher amperage ratings than other blade fuses.

- Regular (ATO, ATC, APR, ATS) fuses
  - Similar to the first type of blade fuse patented by Littelfuse Corp. in 1970s.
  - Several different alternate versions that all fit in the same slots.
  - Found in most modern cars and trucks.
  - Most used: ATO and ATC are actually the same fuse (as size and shape), but, "C" stands for closed (sealed) and the "O" stands for open (exposed).
- Mini
  - Smaller than regular blade fuses, but available across a similar amperage range.
  - Also available in a low-profile mini version.
- Micro
  - The smallest type of blade fuse.
  - Available across the smallest range of amperage ratings.
  - Comes in two versions: two prong micro2 and three prong micro3.

Automobile fuses are available in three types: glass cartridge type, ceramic type, and blade type (Figure 11.12). The latter is the most used. Automotive fuses are rated with continuous current (current that the fuse will carry without risk of failure), and peak current value (peak value is the current that the fuse will carry for a short time without failing). The peak value of an automotive fuse is usually double the continuous value.

For easier identification, automotive fuses are color-coded. This is general rule for all original equipment manufacturer (OEM) products. An original equipment manufacturer is a company that produces parts and equipment that may be marketed

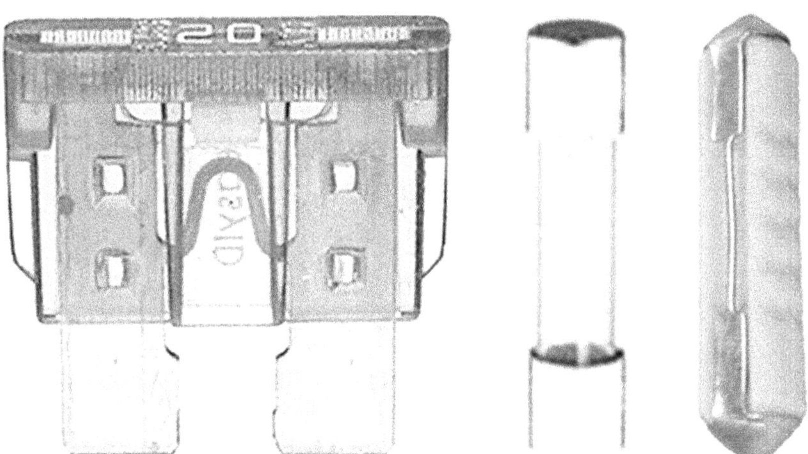

(a) (b) (c)

FIGURE 11.12 Constructive types of automotive fuses: (a) blade, (b) glass, (c) ceramic.

TABLE 11.2
Color Codes for Automotive Fuses

Type	Current [A]	Color
Blade	3	Violet
	4	Pink
	5	Clear/beige
	7.5	Brown
	10	Red
	15	Blue
	20	Yellow
	25	White
	30	Green
Ceramic	5	Yellow
	8	White
	16	Red
	25	Blue

by another manufacturer. Table 11.2 provides examples of color codes for automotive fuses.

Finally, it is worth remembering the previous construction of fuses allow for changing the cooper (Cu) wire. In such a setup, a simple cooper (Cu) wire can be used according to Table 11.3. The drawback with this is that it gives an imprecise control of the blow-out current level. The arrangement was used under the premise that a short-circuit would produce a current higher than the operational current anyways.

TABLE 11.3
Selection of the Cooper Wire for a Self-Manufactured Fuse

Fuse rating [A]	Cu wire diameter [mm]
3	**0.15**
5	**0.20**
10	**0.35**
15	**0.50**
20	**0.60**
25	**0.75**
30	**0.85**
45	**1.25**
60	**1.53**
80	**1.80**
100	**2.00**

FIGURE 11.13 Automotive circuit breakers with their current ratings.

11.5 CIRCUIT BREAKERS

Circuit breakers are used in place of fuses in certain heavy vehicles. A circuit breaker has the same rating and function as a fuse but with the advantage that it can be reset, through a push-button. The similarity with a blade fuse can be seen in Figure 11.13, where the same color coding applies for the current ratings. The disadvantage of the circuit breaker is the much higher cost. This is why circuits with larger current loads, subjected to difficult operating conditions, justify the use of circuit breakers instead of fuses.

Circuit breakers use a bimetallic strip which, when subjected to excessive current, will bend and open a set of contacts. A latch mechanism prevents the contacts from closing again until the reset button is pressed.

11.6 AUTOMOTIVE VARISTOR AND TRANSIENT-VOLTAGE-SUPPRESSION (TVS)

While fuses protect against high currents, a set of devices protect electronic equipment against temporary/accidental overvoltage. Such devices are called transient voltage surge suppression devices, or *varistors*. A combination of the words *VARI-able resi-STOR* is used to define the device. A varistor is an electronic component with an electrical resistance that varies with the applied voltage, as shown in Figure 11.14. Also known as a voltage-dependent resistor, the varistor has a nonlinear, non-ohmic current–voltage characteristic that is similar to that of a diode.

The most common type of varistor is the metal-oxide varistor (MOV), which contains a ceramic mass of zinc oxide grains, in a matrix of other metal oxides (bismuth, cobalt, manganese) sandwiched between two metal plates (electrodes). The boundary between each grain and its neighbor forms a diode junction, which

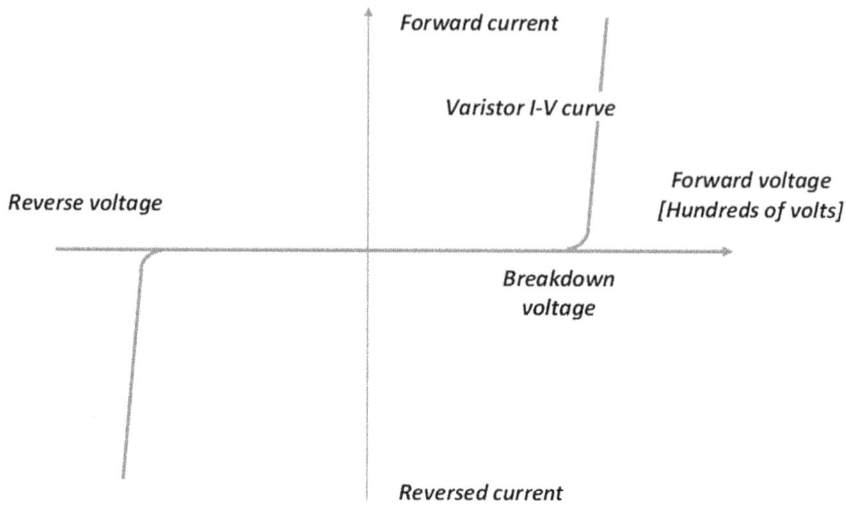

FIGURE 11.14 V-I Characteristics of a varistor.

allows a current to flow in only one direction. When high voltage is applied, the diodes' junctions break, allowing for a large current. The device survives for a certain time under this large current until the problem is cleared or the circuit is disconnected.

Since the role of the varistor is mostly to protect electronics, the varistor comes packaged as a printed-circuit-board device. Its rating is based on a series of parameters related to the damaging energy. When voltage transients occur in rapid succession, the average power dissipation is the energy (watt-seconds) per pulse times the number of pulses per second. For a single pulse, clamped at a certain voltage, the fault source should clear before the device ratings are reached. Datasheet information is defined based on the test shown in Figure 11.15 where the current pulse corresponds to IEEE ANSI standard C62.1 ("*Standard for surge arrestors for ac power circuits*," 1984 Edition, November 24, 1984, and subsequent) waveform. This is an

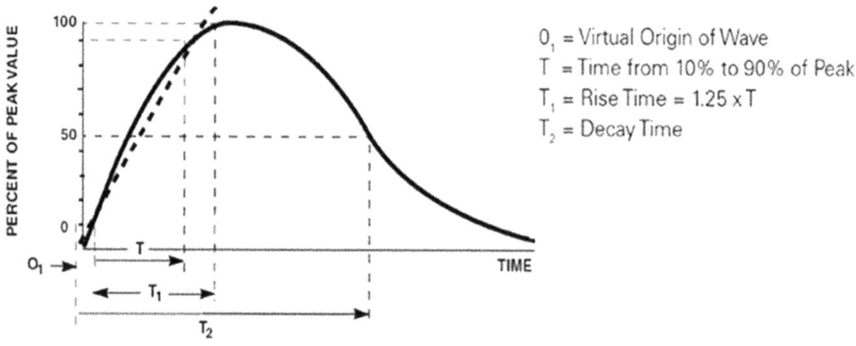

FIGURE 11.15 Rating definition or a varistor. Example for a 8/20 μs current waveform, 8 μs = T_1 = rise time and 20 μs = T_2 = decay time.

exponentially decaying waveform representative of lighting surges and the discharge of stored energy in reactive circuits.

Therein, the rise time is defined as the interval within the positive slope of the response, while the decay time is usually measured down to 50 % of the peak value.

The automotive market allowed designers and manufacturers to come up with numerous alternative solutions. The TVS diode, also called *transil* or *thyrector* is one of these, even if it is rarely used. It has a construction very similar to a Zener diode. A TVS diode can respond to overvoltages faster than other common overvoltage protection components, such as varistors or *gas discharge tubes* (GDT). The actual clamping occurs in roughly one picosecond, but in a practical circuit the inductance of the wires leading to the device imposes a higher limit. A transient-voltage-suppression diode may be either unidirectional or bidirectional.

11.7 SOLENOIDS

11.7.1 Electromechanical Solenoid

Electromechanical solenoids are highly used in automotive applications. Solenoids in automotive applications include: engine control valves, advanced suspension controls, coolant control valves, thermal management control valves, variable water pump controls, valve deactivation controls, high-pressure fuel inlet valves, air intake controls, door locks and latches, air bag inflation control valves, urea/ad blue control valves, discrete proportional control valves, transmission fluid controls, power steering valves, compressor clutches and valves, and possibly others. It can be seen that some of these applications assume an on/off control, while others a linear control.

These devices consist of an inductive coil wound around a movable steel or iron plunger. The coil is shaped such that the movable piece can be moved in and out of the space in the center of the coil. Such displacement alters the inductance of the coil, transforming it into an electromagnet.

The solenoid operation is based on Faraday's law of induction, which predicts how a magnetic field will interact with an electric circuit to produce an electromotive force. Herein, a current through the coil produces movement of the movable piece, also called the armature. This movement of the armature is further used to provide a mechanical force to some mechanism.

A simple experiment is shown in Figure 11.16. A soft-iron piece is placed along the axis of a coil. Initially (Figure 11.16a), there is no current applied to the coil. A dc current is applied suddenly to the coil producing a magnetic field. The soft-iron piece snaps into the center of the coil (Figure 11.16b). An end-trip stopper can define the final position of the moving part. If the dc battery reverses its polarity, the magnetic field reverses its direction and the soft-iron piece snaps out of the coil (Figure 11.16c). Attention should herein be paid to the direction of the winding in respect to the desired effect of the current.

Solenoids are controlled directly by a controller circuit through the current applied to the coil, and thus have very quick response times. The force applied to the armature is proportional to the change in inductance of the coil due to the change in

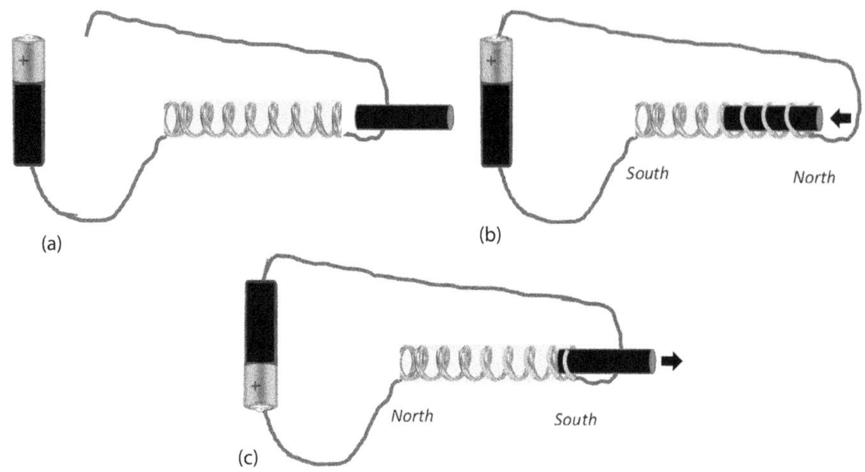

FIGURE 11.16 Solenoid principle: (a) before supplying current; (b) after a current is applied to the coil; (c) after a negative current is applied to the coil.

its position and the current flowing through the coil. The force applied to the armature will always move the armature in a direction that increases the coil's inductance.

Special classes of solenoids can be identified in proportional and rotary solenoids.

The proportional solenoids have an analog positioning of the solenoid plunger or armature as a function of coil current. The flux carrying geometry of the solenoid produces a high starting force (torque), followed with a section that quickly begins to saturate magnetically. The resulting force (torque) profile is nearly flat or descends from a high to a lower value. This concept is fully defined in SAE publication 860759 from 1986. That publication presents the fundamental contrasts between snap-acting and proportional solenoids.

The rotary solenoid uses a special construction able to rotate a ratcheting mechanism when power is applied. The ratchet mechanical device allows continuous rotary motion in only one direction while preventing motion in the opposite direction. When the solenoid is activated, the armature core is magnetically attracted toward the stator pole, and the disk rotates on the ball bearings as it moves toward the coil body. When power is removed, a spring on the disk rotates it back to its starting position.

Electromagnetic solenoids can help transmission shifting, vehicle starting, activating/deactivating of the four-wheel drive, fuel-injections systems and locking/unlocking of cars. Solenoids are also able to help vehicles achieve both higher and better gas mileage. A solenoid can control the lock-up of the torque converter inside of a vehicle. Without the solenoid, the torque converter might slip when in motion, causing excess use of gasoline.

11.7.2 Solenoid Valve

Fluid power deals with the use of fluids under pressure to generate, control, and transmit power. Fluid power is also known as hydraulics when a liquid such as mineral oil or water is used, and pneumatics when a gas such as air is used. Compressed-air and water-pressure systems are the most known systems for fluid power.

A solenoid valve is the device used to control the fluid power. These devices are often simply called solenoids. Some people call the solenoid both the solenoid valve and the electromagnetic relay as a reference to the coil in their construction.

The *solenoid valve* construction is not unlike the electromagnetic relay in Figure 11.3. A coil is used to generate a magnetic field when supplied with a dc current. The magnetic field acts upon the actual valve, allowing or blocking the flow. The major difference from electromagnetic relay in Figure 11.3 is that the relay is manufactured and packaged as an universal component designated to make or break any electric circuit, while the solenoid valve has a construction depending on application, closely related to the mechanics of the fluid power.

Similar to electromagnetic relays, the solenoid valve can be normally open and normally closed, meaning that controls whether or not the fluid passes through. For a simple example, Figure 11.17 reviews the cruise control system based on manifold power, previously shown in Chapter 4.

Starting from the on/off solenoid valve, a more precise device can be designed to linearly control the position for the opening and, accordingly, the flow of fluid. Such

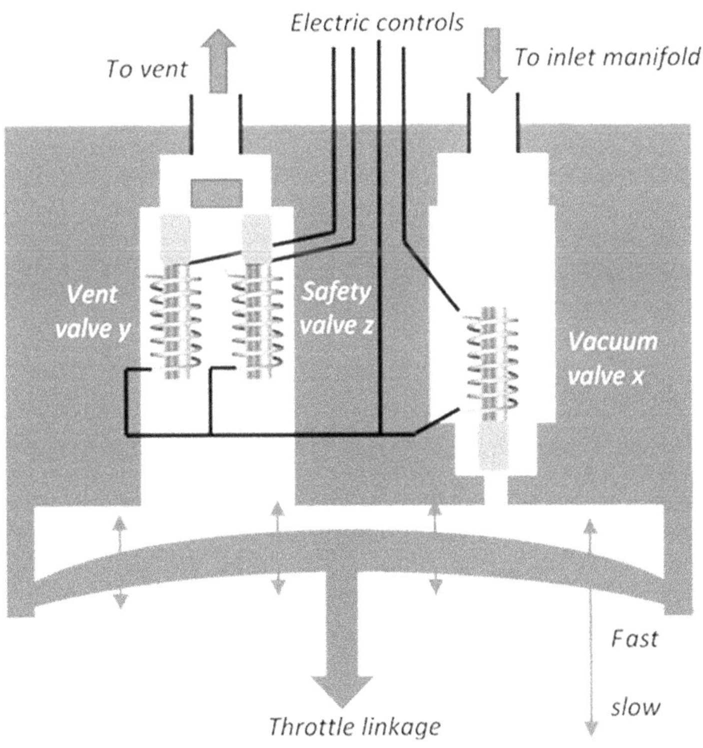

FIGURE 11.17 Three solenoid valves are part of the cruise control system using the manifold power.

devices are properly called solenoids. They are used to operate pistons and valves for accurate control of fluid pressure or flow in applications such as transmissions and fuel injection. Solenoids can be seen as linear motors with a fixed range of travel since they produce displacement.

The mechanism of a solenoid varies from linear action, plunger-type actuators, pivoted-armature actuators, and rocker actuators. The valve can use a two-port design to regulate a flow (turn on or off) or use a three or more port design to switch flows between ports.

11.7.3 Power Converter Drive

The more exciting linear solenoid offers more opportunities for power electronic control. The position of the linear solenoid is controlled in a feedback loop. The simpler solution consists of relating the position to the actual fluid pressure. A precise measurement of the downstream (after the valve) pressure will allow for precise control. This solution is very expensive.

Since the force imposed by the mechanical load on the solenoid is directly proportional to the magnetic field, which, in turn, is directly proportional to the current through the coil, a practical alternative is to estimate the position of the solenoid by measuring the current through the solenoid in a steady-state. When the balance of the forces between the spring-type load and the electromagnetic force is achieved, the average current through the solenoid is proportional to the flow.

The current through the solenoid is controlled with a switched-mode MOSFET. A microcontroller-generated pulse-width modulated signal controls the MOSFET to turn on and off at a high frequency. The MOSFET is in a series with the solenoid and the battery. The average voltage is determined by the ratio of the operation time to the pulse period.

Due to small nonlinearities in the system, the actual characteristics need to be measured and stored for a proper control.

11.8 CONCLUSION

An automotive power system is based on a multitude of fuses and relays able to protect and connect various motors and electromagnetic actuators.

Despite being based on the same power MOSFET devices, the intelligent switch and the relays are different in construction, since the latter can provide galvanic separation.

Due to the large OEM market for automotive components, fuses are highly standardized and available in a series of packages and nominations.

Along with fuses, varistors are used to protect electronics equipment from accidental overvoltages.

REFERENCES

Anon. 2015. *Automotive Power Distribution*. TE Connectivity Corporation Documentation.

Anon. 2017. *Automotive Relay Replacement Reliability Meets Space Savings*. Nexperia Corporation Documentation.

Anon. 2017. *MLA Automotive Varistor Series*. Littelfuse Corporation Documentation.

Anon. 2017. *TI Designs: TIDA-01506 Reference Design for Automotive, Proportional-Solenoid Current Sensor.* Texas Instruments Reference Design.

Anon. 2018. *Power MOSFET Selecting MOSFFETs and Consideration for Circuit Design Application Note*. Toshiba Corporation Documentation.

Denton, T. 2017. *Automobile Electrical and Electronic Systems*. 5th edition, Abingdon-on-Thames: Routledge.

Mohler, D. 1986. *Rotary and Linear DC Proportional Solenoids*. SAE Technical Paper 860759, 1986. doi:10.4271/860759.

Neacşu, D. 2004. *Power Semiconductor and Control for Automotive Applications*. Tutorial Presented at IEEE APEC, Anaheim, CA, USA.

12 Small Motors

HISTORICAL MILESTONES

1827: Georg Ohm discovered Ohm's Law

1831: Faraday discovered electromagnetic induction, and Joseph Henry and Michael Faraday created early motion devices using electromagnetic fields.

1834: Thomas Davenport of Vermont, U.S.A., developed the first real electric motor ("real" meaning powerful enough to do something). His dc commutator motor ran at up to 600 RPM, and could power machine tools or a printing press.

1867: A benefit to dc machines came from the discovery of the reversibility of the electric machine, which was announced by Siemens in 1867.

1879: Walter Baily demonstrated the first primitive induction motor by manually turning switches on and off. This is known as Arago's principle.

1887: Tesla invented his electrical motor and demonstrated their industrial use.

1889: Mikhail Dolivo-Dobrovolsky invented the three-phase induction motor and started industrial production of large motors, a 20-hp squirrel cage and a 100-hp wound rotor in the 1890s.

1965: Ultrasonic motors were invented by V.V. Lavrinko.

12.1 PRINCIPLE OF ELECTRICAL MOTORS

12.1.1 Permanent Magnet Motors

Because small electrical motors, with a simplified construction and control, are present throughout motor vehicles, their operation principles are detailed herein.

The simplest possible motor can be considered as starting from a magnetic needle, similar to the one used for a compass. If the Earth's magnetic field were replaced with a strong local magnetic field produced with a coil, the needle would point to this magnetic field, rotating and settling at 90° from the coil. This magnetic field can be achieved with an electric circuit powered by a battery and coil, as shown in Figure 12.1. Overall, this is still a static setup.

The previous experiment can be enhanced with a set of three similar coils, placed at 120° from each other, as sketched in Figure 12.2. The same battery is used with three switches to apply a dc current alternatively to the three coils. Each time a coil is switched on, the needle revolves 120°. This is the principle of a motor with a *permanent magnet as a rotor.* The moving part is called a rotor and this is also the magnetic needle in our experiment. The stationary part with the three-coil setup is called a stator.

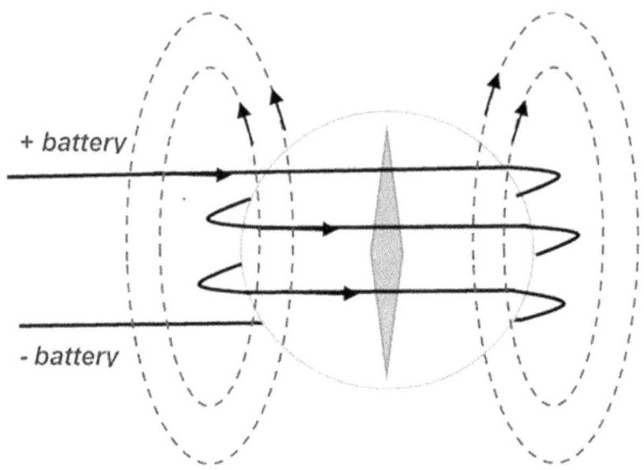

FIGURE 12.1 Magnetic needle (compass) orientation due to an electric current.

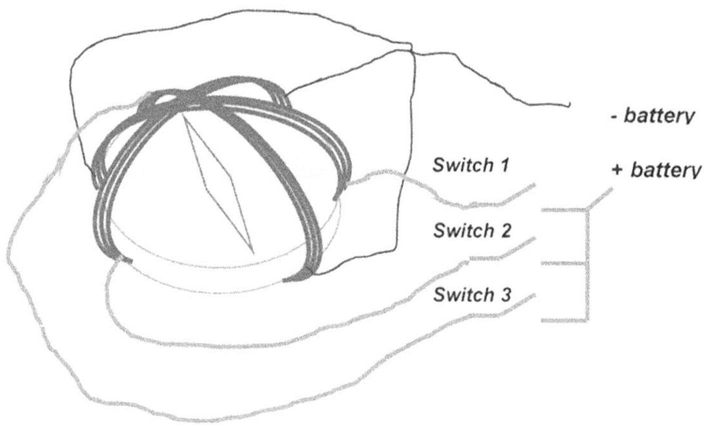

FIGURE 12.2 Using three coils for creation of the magnetic field.

An improvement can be further made by replacing the needle with a cylindrical magnet, and using four coils instead of three. This is called a *stepping motor.*

12.1.2 Variable Reluctance Motor

The arrangement with a magnetic needle or cylinder may be somewhat expensive. The experiment in Figure 12.3 illustrates a rotating magnetic field applied to an iron core. The stator can be made with six coils placed on six teeth. These teeth are called *salient poles* and they are built on an iron core support. Any coil and its opposite are both connected in a series, together forming a magnetic field, and this arrangement is called a *phase*. Hence, the motor in the experiment has three phases. The rotor can be made of an iron core with four teeth, which are also called poles. It is noteworthy that the number of poles in the rotor arrangement must be different from the number of phases in the stator coil system. Finally, the current is supplied

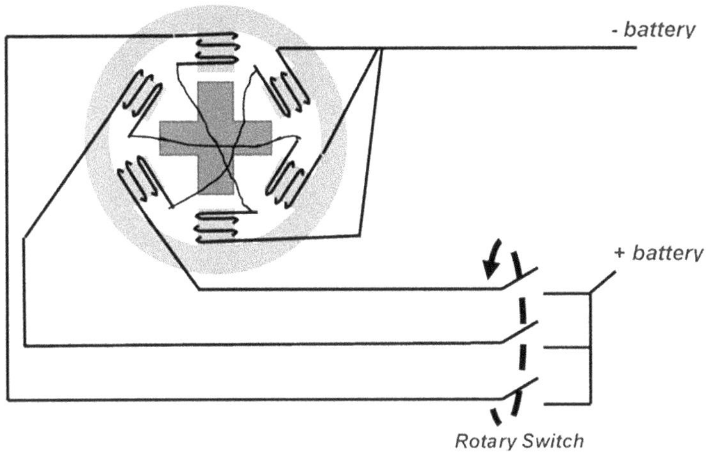

FIGURE 12.3 Using a rotary switch with a variable reluctance motor.

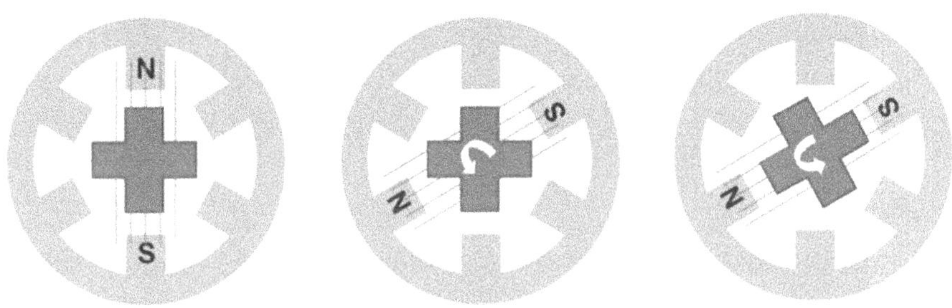

FIGURE 12.4 Instances in the operation of a variable reluctance motor.

through switches to the motor, as illustrated in Figure 12.3. Figure 12.4 illustrates instances in operation.

A lever provides a successive supply to the coils on the stator. The operation is shown with a counterclockwise rotation, in steps of 30°. This principle of a motor without a magnet is called a *variable reluctance motor.*

Using this principle within an actual application implies the use of an inverter able to create alternative voltages on the three phases. On each phase, the Kirchhoff equation yields:

$$\begin{aligned} v(t) &= i \cdot R + \frac{d(L \cdot i)}{dt} = i \cdot R + i \cdot \frac{dL}{dt} + L \cdot \frac{di}{dt} = i \cdot R + i \cdot \frac{dL}{d} \cdot \frac{d}{dt} + L \cdot \frac{di}{dt} \\ &= i \cdot R + L \cdot \frac{di}{dt} + i \cdot \omega \cdot \frac{dL}{d} = i \cdot R + L \cdot \frac{di}{dt} + e \end{aligned} \tag{12.1}$$

where the dynamic dependency of both current and inductance on time is observed, and "*e*" represents the self-induced emf voltage. This self-induced emf (back-electromotive-force, BEMF) is proportional to the rate of change in inductance and the rotational speed.

Mechanical power produced by the reluctance motor yields as follows.

$$P_m = \frac{1}{2} \cdot e \cdot i = \frac{1}{2} \cdot i^2 \cdot \omega \cdot \frac{dL}{d} \tag{12.2}$$

As with any motor, this ideally equals the rotational power.

$$P_m = \omega \cdot T \tag{12.3}$$

This derives the electromagnetic torque within a reluctance motor as follows.

$$T = \frac{1}{2} \cdot i^2 \cdot \frac{dL}{d} \tag{12.4}$$

12.1.3 DC Motors

The two experiments so far used a set of switches to supply the coils with the alternative current. Since a motor vehicle benefits from a dc distribution bus and a battery, electrical motors able to use the dc voltage directly are of interest. Figure 12.5 illustrates the principles of a dc motor.

This is the most used type of motor. The stator has a housing, and two permanent magnets. The rotor has copper windings uniformly placed on an iron core. These coils are supplied through a commutator and two brushes are mounted on the stator. Each rotor coil is set at 90° of the magnetic field produced by the magnets on the stator. Such an arrangement produces a force tangential to the rotor, according to Fleming's left-hand rule (Figure 12.5a). If the middle finger is the current, and the index finger the magnetic field, the force direction is given by the thumb. If both the field and the current changes direction at the same time, the force direction stays the same.

If we reverse the roles and consider first the mechanical rotation of the rotor within the magnetic field applied from the stator magnets, a voltage appears on the rotor, whichcan be measured at the rotor terminals. This voltage is proportional to the rotational speed, the induction of the magnetic field, and the length of the conductor ($e = v \cdot B \cdot l$). This operation mode has the same principle as the bicycle dynamo, and it is called a *generator mode.*

If you reverse the polarity of the voltage applied to a dc motor drive, the motor will rotate in the opposite direction. Analogously, if we force a rotation in the opposite direction, the generated voltage will yield negatively. It can be seen that the same dc electrical machine can work on four quadrants of the current–voltage plane.

The dc motor is supplied with a dc current and the Kirchhoff equation for the electrical supply circuit yields as follows.

$$V = R \cdot i + e \tag{12.5}$$

where "e" represents the *counter-emf voltage* (or *back-electromotive-force*), that is proportional with rotational speed (angular velocity of the motor).

$$e = k_w \cdot \omega \tag{12.6}$$

$$V = R \cdot i + k_w \cdot \omega \tag{12.7}$$

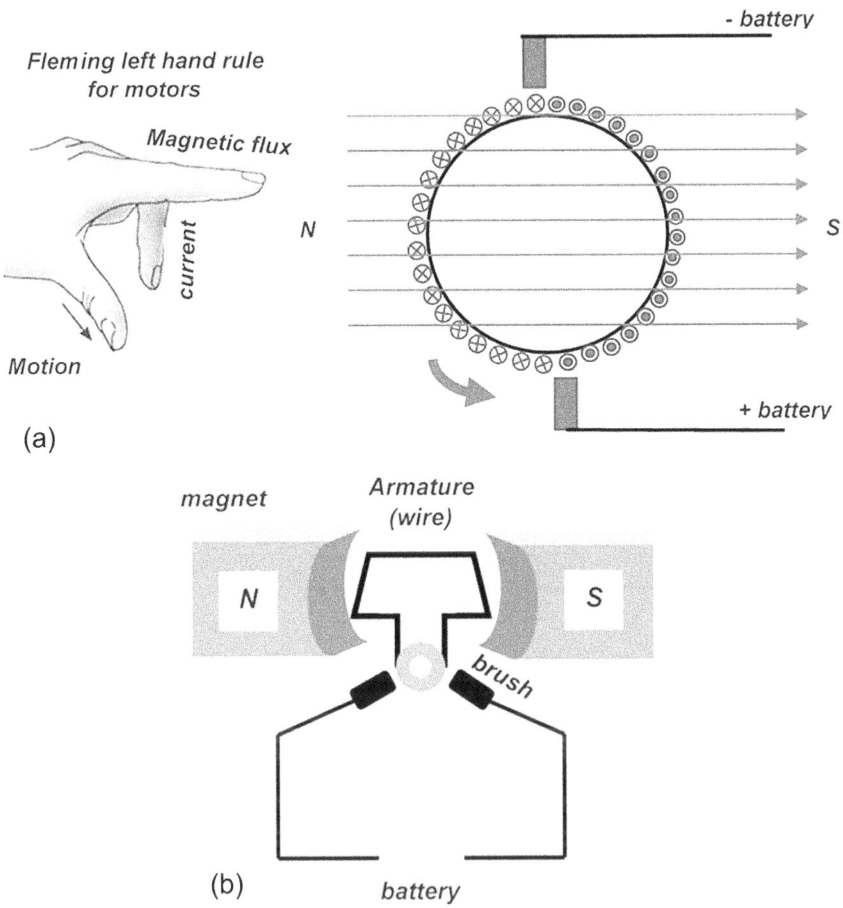

FIGURE 12.5 Principle of a dc motor. (a) Physics; (b) Actual setup.

The torque produced by the rotor is directly proportional to the current in the armature windings. The proportionality constant is the torque constant of the motor.

$$T = k_T \cdot i = k_T \cdot \frac{V - k_w \cdot \omega}{R} \tag{12.8}$$

where T represents the torque developed at the rotor, and k_T represents the motor torque constant.

12.1.4 Induction Motor

Another constructive version of an electrical motor can be achieved when a cylindrical iron rotor is used instead of the four-pole iron rotor and the stator is supplied with similar three-phase currents. Thus, a rotational movement is produced. The physical reasons are different from the salient rotor shown in Figures 12.3–4, even if the experiment seems similar.

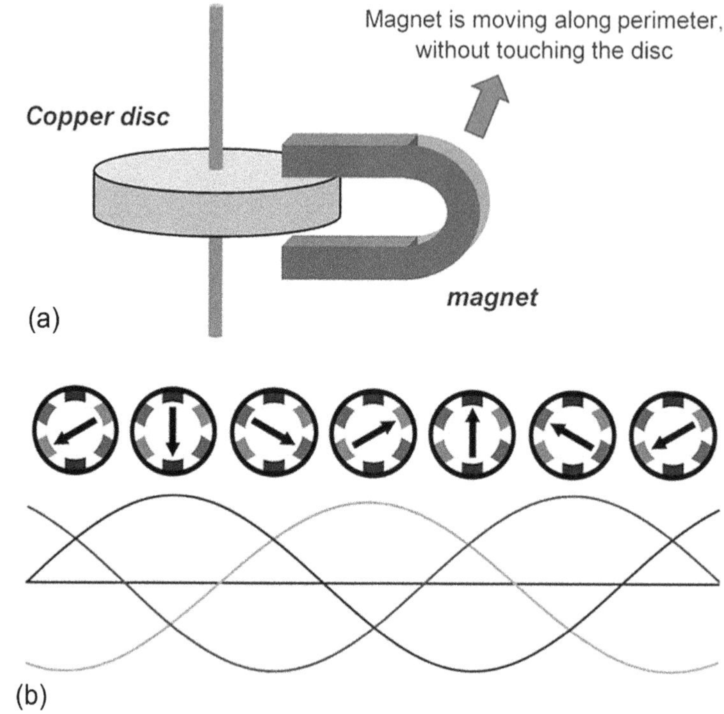

FIGURE 12.6 Arago's experiment for a simplified induction motor.

This time, the magnetic field travels through the iron (or copper) rotor, or this can be seen as the rotor material moving across the field, creating an electromotive force. Such generated voltage produces a current, which in turn interacts with the magnetic field created by the second coil (second phase in stator).

An explanatory experiment is suggested by Arogo's disk experiment (Faraday), Figure 12.6a.

A copper disk is placed under the influence of a permanent magnet. If the magnet is moved around the circumference of the disk, the disk will try to follow. The difference between the rotational speed of the magnetic field and the mechanical rotation of the copper disk is called *slip frequency*.

Performance can be improved if the magnetic field is produced with a system of coils similar to Figure 12.3. However, this new motor setup does not benefit from sudden changes in the magnetic field, or sudden changes in the current through the set of coils. If voltages applied to the three phases vary sinusoidally, as shown with Figure 12.6b, the magnetic field moves uniformly instead of switching currents quickly and this is the benefit of using an *induction motor*.

12.1.5 Brushless Motors

The major advantages of dc motors are that they can operate from a battery and the torque is proportional to the current. On the downside, a conventional dc motor uses the brushes and commutator assembly, which produces poor lifetime and efficiency. Despite all these, dc motors are still the most used components, since they are preferred in applications with limited lifetime requirements, such as wiper motors which are only used when it rains.

Conversely, ac motors do not have brushes and commutators since they do not need to supply the rotor. As a trade-off solution, a motor was built to replace the brushes–commutator assembly (from the construction of a conventional dc motor) with an electronic switching circuit made of two transistors and two diodes for each phase. Diodes are required for protection when the phase is turned off. The number of phases is limited to three to reduce the number of transistors and keep the costs lower. A synchronization of the motor's movement with voltages applied to poles is achieved with one or more Hall effect sensors. The principle for a *brushless dc motor* is sketched in Figure 12.7.

Brushless dc motors (BLDC) and the *permanent magnet synchronous motors* (PMSM) can be the same or different depending on people's views. Both are synchronous motors with permanent magnets attached to their rotors.

A possible definition assumes the BLDC motor has trapezoidal back-electromotive-force (BEMF), while PMSM has a sinusoidal BEMF. Figure 12.8 illustrates possible waveforms for BEMF. The movement of the rotor magnets past the stator teeth induces BEMF into the windings of the armature. When the phase–phase BEMF is plotted with an electrical angle (or time), it may look sinusoidal, trapezoidal, or somewhere in between. The electrical angle is equal to the mechanical angle divided by the number of pole pairs. The BEMF can also be seen on an oscilloscope by measuring phase–phase voltage when the motor rotates. A sinusoidal BEMF typically means a motor has been wound with *distributed windings* (instead of *concentrated windings*), where the windings are distributed over many slots, which is more common for large electric motors. Distributed windings are easily seen by the overlap at the end of the motor. Conversely, concentrated windings are seen as separated from each other.

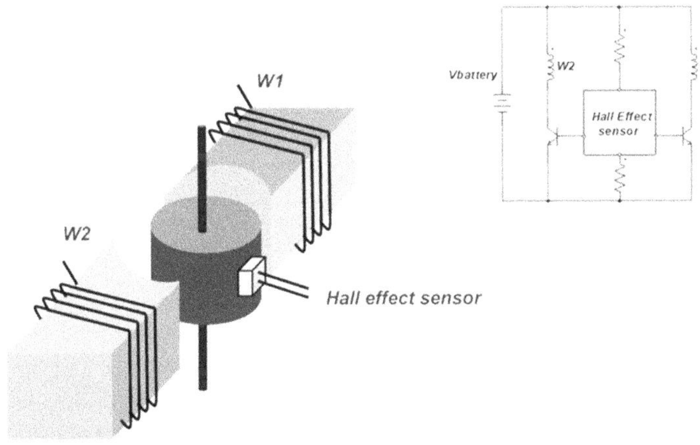

FIGURE 12.7 Magnetic principle of a brushless dc motor.

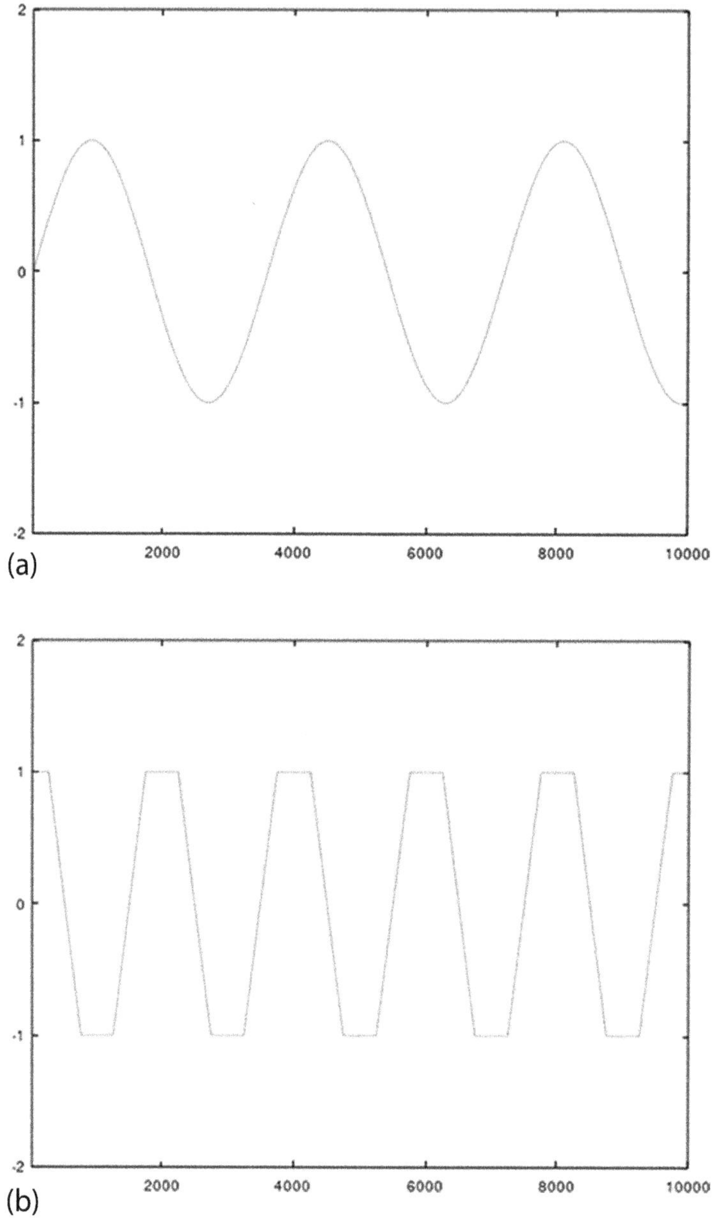

FIGURE 12.8 Sinusoidal versus trapezoidal BEMF.

The brushless dc motors differ in construction as *in-runners* and *out-runners* (Figure 12.9) based on the location of the magnets with respect to the stator windings. While the in-runner BLDC motors have a more conventional description—with the rotor inside the stator enclosure—the out-runners have the magnets on a can outside the stator windings. At the same power level, the out-runners generally produce

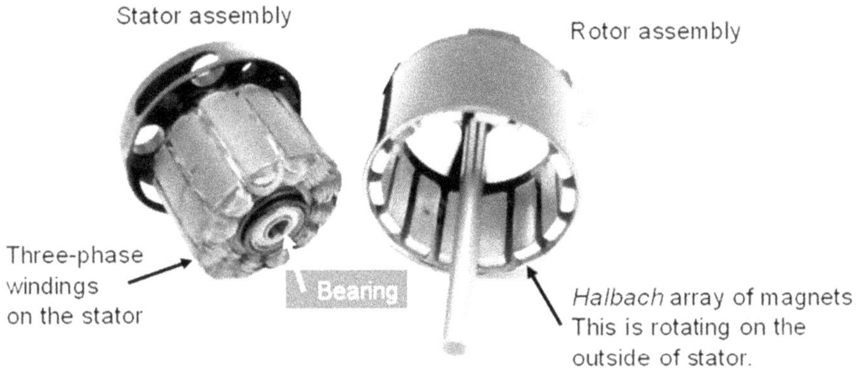

FIGURE 12.9 Out-runner BLDC motor.

a lower speed and a higher torque than the more conventional in-runners since the larger-can diameter of the out-runner allows for a higher quantity of magnets to be used in the outer case, creating more magnetic poles. More magnets alternating magnetic poles forces the controller to switch more rapidly, slowing down the overall rotational speed. An additional advantage is that the out-runners produce nearly sinusoidal BEMF even if they have concentrated windings on the stator. This is due to a larger number of magnetic poles (for instance, 12 or 14).

As an example, the Mercedes Benz S-class W220 (1998–2005) engine cooling fan uses an 850 W out-runner BLDC.

BLDC motors are often rated with “Kv.” This should not be confused with “kV,” the abbreviation for kilovolt. It means the number of revolutions per minute (rpm) that a motor turns when 1 V (1 volt) is applied with no load attached.

Torque-speed characteristic of the BLDC motor is similar to that of a dc motor where torque is proportional to the supplied current. However, the trapezoidal BEMF produces more torque ripple than the sinusoidal BEMF. For most low-power automotive applications, using a motor with a conventional six-step control or trapezoidal BEMF may not lead to any noticeable problems. High-performance applications requiring reduced noise, reduced vibration, and increased efficiency may require a sinusoidal BEMF.

The electronic control is mandatory for a BLDC motor and this is sketched in Figure 12.10. Without a three-phase inverter properly controlled, the brushless motors are useless devices. The circuit in Figure 12.10 takes advantage of six transistors used as electronic switches. The operation of this circuit identifies six states, each characterized by three MOSFET transistors in conduction.

During rotation of the rotor, a special device based on a sensor secures synchronization of the control with the actual rotation. For example, here, a set of optocouplers can be used along with a slotted disk. The rotation of the motor allows the light to pass through the slots and toward three optocouplers only. These optocouplers provide signals that are used to control three transistors at a given moment. Therefore, three switches are in conduction simultaneously, and all valid combinations define the six states. The individual circuits described by the six states are drawn in Figure 12.11. This sequence of states produces a rotating magnetic field inside the motor.

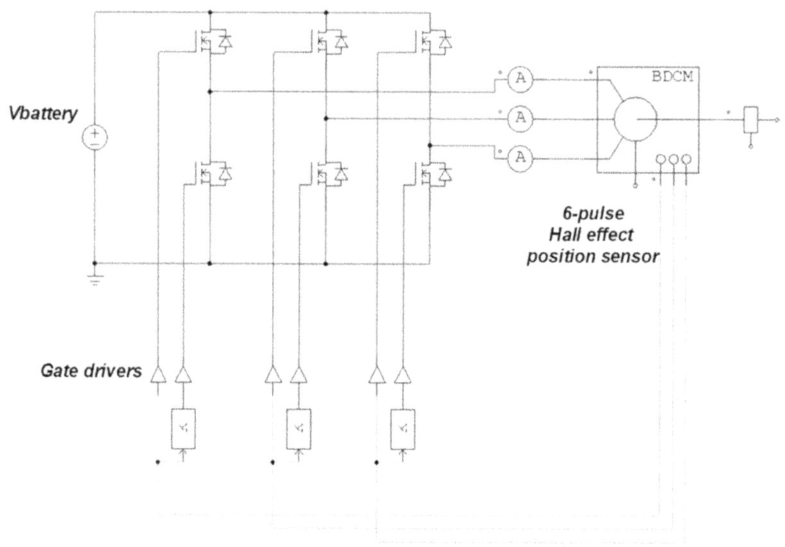

FIGURE 12.10 Open-loop control principle for a brushless dc motor.

12.1.6 Comparison between Brushless dc Motor and the Induction Motor

Since both the brushless dc motor and the induction motor use a similar power electronic converter (Figure 12.10) for control, it is worthwhile making a direct comparison between the two.

When a three-phase induction motor is connected directly to a three-phase utility power, torque is produced immediately and the motor has the ability to start under load without any special electronic circuit. From Tesla's invention of the induction motor in 1889 and until the 1960s, there were no electronic inverters used for supplying induction motors. This direct compatibility with conventional utility power is the main reason for their success. In contrast, a brushless dc motor produces no starting torque when directly connected to a fixed-frequency utility power. They work in association with electronic circuits only. This generic advantage of induction motors is not practical in automotive applications since the primary source of energy is the dc battery. Hence, the induction motor also needs a three-phase inverter as a mandatory condition for operation.

The advantages of brushless dc motors are:

- Much less rotor heat is generated with the brushless dc motors since there are no windings and no calorimetric loss on those windings. Rotor cooling is therefore easier and peak point efficiency is generally higher for the brushless dc motor drive.
- The brushless dc motor can also operate at a unity power factor, whereas the best power factor for the induction drive is about 85%. This also contributes to better efficiency.

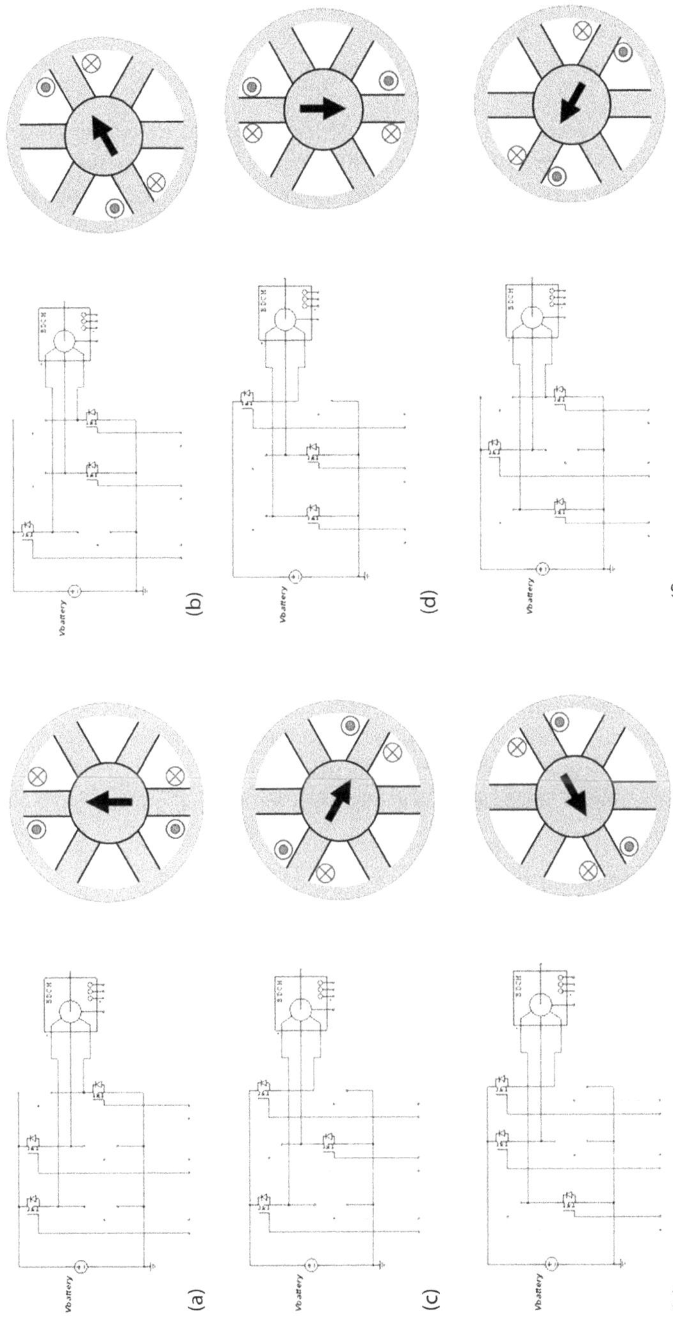

FIGURE 12.11 The six states (a–f) during operation of a three-phase brushless dc motor.

- Apart from synchronization, the actual control of the brushless dc motor is simpler than the control of an induction machine, where the designer needs to account for the variable slip frequency.

The advantages of the induction motor drives are:

- The brushless dc motors cannot adjust the magnetic induction since the magnets provide the same magnetic field either when the motor is used at low power or high power. The induction motor can reduce the voltage at light loads such that magnetic losses are reduced and efficiency is maximized. Induction drives may be the favored approach where high-performance is desired since peak efficiency will be a little less than with brushless dc motors, but average (or work profile) efficiency may actually be better.
- Generally, induction motors are easier to protect.
- Induction motors are less costly than brushless dc motors since permanent magnets are expensive.

Today, most electric and hybrid vehicles use brushless dc motors. However, the electrification process and the transition toward more electrical power may favor induction motors for the increased performance opportunity.

12.2 DESIGN OF LOW-POWER DC MOTORS

Since the conventional dc motors are very used in automotive applications, a look into their application circuit is required. In practice, the motor is set within an electrical circuit, supplying a stator winding to create the magnetic field. The design of a motor is made with task-dependent characteristics.

Circuit connection of stator and rotor identifies four types of circuits.

12.2.1 SHUNT-WOUND DC MOTORS

The field winding is connected in parallel with the armature, as shown in Figure 12.12. Due to the constant excitation of the fields, the speed of this motor remains constant, virtually independent of torque.

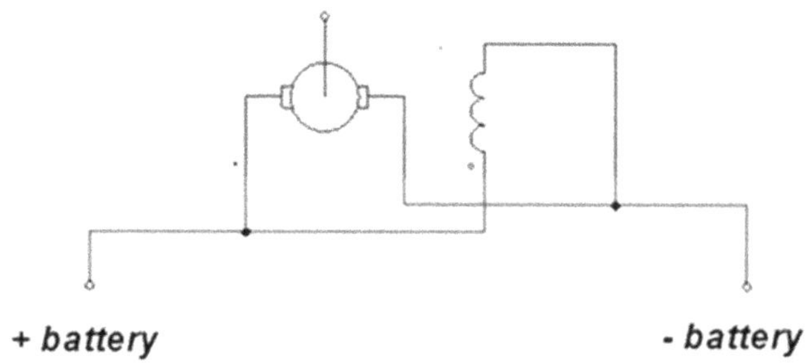

FIGURE 12.12 Shunt-wound dc motor.

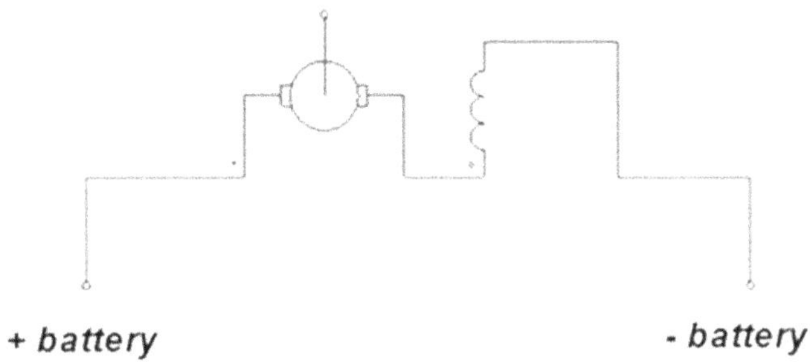

FIGURE 12.13 Series-wound dc motor.

12.2.2 Series-Wound dc Motors

The field and armature windings are connected in a series, as shown in Figure 12.13.

The armature current passes through the field windings, requiring the field windings to consist of only a few turns of heavy wire. When this motor starts under load, a high initial current is generated due to low resistance and no BEMF. This current produces a very strong magnetic field and therefore high initial torque. This characteristic makes the series-wound motor ideal as a starter motor.

12.2.3 Compound Wound dc Motor

A combination of shunt and series-wound circuits are shown in Figure 12.14. Characteristics can vary depending on how the shunt winding is connected, which is either across the armature or across the armature and series winding.

Larger-power starter motors are often compound wound and can be operated in two stages. The first stage involves the shunt winding being connected in series with the armature. This unusual connection allows for low meshing torque due to the resistance of the shunt winding. When the pinion of the starter is fully meshed with the ring gear, a set of contacts causes the main supply to be passed through the series winding and armature, giving full torque. The shunt winding will now be connected in parallel and will act in such a way as to limit the maximum speed of the motor.

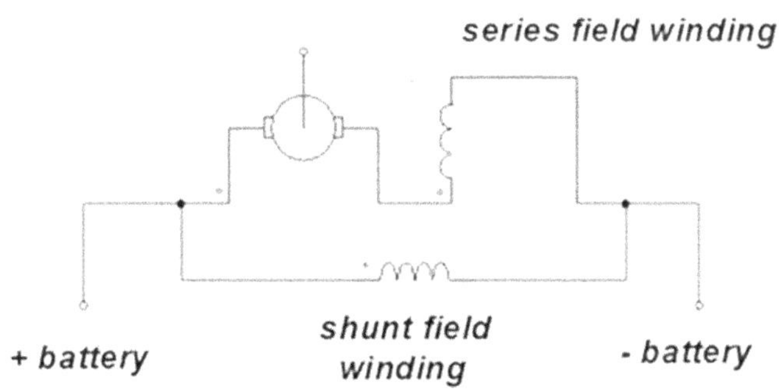

FIGURE 12.14 Compound wound dc motor.

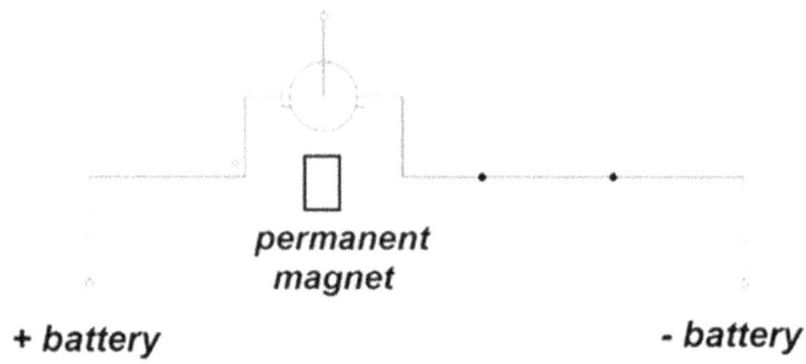

FIGURE 12.15 Permanent magnet motors.

12.2.4 Permanent Magnet dc Motors

Permanent magnet motors are smaller and simpler compared with the other three types, previously discussed. The electrical circuit is shown in Figure 12.15. The field excitation is herein by a permanent magnet, which means the excitation will remain constant under all operating conditions. Characteristics of this type of motor are broadly similar to the shunt-wound motors. When one of these types is used as a starter motor, the drop-in battery voltage tends to cause the motor to behave in a similar way to a series-wound machine.

Characteristics of the four types of small motors are generally reported as speed-torque characteristics, as illustrated in Figure 12.16. Whereas the actual numerical values depend on the actual motor, the shape of each characteristic is herein representative.

While all these types are available on the market, the most frequent choice is the *permanent magnet dc commutator motors* (Figure 12.17). This is mostly because of

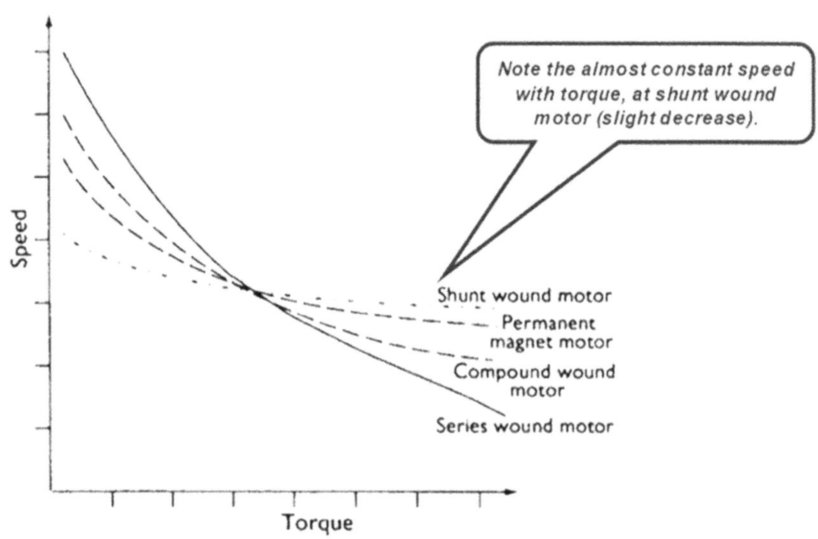

FIGURE 12.16 Characteristics of various constructive types of dc machines.

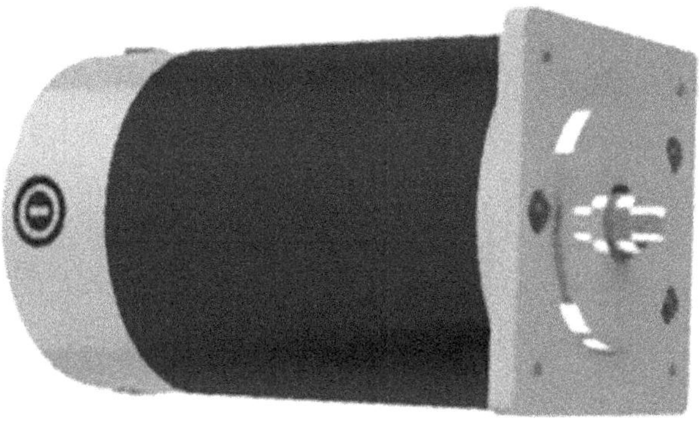

FIGURE 12.17 Generic example for a 200 W automotive *Permanent Magnet dc Commutator Motor.*

the smaller package and simpler construction. The magnets are fixed on the outside shell while brushes allow current to flow into a commutator placed on the rotor. The rotor also accommodates the windings on an iron core. A comparison between typical weight ratios for various components of permanent magnet motor and wound field motor are suggested next.

- Cooper, at 1:10
- Rotor, at 1:2.5
- Magnets, at 1:7
- Case, at 1:1

The size and weight advantage of the *permanent magnet motor* are obvious.

What concerns the control, despite being possible to have a simple supply from the dc battery, more and more applications use a microcontroller-based control of a power converter for motor drive control. The dc motor drives can be unidirectional or bidirectional. For the bidirectional motor drive, the power converter should be able to reverse the supply voltage.

The unidirectional motor drives for the permanent magnet or wound field commutator motors can be controlled by an electronic switch in series with the power supply (Figure 12.18a), whereas the bidirectional (reversible) motor drives require an H-bridge converter (Figure 12.18b). The H-bridge converter is used here as a switch able to deliver both voltage polarities to the motor, even if it can be used with PWM control of applied voltage.

12.3 APPLICATIONS: FANS, BLOWERS, PUMPS

Many electrical machine drives are used as fans, blowers, or pumps. The difference between fans and pumps is in the fluid being moved; in the case of a fan, it is air or some gas, while the pump moves liquid. A second, but less clear, distinction is that pumps are usually coupled directly to the drive motor while the fan uses a system of pulleys. The term *blower* is used sometimes for a machine used for moving gas with a moderate increase of pressure, like a more powerful fan.

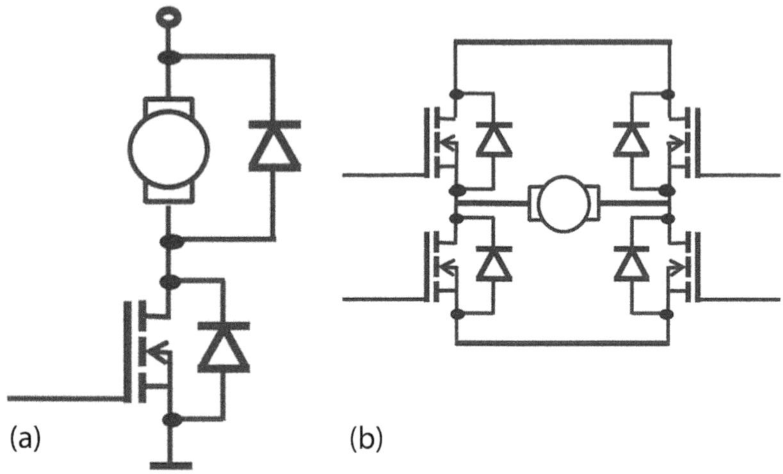

FIGURE 12.18 Various topologies for motor drives: (a) unidirectional, (b) bidirectional.

The use of an electric motor drive for pumps or fans constitutes a special category in terms of both the construction and use of the electric motor. Such motors are mostly supplied directly from the battery, but tentatively replaced by electronic controlled drives.

Historically, pumps in motor vehicle applications started as being powered from mechanical ancillary energy sources. Whether the pump is pushing water, coolants, or fuel, the current trend with automotive electrification consists of replacing mechanical pumps with electrical drives.

Mechanical pumps—starting the combustion engine operation—have a high hydraulic loss and they are therefore less energy efficient than their low-friction electric counterparts. Conversely, electric pumps work independently and they can be controlled with much greater precision. The control includes the option to use the electrical drive and pump on demand, only when needed.

For instance, an electric *fuel pump* offers a substantial reduction in fuel consumption and lower emissions result. A *water or coolant pump* with electric motor drive benefits from efficiency and controllability in order to reduce power consumption and to improve thermal control of the engine.

The semiconductor device manufacturers respond to the automotive electrification process with a set of hybrid integrated circuits able to control the electric drives used in fans and pumps. Such multifunction and scalable solutions usually require brushless dc motor control.

Consider the application details of a cooling system for the front-wheel-drive of a modern car with dc motor electric fans. For an actual example, the *1968–1977 Chevrolet Chevelle aluminum radiator dual fans* were able to cool engines up to 700 HP with dual 11" fans giving a total of 3,000 CFM (that is the measurement of the velocity at which air flows into or out of a space, in cubic feet per minute or cu-ft/min). These fans are controlled with a thermostatic switch by some engine computer. The fans are turned on when the temperature of the coolant goes above a

TABLE 12.1
Typical Motor and Switch Requirements: [After *Philips Semiconductor Application Note*]

Application	Power [W]	Current [A]	Type of Drive	MOSFETs
Air conditioning	300	25	1 motor Unidirectional, variable speed	1
Active suspension	300	25	1 motor Unidirectional, variable speed	1
Radiator Fan	120–240	10–20	1 motor Unidirectional, variable speed	1
Fuel pump	100	8	1 motor Unidirectional, variable speed	1
Wipers	60–100	5–8	1 motor Unidirectional, variable speed	1
Sunroof	40–100	3.5–8	1 motor Reversible	4
Power Seat	50	4	4–16 motors Reversible	4
Power window	25–120	2–10	2–4 motors Reversible	4
Power door lock	12–36	1–3	6–9 motors Reversible	4
Radio antenna	25	2	1 motor Reversible	4
Power mirror	12	1	2 motors Reversible	4
Seatbelt	50	4	2–4 motors Reversible	4

set point, and they turn back off when the temperature drops below that point. The drawback here consists of the high loss in brushes with a conventional 12 V battery due to the inherent high current, like 25 A in the example. For the same power level, a lower current can be achieved with a higher voltage on the dc distribution bus. Hence, the transition to a 48 V dc bus can help with the reduction of caloric (resistive) loss.

For completeness of information, a comprehensive list of the motors used motor vehicles is included in Table 12.1, which follows an elderly *Philips Semiconductor Company* classification. A very similar classification can be found on the webpages of similar semiconductor companies.

12.4 DESIGN ISSUES RELATED TO THE DC DISTRIBUTION BUS

A automotive power distribution bus constructed on a low dc voltage, like 14 V or 28 V, brings certain limitations. These include a large voltage drop on brushes (percent-wise), and the inherent use of low inductance machines. Both lead to complex control systems.

In a different line of thought, automotive applications provide a difficult heat removal in high current operation, due to limited space. This reflects onto the tough packaging requirements for size, cost, and cooling.

Multiple motor drives are supplied from the same dc voltage bus, and their individual transients bring large current variations into the dc bus. This may produce instability, and requires a power management system able to follow an analysis of bus stability.

12.5 MOTOR DESIGN: INERTIA MATCHING

It is understandable that a motor needs to meet the demands of its mechanical load during any operational conditions. Therefore, a procedure for *inertia matching* is herein followed. Inertia is best described as an object's resistance to a change in speed, and it is related to the object's mass and distance from the axis of rotation. Since belts stretch, gears have backlash, and couplings are elastic, one can consider that mechanical components are not infinitely rigid. Hence, designers must decide what inertia ratio is acceptable for each specific application.

For the electrical engineering graduate, *inertia matching* can be seen as being similar to the concept of impedance matching from electronics classes.

The following equations consider ω = angular velocity, α = angular acceleration, I_L = moment of inertia of the motor, I_M = moment of inertia of the load, T = torque, and n = gear ratio.

In the absence of gears, the power needed to accelerate the system yields as follows.

$$P = (I_M + I_L) \cdot \alpha \cdot \omega \tag{12.9}$$

This power is produced by the electric motor as

$$P = T \cdot \omega \tag{12.10}$$

Equaling the two forms of power yields next.

$$T = (I_M + I_L) \cdot \alpha \tag{12.11}$$

Observing this calculation calls for some comments.

- This calculation is very important in the case of small motors with fast positioning, where we need the acceleration α to get to the highest value possible from the available torque.
- Usually, motor vehicle load requires large torque and low speed. Hence, there is always a need for gears to accommodate a conventional electric motor with high rotational speed and lower torque.

For optimization, in order to minimize the torque required to achieve a certain acceleration, the motor inertia must equal the load inertia: $I_L = I_M$. Such a "perfect" inertia match of 1:1 (load inertia equals motor inertia) is difficult to achieve in the real world where a higher ratio is usually the case. However, the higher the performance demanded by the system requirements, the lower the inertia ratio should be (meaning closer to 1:1) in order for the motor to effectively control the load and minimize overshooting.

A realistic goal is to attempt an inertia ratio of 10:1 or less. A higher ratio than this increases system response time and resonance, causing the system to overshoot the target velocity and position, and this makes the task of the control system difficult or impossible. If the control system is not enough, the designer can reduce system inertia with two mechanical options:

- decreasing the inertia ratio by adding a gearbox
- using a larger motor

If the motor rotates the mechanical load through some of the gears, the condition of maximum power transfer is when the moment of inertia of the motor equals the reflected moment of inertia of the load, which is $(n^2 \cdot I_L)$.

Belt driven systems often use a gearbox to optimize the motor torque and speed, and decreasing the inertia ratio via gear reduction is possible. But adding a gearbox to a system just for the purpose of reducing the load inertia is not always the best solution since a gearbox adds inefficiency to the system. Gearboxes also increase cost both in terms of added components and additional energy consumption.

Using a larger motor with a higher inertia can also help. This can also be a costly solution since a larger motor is typically more expensive and consumes more energy. A motor with higher inertia also uses more of its available torque to overcome its own inertia, making it less efficient.

12.6 MOTOR DESIGN: TORQUE REQUIREMENTS

Torque is a measure of the force that can cause an object to rotate around an axis. Torque is a vector quantity. The direction of the torque vector depends on the direction of the force on the axis. Hence, torque is often defined as the product of the magnitude of the force and the perpendicular distance of the line of action of a force from the axis of rotation. More analytically, torque equals the vector product between force and distance vector.

Torque can be either static or dynamic. A static torque is one that does not produce an angular acceleration. Someone pushing on a closed door is applying a static torque to the door because the door is not rotating on its hinges, despite the force applied. An engine maintaining a motor vehicle at a constant speed is also applying a static torque because there is no acceleration or movement. Conversely, the drive shaft in a racing car accelerating from the start line is carrying a dynamic torque because it produces an angular acceleration in the wheels and therefore movement.

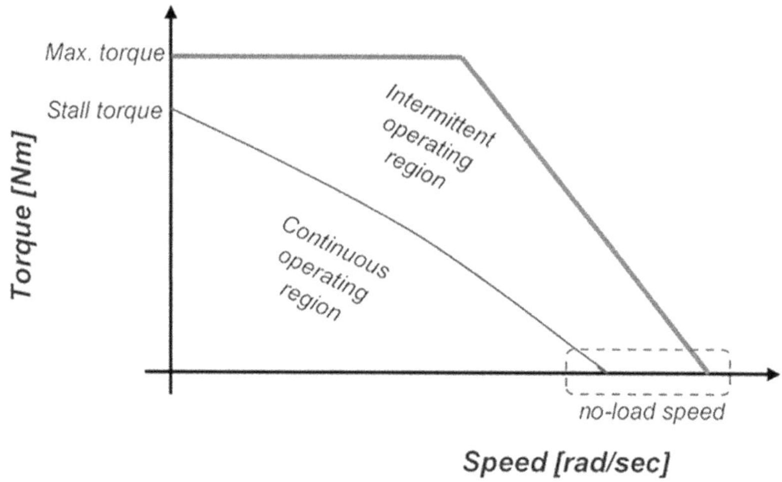

FIGURE 12.19 Generic torque-speed characteristics.

It can be seen from these definitions that torque needs to be analyzed with respect to speed. The *mechanical power* equals the product of torque and speed. Hence, the electric motors are always reporting the *torque-speed characteristics,* which allow the designer to identify the extreme operating points.

Figure 12.19 illustrates this characteristic with an example. For instance, *stall torque* is the maximum torque that can be externally applied to the shaft and cause the motor to stop rotating. When there is no rotation in the motor, the wound coils are acting as a fixed resistance. For continuous running, the stall torque should not be exceeded since this is the maximum torque at which overheating will not occur. For intermittent use, higher torque is possible for short intervals of time, during transients. The torque-speed characteristics also show that the available torque decreases at higher speed with the maximum available speed at no-load operation. The actual curve for the maximum torque at each speed depends on the actual motor technology, which is sometimes linear, sometimes slightly curved, as suggested in Figure 12.19.

Algebraically, the required motor torque needs to satisfy both the continuous load requirement and support the possible acceleration.

$$T_m = \frac{T_L}{n} + I_m \cdot \alpha_m = \frac{T_L}{n} + I_m \cdot (n \cdot \alpha_L) = \frac{T_f + I_L \cdot \alpha_L}{n} + I_m \cdot (n \cdot \alpha_L)$$
$$= \frac{1}{n} \cdot \left[T_L + \alpha_L \cdot \left(I_L + n^2 \cdot I_m \right) \right] \tag{12.12}$$

where T_f represents the friction torque, all "L" indices refer to the load and "m" indices refer to the motor.

The motor needs to be able to run at maximum required speed (velocity) without overheating.

The required power is the power needed to overcome friction (which means to provide static torque) and to accelerate the load (then add some dynamic torque). Algebraically

$$P = T_f \cdot \omega + I_L \cdot \alpha \cdot \omega \tag{12.13}$$

12.7 ULTRASONIC MOTORS (PIEZOELECTRIC MOTORS)

12.7.1 Principle

The automotive market has overwhelmingly encouraged research on other technologies for motor drives. Among these an attractive yet not fully explored alternative is offered with *ultrasonic motors.*

An ultrasonic motor is an electric motor powered by the ultrasonic vibration of a component identified as the stator. This stator is placed against another component. If the movement is a rotation, the second component is called the rotor. If it is a linear translation, it is called the *slider.*

Ultrasonic motors are not unlike piezoelectric actuators since both use piezoelectric material, most often *lead zirconate titanate*, *lithium niobite*, or other single-crystal materials. The most obvious difference is the use of resonance in ultrasonic motors to amplify the vibration of the stator in proximity to the rotor. Ultrasonic motors also create large rotation or sliding distances, while piezoelectric actuators are limited by the static strain that may be induced in the piezoelectric element.

Two different ways are generally available to control the friction along with the stator–rotor contact interface: *traveling wave vibration* and *standing-wave vibration*. The *ring-type traveling wave ultrasonic motor* operates based on standing and traveling waves.

In this respect, two piezoelectric ceramic elements are used as vibrators and positioned close to each other. The first generates a vibration described with a sine function while the other generates a vibration described as a cosine function.

Figure 12.20 illustrates the physics of the ultrasonic motor.

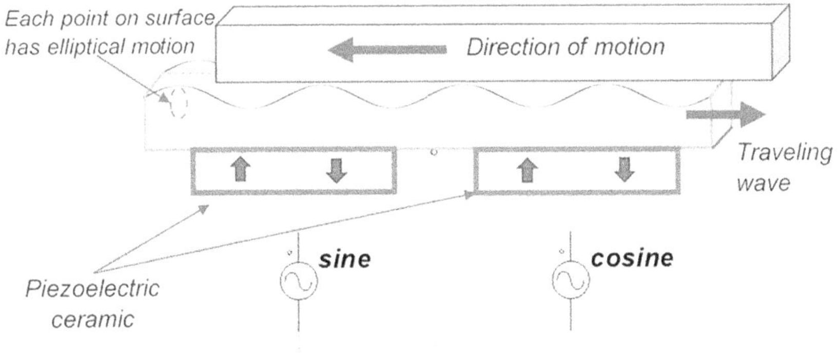

FIGURE 12.20 Principle of an ultrasonic motor with traveling waves.

Two sine/cosine generators apply voltage to the ceramic electrodes, which start generating waves. The rotary ultrasonic motor alternates the pairs of generating elements on a circumference while the linear ultrasonic motor alternates the elements along with a linear design. A traveling wave (vibration) is generated and then propagates to the right in the example shown in Figure 12.19. The metallic stator surface vibrates up and down and each point on the surface follows a small elliptical trajectory. A metal or plastic component is placed atop the metal that generates the waves. This piece becomes the rotor (or slider). If stator and rotor are in rigid contact—eventually helped with an applicator pressure—the rotor/slider will be driven by a tangential force at the contact surface and move in a direction opposing the traveling wave.

Similar to the conventional electric motors, the ultrasonic motor is characterized by a torque-speed curve, as shown in Figure 12.21. The speed-torque characteristics of the ultrasonic motor are provided herein without demonstration.

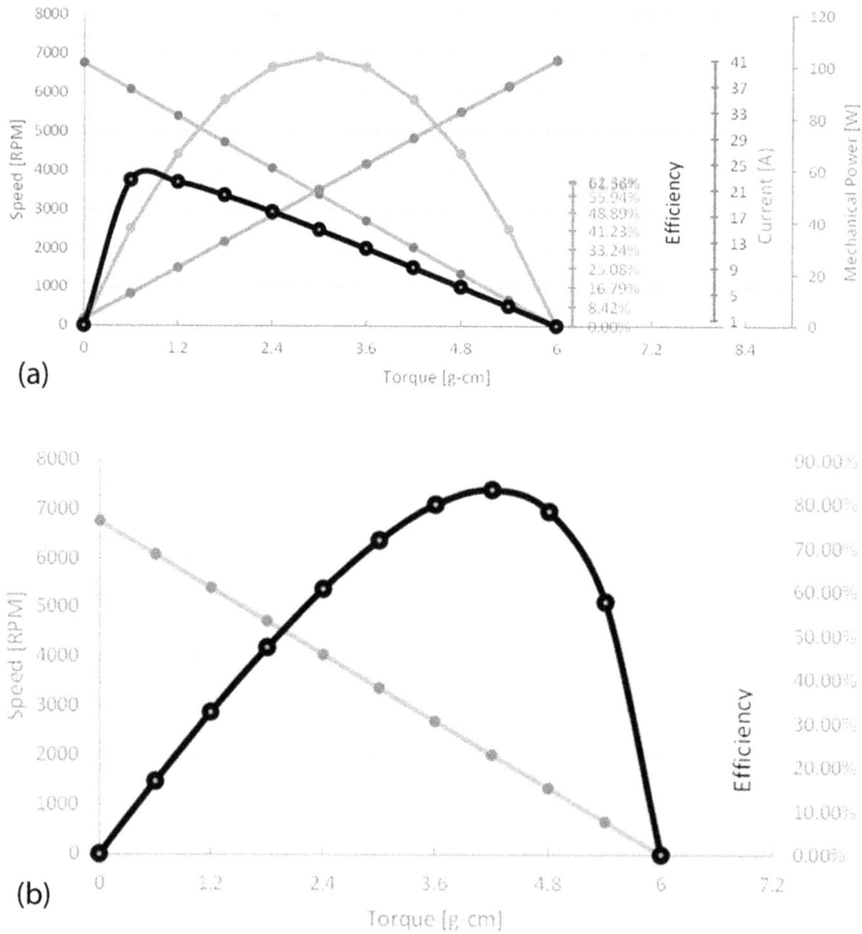

FIGURE 12.21 (a) speed-torque electric motor characteristic (Chapter 6, Figure 6.3); (b) speed-torque ultrasonic motor characteristics.

It can be seen that the optimal conditions for maximum efficiency at dc and ultrasonic motors are different. For the dc motor, the maximum efficiency occurs at near no-load speed, that is also the case for low-torque. The dc motor is well suited for high-speed operation. Conversely, for the ultrasonic motor, the maximum efficiency occurs at low speed and high torque. This motor is well suited for low-speed applications. This means it has potential for being used in automotive applications without a gearbox. Such characteristics recommend the use of ultrasonic motors in applications like power steering, power windows, tilt steering, car seat adjustment and so on.

12.7.2 Control and Optimization

The ultrasonic motor uses a two-phase power electronic converter, as shown in Figure 12.22.

Each phase is supplied from battery and a square wave converter with phase shift control then applies voltage to the transformers. Since the actual phase within the ultrasonic motor is characterized by a capacitance due to the piezoelectric effect, series inductances are used to form resonant circuits. A pair of quasi-square-waves generated by the inverters are supplied to each resonant circuit and sinusoidal waveforms appear on the capacitors.

Due to asymmetries in motor construction (inaccuracies during manufacturing or inhomogeneities when bonding) and cross-coupling terms due to stator vibrations, each inverter is controlled separately with the controllability of phase shift and duty cycle. This outlines the major differences between a high-performance controller and the ideal principle of operation illustrated in the previous section.

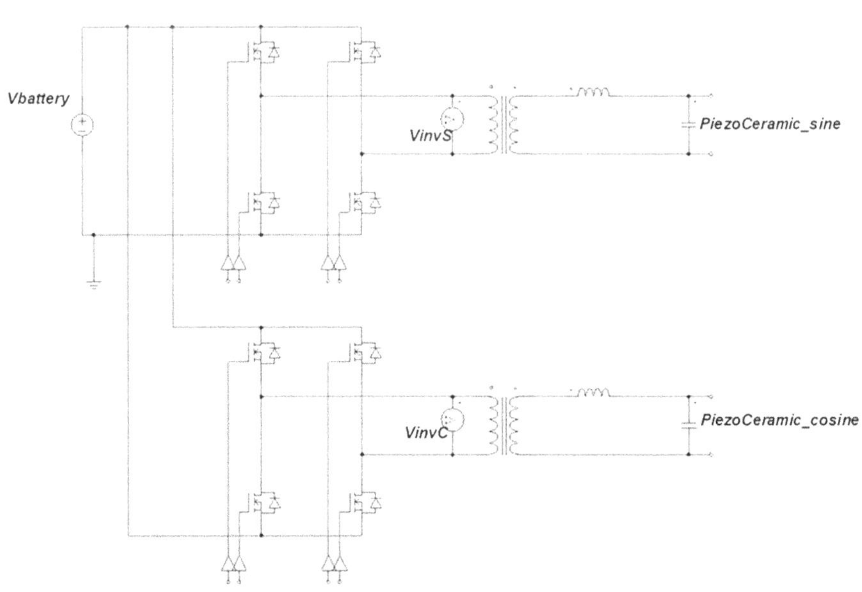

FIGURE 12.22 Power converter system for ultrasonic motor.

The high-performance controller may produce "sine/cosine waves" of different magnitudes from each other and their phase shift may be more of less than theoretical 90°. This makes the modeling and control difficult, and overall limits the actual use of this motor.

12.8 CONCLUSION

Electrical motors are the most important component in automotive mechatronics, and their application is growing with the advent of convenience and safety systems.

However, the most important trend with benefits for electric motors relates to the electrification of transportation. Various ancillary sources of energy derived from the operation of the combustion engine are nowadays replaced with electric motor drives. This offers better efficiency, reduced thermal requirements, and the benefit of advanced digital control. A multitude of new features can be implemented by software around the electric motor drive.

The electric motors used in motor vehicles are brushed or brushless dc motors, with rare use of induction or variable reluctance ac motors. All these varieties have been presented in this chapter, starting from the core physical description and operation.

REFERENCES

Alciatore, D., Histand, M. 2003. *Introduction to Mechatronics*. 2nd edition, New York: McGraw-Hill Education.

Anon. 2006. *PICDEM™ Mechatronics Demonstration Board User's Guide*. Microchip Corporation Documentation DS51557C.

Anon. 2018. *The Future of Automotive Pumps is Electric!* ST Microelectronics blog. Internet reading at https://blog.st.com/the-future-of-automotive-pumps-is-electric/. Last reading May 14, 2018.

Bishop, R. 2006. *Mechatronics – An Introduction*. Boca Raton, FL: Taylor and Francis.

Bolton, W. 2012. *Mechatronics – A Multidisciplinary Approach*. 5th edition, London: Pearson Education.

Denton, T. 2017. *Automobile Electrical and Electronic Systems*. 5th edition, Abingdon-on-Thames: Routledge.

Henderson, D., Schaertl, L. *Mechatronics Meets Miniaturization*. Published in Design World, October 2010 issue.

Kinnaird, C. 2015. *How Many Electric Motors are in Your Car?* Motor and Drive Systems.

Maas, J., Schulte, T., Frohleke, T. 2000. Model based control of ultrasonic motors. *IEEE/ASME Transactions on Mechatronics*, 5(2):165–180.

Neacşu, D. 2004. *Power Semiconductor and Control for Automotive Applications*. Tutorial Presented at IEEE APEC, Anaheim, CA, USA.

Salem, F. 2013. Mechatronics design of small electric vehicles; research and education. *International Journal of Mechanical & Mechatronics Engineering, IJMME-IJENS*, 13(1):23–36.

Sashida, T., Kenjo, T. 1993. *An Introduction to Ultrasonic Motors*. Oxford: Clarendon Press.

Storck, H., Wallaschek, J. 2000. Experimental investigations on modeling assumptions in the stator/rotor contact of travelling-wave ultrasonic motors. *Journal of Vibroengineering*, 4(5).

13 Power Integrated Circuits

HISTORICAL MILESTONES

1970s: Power ICs were realized using pure bipolar technologies (I2L for logic) that obtained a very limited benefit in improving power device performance and increasing logic density as lithography was scaled down.

1980s: CMOS Technology is first used.

1985s: BiCMOS and BCD (BiCMOS+DMOS) are introduced to classify the family of silicon processes that allows mixing of different structures on the same chip.

2000s: Three different roadmap directions can be seen:

- *High-voltage-BCD (500–700 V) = A BCD process that satisfies a high-voltage capability and exploits the "resurf" principle when applied to the Lateral DMOS. This is the so-called "BCDoffline," at 2 μm, utilized mainly in the field of electronic lamp ballasts and power supplies for industrial applications.*
- *High-Power-BCD (40–90 V, high current) = A junction isolation approach with power DMOS, both in lateral and vertical fashion, implementing highly doped N buried layers. The N epitaxial layer, grown on a P-type substrate, is designed to support VDMOS devices from 70 V up to 90 V.*
- *High-Density-BCD (5–50 V).*

2010s: Smart power system-on-chip (SoC) products are almost exclusively designed on Bipolar-CMOS-DMOS (BCD) technologies. These products address applications as diverse as power management for mobile phones, motor drivers, printer head drivers, and so on.

≈2005: The formation of vertical trench DMOS devices can be added to existing integrated BCD process flows in order to improve the efficiency of the BCD devices. Utilizing an integrated trench device in a BCD process can offer at least a factor-of-two $R_{ds(on)}$ area advantage over a planar counterpart. Original solution in 0.5-micron (BCD05) and 0.25-micron (BCD25) processes. For instance, the 0.5 μm, 40 V BCD process features dual gate oxides (5.5 V and 16 V), complementary N- and P-channel MOSFETs with 5 V, 7 V, 16 V, and 30 V capabilities, vertical NPN (VNPN) and lateral PNP (LPNP) bipolar transistors, a variety of passive elements, a NMOS device rated for 40 V operation.

≈2005: 130 nm HV-LDMOS-based technology.

≈2005: Come-back of High-voltage CMOS (HV-CMOS) processes to replace BCD processes in some power applications. LCD driver ICs supporting voltages up to 40 V were the first large volume application for HV-CMOS processes. Other applications such as power management, bus transceivers, printer head drivers. HV-CMOS processes scaled down to feature sizes as low as 0.13 μm.

13.1 INTEGRATED-CIRCUIT TECHNOLOGIES

Due to the advent of mechatronics systems for automotive applications, more and more power electronic circuits are required. Since these are under continuous pressure for volume and weight reduction, electrical loss reduction, and integration of new features, the implementation of integrated circuits are sought. This emerged as a large market and allowed the experimentation of several technologies simultaneously. Different integrated circuits (ICs) currently in production were actually designed at different moments over the last 25 years, and they use either a bipolar transistor, CMOS, BiCMOS, BCD (BiCMOS+DMOS), High-voltage-BCD (500–700 V), High-Power-BCD (40–90 V, high current), High-Density-BCD (5–50 V), or HV-LDMOS. Technology priorities today relate to finer lithography features, wider voltage capability, and a broadening variety of integrable components.

The inherent question of automotive power electronics is how much integration is possible for a power converter. Integrated circuits with various levels of integration are available on the market. A classification that depended on the power level and complexity of control into three levels of function integration (low, medium, high) was detailed in Chapter 5.

For the switching mode power supply, the location of the power switches depends on the load current and bus voltage (Figure 13.1):

- Both the freewheeling diode and MOSFET are inside the integrated circuit.
- The diode is external to the integrated circuit.
- Both the freewheeling diode and MOSFET are external to the integrated circuit.

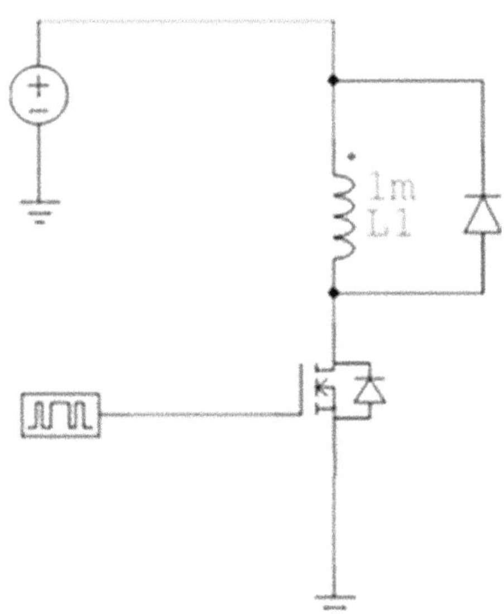

FIGURE 13.1 A simple power circuit.

Interestingly enough, the size of the freewheeling power diode is of more concern than the power MOSFET, which encourages the idea of a synchronous rectification setup where the power diode is replaced with a self-controlled power MOSFET.

The trade-off for the final choice between these three options is more complex. For higher power, the obvious choice is the external MOSFET and/or diode. For a similar power level, the choice depends on the real estate available inside the integrated circuit. As mentioned, the power diode takes more space than the power MOSFET for the same current, and it is the first option when placed externally.

If efficiency is critical and a lower R_{dson} is desired, the transistor should be external in order to accommodate a larger semiconductor area for a lower R_{dson}. The overall real estate on the printed-circuit board (PCB) may be similar since the external device requires additional space but dissipates less power given its lower R_{dson}, while the internal power MOSFET still needs to be cooled with the same PCB surface, using one or more PCB layers.

Maximum supply voltage is most critical in automotive applications since most conventional Power IC technologies are rated up to 40 V. This should be compared for recommending the voltage rating for various automotive applications, which is 40 V for a conventional 12 V battery as the primary voltage source, and 75 V for a newer 48 V bus.

13.2 ARCHITECTURE OF ANALOG OR MIXED-MODE POWER IC

Integrated circuits for automotive power applications follow simple, well-known architecture solutions, which subscribe to the main application topics:

- Non-isolated power supplies or dc/dc converters
- Isolated power supplies
- Motor drives

High-voltage gate drivers for High-voltage propulsion drives.

13.2.1 Example of Disruptive Innovation—PWM Control Chip

The PWM control integrated circuit was invented by Bob Mammano in 1975, and introduced to the market in 1976 by the *Silicon General Company* as SG1524. In the 1970s, PWM Control IC was developed by multiple corporations, with products like the Motorola MC3420, Texas Instruments TL454, Signetics NE5560, and Ferranti ZN1066. The solution uses a mixed-mode integrated-circuit technology with a simple (well-known today) structure. Its invention coincided with the advent of switching power supplies in the late 1970s, and satisfied a clear market need, able to add value to the customers.

The PWM IC allowed for the creation of a new market with a paradigm (*vendors competing customers*), the addition of a new set of customers, enabling new power supply technologies and more incremental development like current-mode ICs, cycle-by-cycle current-limiting protection, single-ended, push-pull supplies, LDOs, hot-swap, soft-switching, and so on.

Figure 13.2 provides the description of a current-controlled buck converter. An architecture for the appropriate controller is shown in Figure 13.3.

A *band gap reference* inside the integrated circuit provides a fixed 1.2 V reference. The reference voltage is compared with the feedback information, either current from a current sensor or voltage from a resistive divider. The resulting error

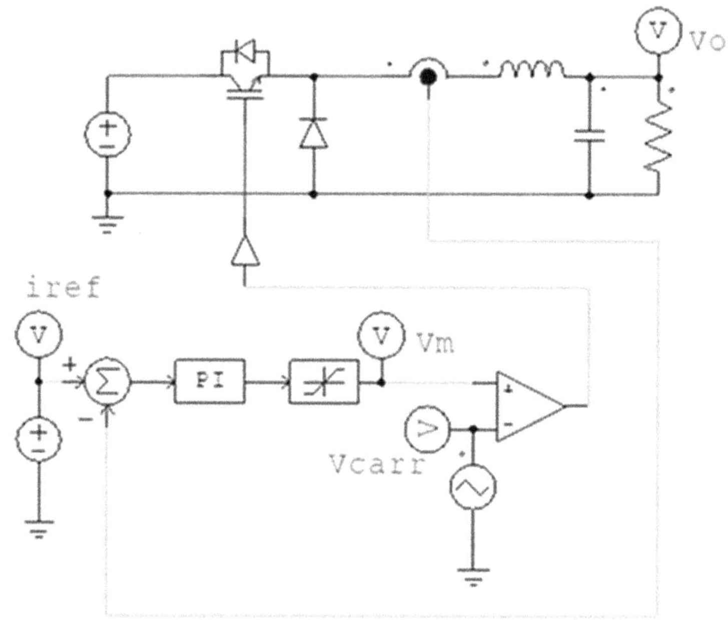

FIGURE 13.2 Current-controlled buck converter.

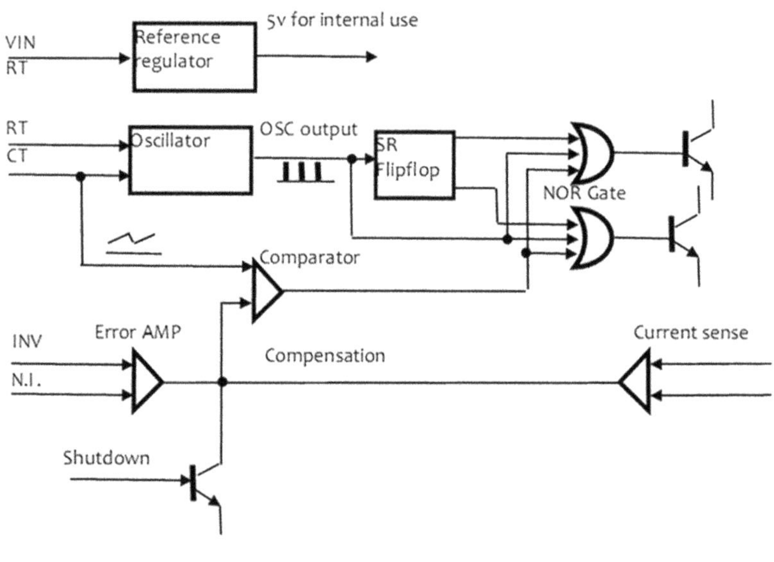

FIGURE 13.3 Principle schematic for the control integrated circuit.

is processed through an *error amplifier*, which is an operational amplifier with a passive network on the feedback path, able to correct (compensate) the dynamic response of the overall system. This passive network classifies the controller as Type I, Type II, or Type III (for details, see Chapter 9).

The output of the error amplifier represents the reference for the PWM generator. The PWM is generated with a comparison between a triangular carrier signal of high frequency (kHz) and the output of the error amplifier. The triangular signal is generated outside the integrated circuit with a resistive–capacitive network, aiming at charging the capacitor with a delay. The linearly increasing voltage is reset when the rising slope reaches a predefined threshold. The capacitor is suddenly discharged and the linear charging starts over. This forms a triangular waveform, called a carrier, produced at a high frequency, in the range of kHz.

When the output of the error amplifier is larger than the triangular carrier, the switch is controlled in an "on" state, but otherwise stays "off".

An auxiliary protection circuit monitors various parameters (current, voltage, temperature); if they go beyond limits of good operation, a shutdown signal is generated to block the further generation of control [PWM] pulses. The protection circuit can either wait for operator reset or work pulse-by-pulse and automatically reset at the end of each PWM sequence.

13.2.2 Flyback PWM Controller IC for Isolated Power Supplies

Most power supplies require isolated power, which means delivering voltage with a different grounding than the supply voltage. Among several topologies available, the flyback converter is the most used solution due to its simplicity. Figure 13.4 gives the topology of a flyback converter.

The operation is somewhat similar to a boost converter. When the transistor is turned on, the primary current stores energy within the magnetic field of the transformer, without transfer to the secondary circuit where the diode has a negative voltage. When the transistor is turned off, the primary current ceases and the stored energy produces a voltage able to turn on the secondary-side diode and to deliver energy to the output capacitor.

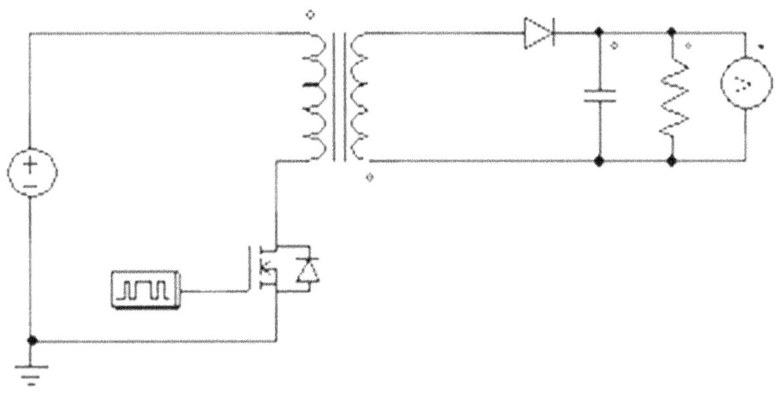

FIGURE 13.4 Flyback converter.

The flyback PWM control IC represents the most used class of ICs for isolated power supplies, the entire controller being delivered within an eight-pin package. The control IC was introduced in the 1980s as UC3842, with a simple structure, 144 transistors, in 7.5 mm technology, sold for $1.75. In the 1990s, an important feature improvement was marketed as UCC3802, with 478 transistors, in a 3.0 mm technology, and sold for $0.85. In the 2000s, more features were added within a device marketed as UCC286xx/287xx, with 1,158 transistors, in a 0.5 mm technology, and sold for $0.45.

These modern features added to the core controller include advanced soft-start, quasi-resonant flyback operation, valley switching for low EMI, stand-by power requirements, no-load power requirements like consumption less than 300 mW at no-load, current-mode control, multimode power saving with automatic switching between operation modes, pulse skipping, or pulse density modulation, burst operation up to hard-switching, green mode status indicator, line and load overvoltage protection, bounded frequency range, with frequency foldback, and so on.

13.2.3 Three-Phase Power MOSFET Controller

The automotive power systems use numerous motors to actuate various mechanical loads. While the initial use was reduced to turning on and stopping these motors, efficiency requirements, speed control, as well as precision positioning, recommend the use of variable-speed motor drives. These require single-phase or three-phase power converters, made up of power semiconductor switches.

Depending on the output current, the power switches can be inside (up to 3–5 A) or outside the integrated circuit (to tens of A). Other features within the integrated circuit include the six gate drivers for the power MOSFETs, which can be setup with a charge pump or LDO voltage regulator, the PWM and PWM control output with adjustable dead-time, the synchronous rectification, the battery overvoltage and undervoltage protection, thermal shutdown with hysteresis, power MOSFET protection, front-end boost converter for dc bus voltage control, and brake/direction I/O control. The control integrated circuits work with a typical 40–45 V supply voltage, and within a temperature range between –40 to 150°C.

A special set of features are included for brushless dc motor control. They include the conventional PWM control of the power converter, as well as a BEMF (back-electro-motive-force) sensing, a tachometer I/O interface, and speed/current control.

Another special case is introduced by the micro-stepping motor driver, able to operate bipolar stepper motors. Thus, by simply inputting one pulse on a designated input, the motor will take one step (full, half, quarter, or eighth depending on programming). Usually, there are no phase-sequence tables, high-frequency control lines, or complex interfaces to program.

13.2.4 High-Voltage Gate Drivers for High-Voltage Propulsion Drives

The advent of electric and hybrid vehicles introduced power converters operated from a higher voltage bus, of 300–400 V dc voltage. Both the single-phase and the three-phase converters using this bus voltage have upper-side switches with a local ground

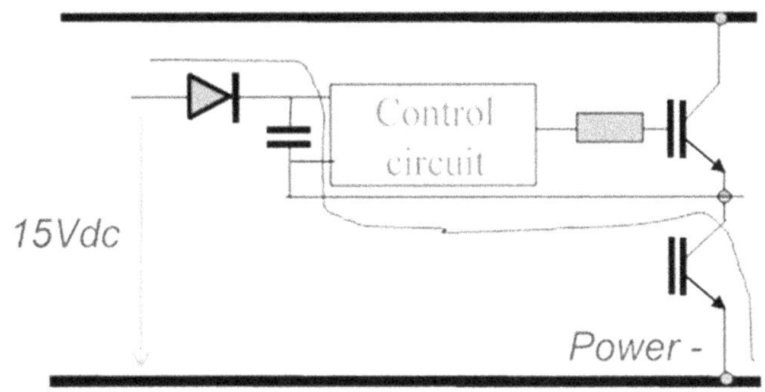

FIGURE 13.5 Bootstrap power supply.

well above the vehicle grounding. Hence, a local power supply is required for each gate driver, which is able to deliver gate power with isolation from the vehicle grounding. While it is possible to use any isolated power converter topology with a transformer, cost reduction pressure recommends the use of a bootstrap or charge pump supply for the high-side switch. Bootstrap type circuitry is used as an energy source for the high-side device, with a floating ground and level shifting for information transmission.

The operation of the bootstrap power supply is very simple, as depicted in Figure 13.5 (a full application circuit is shown with Figure 10.16). When the low-side switch turns on, the bootstrap capacitor is charged with 15 V through the series diode and the power semiconductor switch (IGBT or MOSFET). When the low-side switch turns-off, the voltage across the bootstrap capacitor can supply to gate driver of the high-side IGBT.

Design considerations for the bootstrap power supply follow. Any particular PWM algorithm defines different values of the minimum and maximum pulse-widths. Since the bootstrap capacitor only charges when the low-side device is on, the duration of this interval should be long enough to ensure the rise of capacitor voltage to V_{CC}. Denoted by R, the resistor is on the charging path of the bootstrap capacitor (diode's resistance plus additional current-limiting resistor),; the voltage across the capacitor is given by:

$$v_c(t) = (V_{cc} - V_f - V_{LS}) \cdot \left[1 - e^{-\frac{t}{R \cdot C}}\right] \tag{13.1}$$

The on-time of the low-side switch is considered as charging time and it should be enough for charging-up the capacitor above the minimum allowed voltage for proper operation of the high-side circuitry within the circuit. For instance, for the IR2137 integrated circuit, which is 11.3 V when $V_{CC} = 15$ V.

Considering the possible discharge of the capacitor during the high-side switch conduction, one can restrict the level of the desirable voltage at the end of the charging interval to 90 %. It yields:

$$(V_{cc} - V_f - V_{LS}) \cdot 0.90 = (V_{cc} - V_f - V_{LS}) \cdot \left[1 - e^{-\frac{(t_{on}^L)\min}{R \cdot C}}\right] \tag{13.2}$$

$$\rightarrow t_{on}^L \geq -R \cdot C \cdot \ln(0.1) = 2.3 \cdot R \cdot C$$

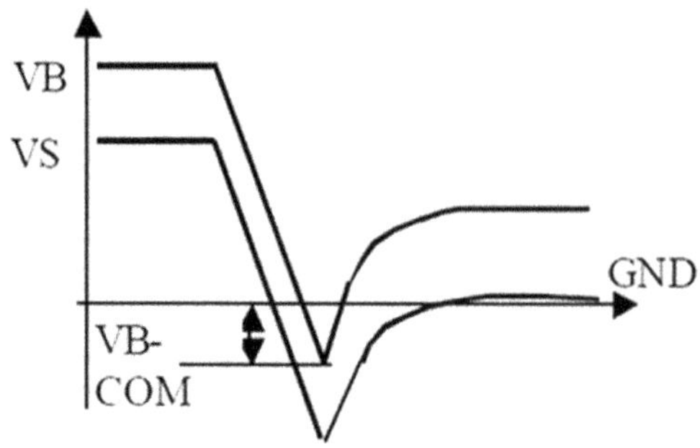

FIGURE 13.6 Negative voltage spike.

The bootstrap diode is designed based on V_{RRM} and I_F. It is also recommended to choose diodes with τ_{RR} = 50 ns or better. A resistor in a series with the bootstrap diode brings benefits for correcting the time constant formed together with the bootstrap capacitor, for limiting the charging current at startup, and for limiting the current through the bootstrap circuit during possible negative V_s spikes.

This negative spike is explained in Figure 13.6, and it happens since both capacitor terminals are trying to follow the V_s negative spike and this can turn on the bootstrap diode allowing a current through *R*.

13.3 IC DESIGN CONSIDERATIONS

13.3.1 Power MOSFET Used within Integrated Circuits

The power MOSFET is used as a switch within a power converter. A buck converter is herein considered as the simplest example, as shown in Figure 13.7. The recovery diode is always replaced with a second MOSFET able to take over the inductive current when the main switch is off. This is called *synchronous rectification.*

For any of the three applications discussed within the previous section (power converters, isolated power supplies, motor drives), both power MOSFET switches

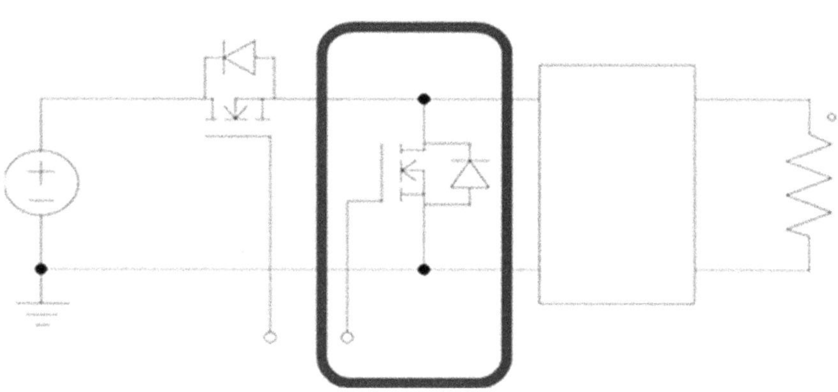

FIGURE 13.7 Buck power converter with synchronous rectification.

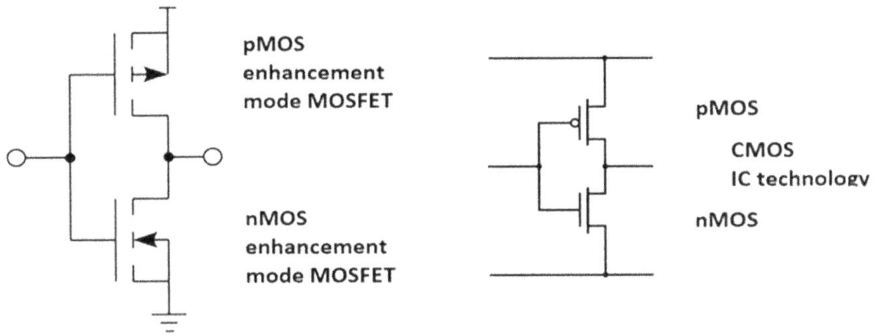

FIGURE 13.8 Symbols for power MOSFET transistor used within integrated-circuit technologies.

can be placed inside the integrated circuit or outside, depending on the maximum load current. Usually, a current under 3 A can accommodate the power MOSFET within the integrated circuit. If a diode is considered for inductive current, it is the first preference to be placed outside the integrated circuit. Some solutions provide a power MOSFET inside the integrated circuit and they expect the diode or equivalent to be outside the integrated circuit.

The design of the power MOSFET is governed mostly by conduction loss, which depends on the R_{dson}. The MOSFET transistor (inside or outside the IC) should be designed with a 1% rule, which is for a load resistance of 1 Ohm, the R_{dson} should be around 10 mOhm. For a buck converter, this design can be softened by considering the duty cycle. For instance, for a duty cycle of 33%, the R_{dson} can be 30 mOhm.

Most integrated circuits follow design procedures for CMOS technology. The symbols used for transistors may be adequate for integrated-circuit technology, as suggested in Figure 13.8.

For either buck or inverter, the transistor is placed before the load, and usually this means it should be of a pMOS type. If a nMOS transistor is used, the gate control voltage needs to be higher than V_{DD} (positive supply voltage) and this may be a problem within the integrated-circuit design. Unfortunately, using a pMOS transistor inside the integrated circuit takes lots of space. Alternatively, it is possible to add circuitry in order to use a nMOS transistor for the high-side switch. This is illustrated in Figure 13.9.

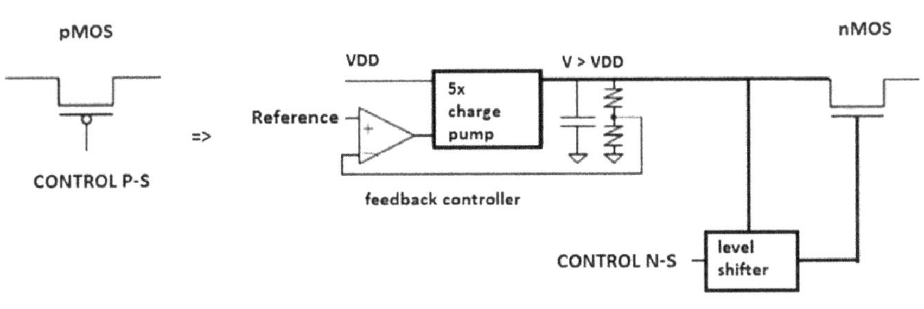

FIGURE 13.9 Using nMOS for the high-side switch.

A calculation example is suggested herein, using an input voltage of $V_{DD} = 1.2$ V, a gate supply voltage of $V = 3.8$V, and a 50 mOhm switch.

- Using pMOS transistor: $(W/L)_{PS} = 1/(mC_{ox} \times (V_{dd} - |V_{tp}|) \times R_{PS}) = 500{,}000$ m/ 0.25 m
- Using nMOS transistor: $(W/L)_{NS} = 1/(mC_{ox} \times (V - V_{tn}) \times R_{NS}) = 33{,}300$ m/ 0.25 m
- A charge pump with 1 pF capacitance is the same size with nMOS
- Hence, the size of pMOS : size of nMOS equivalent = 7.5 : 1 (500 : 66.6)

13.3.2 Power Diode

While the power diode is widely used in a power converter, it presents a series of issues when considered for implementation within an integrated circuit because a diode in integrated-circuit technology would occupy lots of space inside the circuit.

Replacement with a second MOSFET and the use of synchronous MOSFET concepts would again require lots of space for the power MOSFET, gate drive, and control logic. A more economical solution consists of its replacement with an *active diode.*

Alternatively, the designer can opt for an active diode, as shown in Figure 13.10. An active diode is a power MOSFET device controlled by a comparator. An anode potential larger than the cathode potential turns on the power MOSFET. Various implementation solutions are possible, and two examples are shown in Figure 13.11.

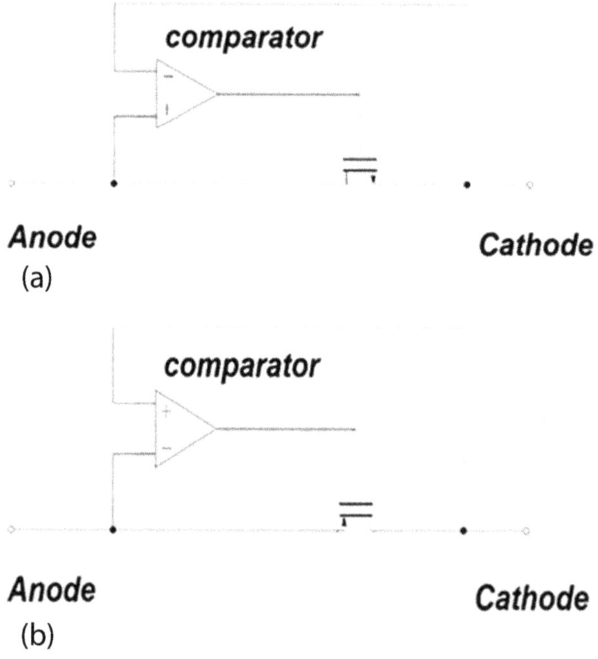

FIGURE 13.10 Active diode circuitry for n-type and *P*-type MOS devices.

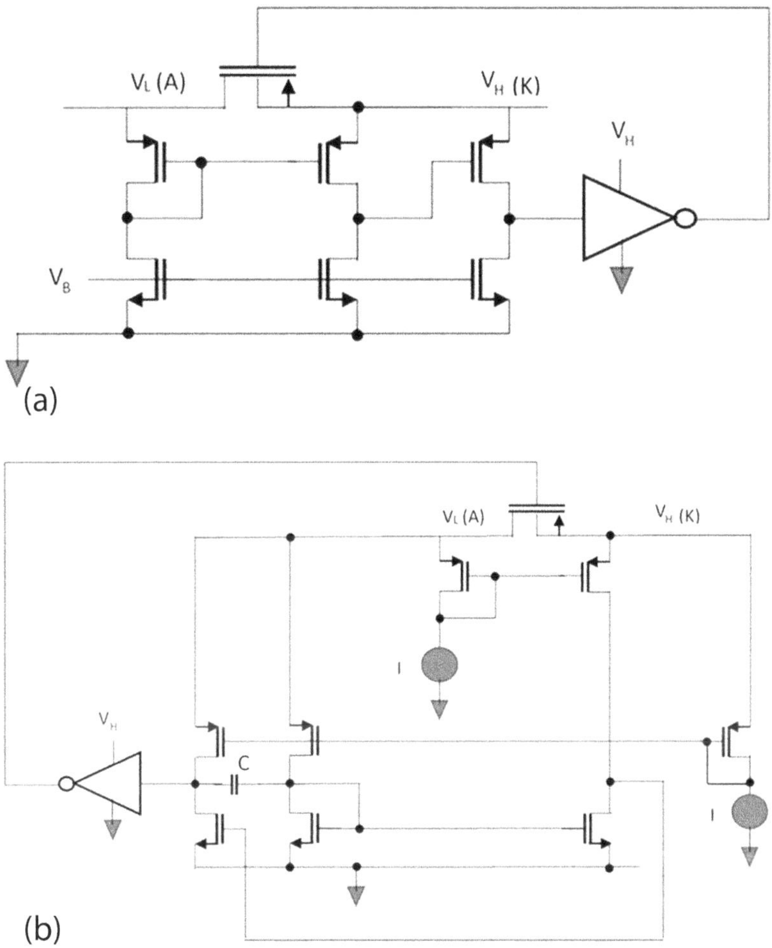

FIGURE 13.11 Complete possible circuits for the active diode.

13.3.3 Gate Driver

"*Logic level MOSFETs*" are designed for operation at 5 V or lower to 2.5 V and have guaranteed low "on" resistance at these gate voltages. The control voltage is zero for the "off" state, which means there is no need for negative voltages.

The power MOSFET switch needs to be driven by a higher power circuitry. Inside the integrated circuit, this means the use of digital buffers. Digital buffers can be designed at different power levels within the same platform, or eventually be cascaded, as suggested in Figure 13.12.

Using larger transistors increase loss due to both switching and conduction components. Each buffer stage introduces a delay. A minimal delay can be achieved for a current ratio of "*e*" (= 2.7), but this is not enough. Exponential sub-sequential ratios would work better.

Finally, the gate driver should also include additional protection, shutdown, and dead-time generator features that are added into the digital/logic side of this design.

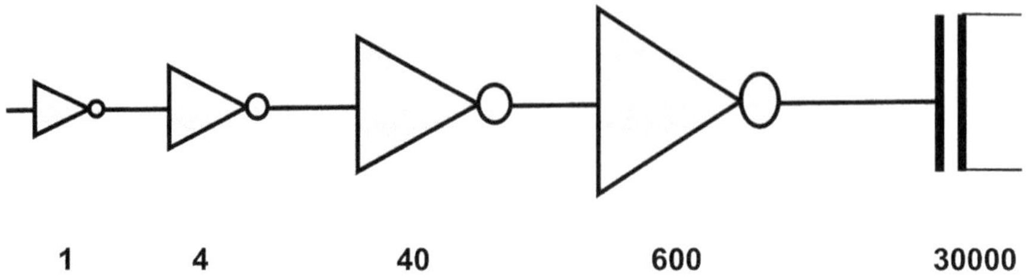

FIGURE 13.12 Cascaded buffers for gate driver control, with their size ratios.

13.3.4 Band Gap Reference

Control circuitry is based on a voltage reference generated inside the integrated circuit.

A *band gap voltage reference* is a temperature-independent voltage reference circuit widely used in integrated circuits. It produces a fixed voltage regardless of power supply variations, temperature changes, and circuit loading from a device.

For a low-voltage integrated circuit ($V_{dd} < 5$ V), there is only one trimmed voltage reference. A trimmed voltage reference involves the design of a precision voltage reference. Using IC technology, invariably this is achieved with a 1.25 V band gap reference.

For higher voltage integrated circuits (V_{dd} ~15 V), as is mostly the case in automotive electronics, there are usually two voltage references: one precision trimmed reference (which may be 1.25 V) used for control, and a second one, untrimmed, used for protection circuits like undervoltage (UVP) or overvoltage protection (OVP) (Figure 13.13).

13.3.5 PWM Generator

The error amplifier follows the structure of an operational amplifier.

The PWM generator inside the integrated circuit is based on a triangular signal generator that uses an external capacitor. Using the internal voltage reference, a constant current generator is created. Sometimes, this is using an external resistor R_T for adjustment.

In most cases, the ramp is generated just inside the integrated circuit. The constant current produced by the generator is mirrored into a capacitor which charges linearly. The capacitor voltage is compared with both a lower and an upper threshold. When the voltage reaches the top threshold, the comparator triggers a *R–S* flip-flop that changes its state. The output of the flip-flop briefly controls a transistor able to suddenly discharge the capacitor. When the capacitor voltage reaches the lower threshold, this produces a change of state of the *R–S* flip-flop and the charging process restarts (Figure 13.14).

13.3.6 Current Sensor

For low current applications, the current through the power MOSFET switch transistor can be detected inside the integrated circuit with a second MOSFET controlled

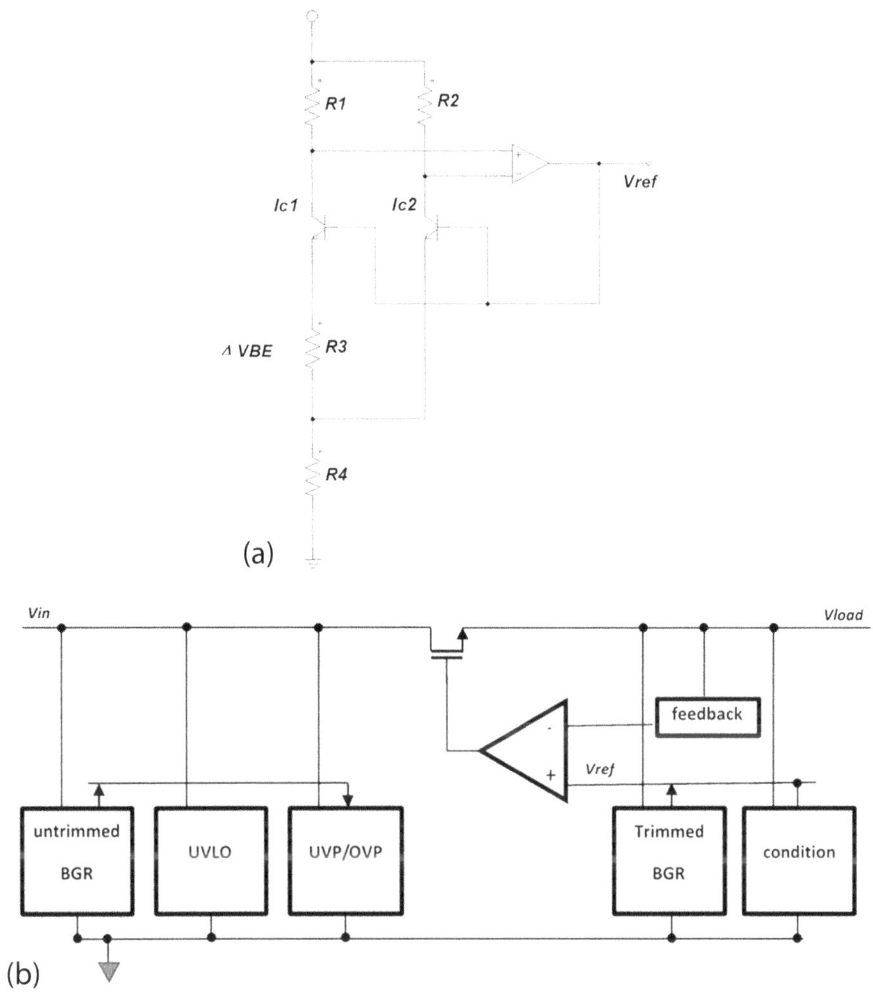

FIGURE 13.13 Voltage references: (a) Circuit solution; (b) Use of various references.

in the same conditions as the power MOSFET switch, and with the same voltages. A current mirror is created and the load current is reflected within the integrated circuit. In order to avoid processing large currents inside the integrated circuit, the current mirror is built asymmetrically with a large ratio (Figure 13.15).

For example, Figure 13.15 shows such a solution applied to a buck converter, with a reflection of the current under a ratio of 1,000:1, so that 1 A is seen as 1 mA. The sensed current is converted into a voltage inside the integrated circuit, through a resistor herein denoted with R_T. This voltage is used inside the PWM controller to setup the width of the control pulse and to thus regulate the power converter current. Whenever, the MOSFET within the buck converter is in the "off" state; the current is switched to another circuit inside the integrated circuit with an assembly M_{sw1} and M_{sw2}.

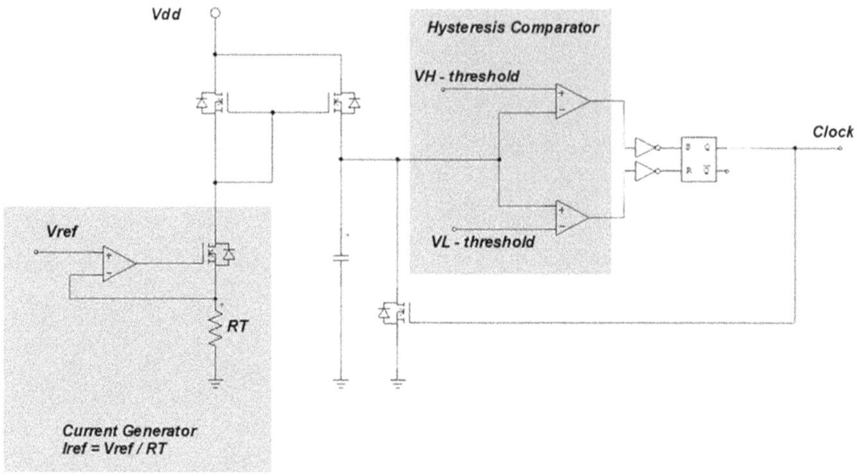

FIGURE 13.14 PWM generator.

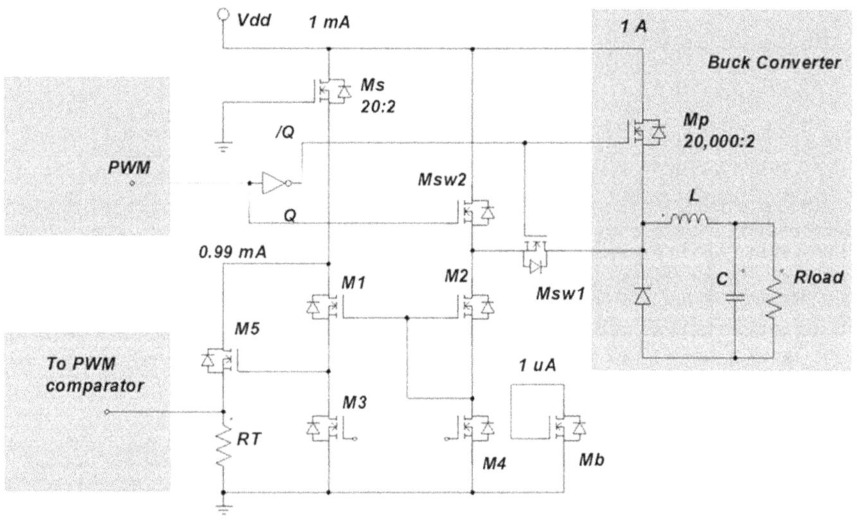

FIGURE 13.15 Currents sensor inside the integrated circuit.

13.3.7 Auxiliary Protection Circuitry

The integrated circuit is enhanced with a series of protection features, which are simple comparator circuits that follow a sensor. Possible protection features include

- undervoltage lock-out for lower input dc bus voltages
- overtemperature
- load overcurrent
- overvoltage on the input dc bus

The undervoltage lock-out protection monitors the supply voltage and shuts the entire IC down if the supply falls under a certain minimal accepted level. The overtemperature protection monitors the chip temperature and shuts the integrated circuit down when a temperature rise is detected. The load overcurrent protection monitors the load current or the current through the MOSFETs and shuts the control gate signals down when a large current is detected. The load overvoltage protection monitors the voltage in the power circuitry and shuts the control signals down when a larger than expected voltage occurs.

13.3.8 Soft-Start Circuitry

The need for a soft-start circuitry is explained with the help of a buck converter. When the output voltage is zero, at startup, the controller determines a duty cycle of $D = 1$, as required to charge the output capacitor. This produces a very large inrush current.

Conversely, a soft-start circuit causes a small current to charge a capacitor, such that the reference for the PWM rises slowly, and limits the duty cycle. A solution for the circuit within the integrated circuit is shown in Figure 13.16, where M_2 is in the subthreshold region and sources a current in the range of nA.

13.3.9 I/O Connections

All I/O connections need to be protected from overvoltage and ESD (electrostatic discharge). Such protection is achieved with various diode solutions: Schottky diodes, large (overrated diodes), or transistors in diode connection (can be MOSFET transistors, depending on main IC technology). Figure 13.17 illustrates these various protection circuits inside the integrated circuit, near the I/O pins.

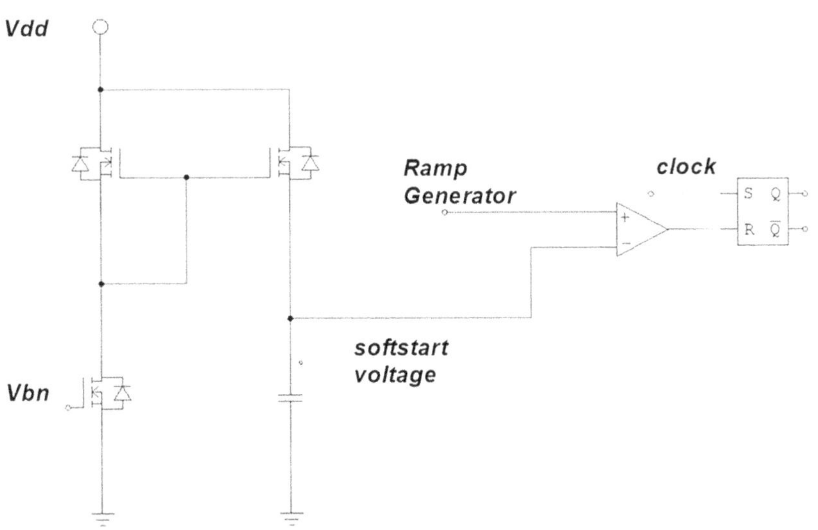

FIGURE 13.16 Possible solution for the soft-start circuit.

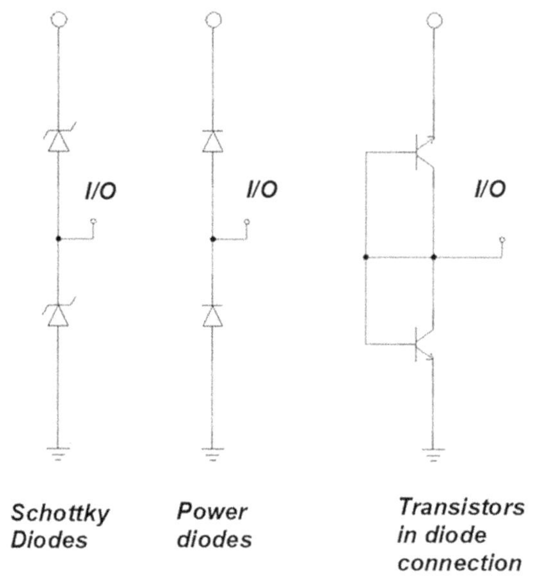

FIGURE 13.17 I/O connections.

13.4 DIGITAL IC SOLUTIONS

While 90% of the market is on mixed-mode or analog type integrated-circuit technology, recent efforts dwell on digital platforms for the control of power supplies. Architecture follows the emulation of the previous circuits and can either benefit from FPGA design for the logic circuitry or microcontroller. Usually these solutions are more expensive. They add value with protection and diagnosis functions.

13.5 CONCLUSION

Due to the advent of mechatronics systems for automotive applications, more and more power electronic circuits are required. Since these circuits target volume and weight reduction, electrical loss reduction, and integration of new features, implementation of integrated circuits is sought for. Various integrated circuits technologies have emerged over the last 40 years (1980–2020).

Without entering the actual design of integrated circuits, this chapter illustrates the topic of building power converter stages within integrated circuits with several examples. A possible classification of the types of applications targeted by power integrated circuits was included.

The circuitry for the main blocks used inside the integrated circuits for automotive power electronics applications were discussed with the goal to familiarize the application engineer with the design aspects derived from the internal structure of each integrated circuit.

REFERENCES

Blokdyk, G. 2019. *Power Management IC A Complete Guide - 2019 Edition.* Warshaw: 5STARCooks.

Chen, K.H. 2016. *Power Management Techniques for Integrated Circuit Design.* 1st edition, New York: Wiley-IEEE Press.

Denton, T. 2017. *Automobile Electrical and Electronic Systems.* 5th edition, Abingdon-on-Thames: Routledge.

Dyer, T., McGinty, J., Strachan, A., Bulucea, C. 2005. *Monolithic Integration of Trench Vertical DMOS (VDMOS) Power Transistors Into a BCD Process.* Presented at the 17th International Symposium on Power Semiconductor Devices and ICs, Santa Barbara, CA, USA, pp. 47–50.

Ki, Wing Hung. 2009. *IC Design of Power Management Circuits.* Tutorial at IEEE International Symposium on Integrated Circuits, Singapore, December 14, 2009.

Neacşu, D. 2004. *Power Semiconductor and Control for Automotive Applications.* Tutorial Presented at IEEE APEC, Anaheim, CA, USA.

Rincón-Mora, G.A. 2015. *Power IC Design - From the Ground Up.* 6th edition, Morrisville, NC: Lulu Press.

Sze, S.M., Ng, K.K. 2006. *Physics of Semiconductor Devices.* 3rd edition, Hoboken, NJ: Wiley.

14 Propulsion Systems

HISTORICAL MILESTONES

1830s: First electric cars in use—R. Anderson (Scotland); Stratingh & Becker (Holland).

1842: Davenport (U.S.) uses the first non-rechargeable electric cells.

1899–1902: The highest point in electric cars—followed by decreased interest.

1912: C. Kettering (G.M.) invented the electric starter—no need for the hand crank.

1920s: Use of the battery charger in engine cars.

Late 1960s and early 1970s: Interest revived with soaring oil prices and gasoline shortages, peaking with the 1973 Arab Oil Embargo.

1976: In the U.S., Electric & Hybrid Vehicle Research, Development, and Demonstration Act was published. However, research efforts didn't take off until the 1990s.

1990: Clean Air Act Amendment in the USA and the 1992 Energy Policy Act In Europe: late 1993, decision to form and locate the European Environmental Agency in Copenhagen.

1992: United Nations Framework Convention on Climate Change (UNFCCC) treaty was adopted on May 9, 1992, and opened for signature at the Earth Summit in Rio de Janeiro from June 3 to 14, 1992.

Rejection again in the late 1990s and early 2000s: Many consumers didn't worry about fuel-efficient vehicles due to a booming economy, a growing middle class and low gas prices in the late 1990s. Development in engine technology and catalytic converters for pollution alleviation.

After 2006: Previous investments started to pay off and production series vehicles are seen on the roads.

In 2014: there were 23 plug-in electric and 36 hybrid models available in a variety of sizes.

14.1 PROPULSION ARCHITECTURE

A multitude of solutions are possible for the use of electric machines in vehicle propulsion, depending on the involvement (or not) of the combustion engine. Additional to the propulsion itself, a solution should be made available for auxiliary and ancillary systems.

Several terms are used for modern hybrid electric vehicles. A *hybrid electric vehicle* (HEV) is a vehicle without the capacity to plug in, despite having an electric drive system and battery. The driving energy comes only from liquid fuel. Two topologies are defined based on the relationship between the combustion engine and the electrical motor: a *series hybrid* and a *parallel hybrid* configuration. A series hybrid allows the engine to generate electrical energy through a generator into the battery bank. The stored electrical energy is used to supply an electric motor that actuates the actual vehicle's axle. The parallel hybrid topology connects both the combustion engine and the electric motor on the same axle in order to actuate the vehicle. During a highly dynamic regime, the electric motor augments the power delivered by the combustion engine and allows the production of a larger torque. The charging of the battery bank is also provided with a generator turned by the combustion engine.

Conversely, a *plug-in hybrid electric vehicle* (PHEV) is a vehicle with plug-in capability, which means it can use energy for driving from either its battery or liquid fuel. An *all-electric vehicle* or battery-electric vehicle is a vehicle that gets its energy for driving entirely from its battery. Hence, it must be plugged in to be recharged. A plug-in electric vehicle (or PEV) is any vehicle that can be plugged in, either a plug-in hybrid or an all-electric vehicle.

Figure 14.1 explains these definitions, along with their architecture, while Figure 14.2 presents a different view on the classification of hybrid and electric vehicle systems.

(a) Pure electric
(b) Series hybrid
(c) Parallel hybrid
(d) Plug-in parallel hybrid

Since the actual propulsion involves an electric motor, the selection and design of the electric motor drive is very important. Depending on the size of the vehicle, the electrical motor generally needs to deliver in between 45 kW (60 HP) and 150 kW (200 HP). Due to reliability and efficiency concerns, a *brushed dc motor* is rarely used. Furthermore, such high-power levels can be achieved easier with three-phase ac machines, which can be induction motors, synchronous motors, permanent magnet (brushless dc motors), or switched reluctance motors. Any of these motors can be a possible solution and actually used in vehicle propulsion applications. Using an ac motor requires a power converter for supplying sinusoidal currents out of the dc battery and distribution bus.

Given the mobility related restrictions, the physical size of the electrical machine is very important, in both volume and weight. In this respect, the nominal frequency is often increased to ~200 Hz instead of the industrial motors' frequency at 50/60 Hz. The power converter may be integrated with the electrical motor for further size reduction.

However, it is noteworthy herein that there is a major difference from *railway propulsion*. In order to achieve better operating conditions, ac railways were supplied with current at a lower frequency than the 50/60 Hz commercial supply. Hence, special traction current power stations were used for the propulsion of trains, or rotary converters transform 50 or 60 Hz commercial power to the 25 Hz or 16 2/3 Hz frequency used for ac traction motors. Such ac systems allowed the efficient distribution of power down the length of a rail line, and also permitted speed control with switchgear on the train.

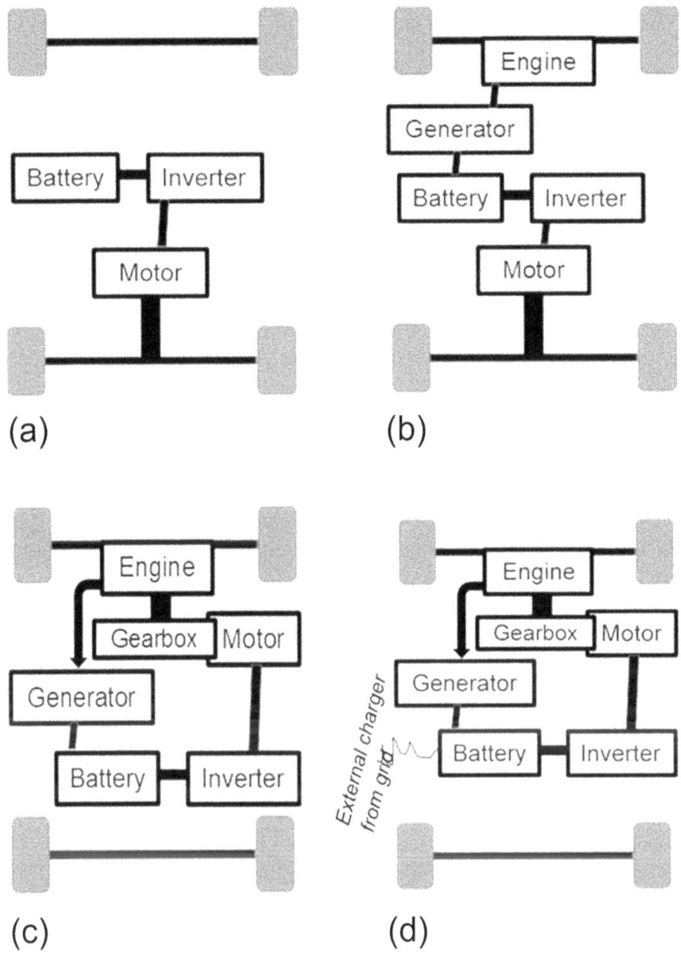

FIGURE 14.1 Various topologies for electric and hybrid vehicles.

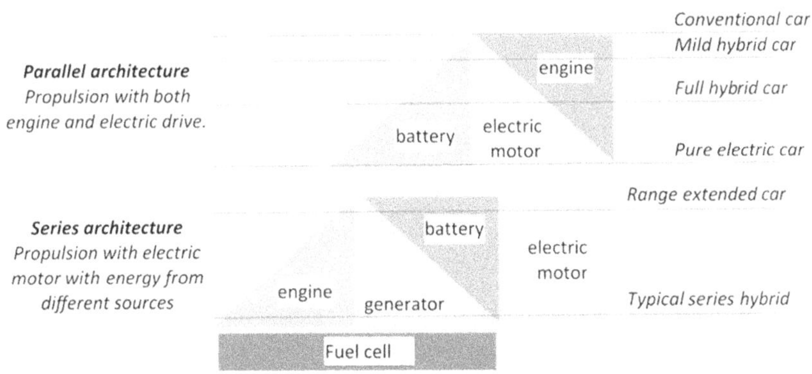

FIGURE 14.2 Definitions.

14.2 INDUCTION MOTOR DRIVE—CONVERTER SYSTEM

This is an easier solution, with lots of previous industrial applications, going back to the 1800s. It has a very big advantage since the induction motor drive starts immediately after a simple application of ac voltage. The induction motor remains the preferred choice for vehicle propulsion in academic development. The only notable production use for vehicle propulsion were with

- *General Motors'* EV-1 vehicle (produced and leased by General Motors from 1996 to 1999).
- AC Propulsion Company vehicles, including tzero (launched in January 1997 by designer Alan Cocconi, as alternating current-based drivetrain systems for electric vehicles).
- *Tesla's* Roadster (introduced in 2006 for $109,000, and in production from 2008 to 2012).

The induction motor is supplied with a three-phase power converter that is called an inverter. The inverter is controlled with PWM (pulse width modulation) to provide a set of quasi-sinusoidal currents through the machine's inductance. Figure 14.3

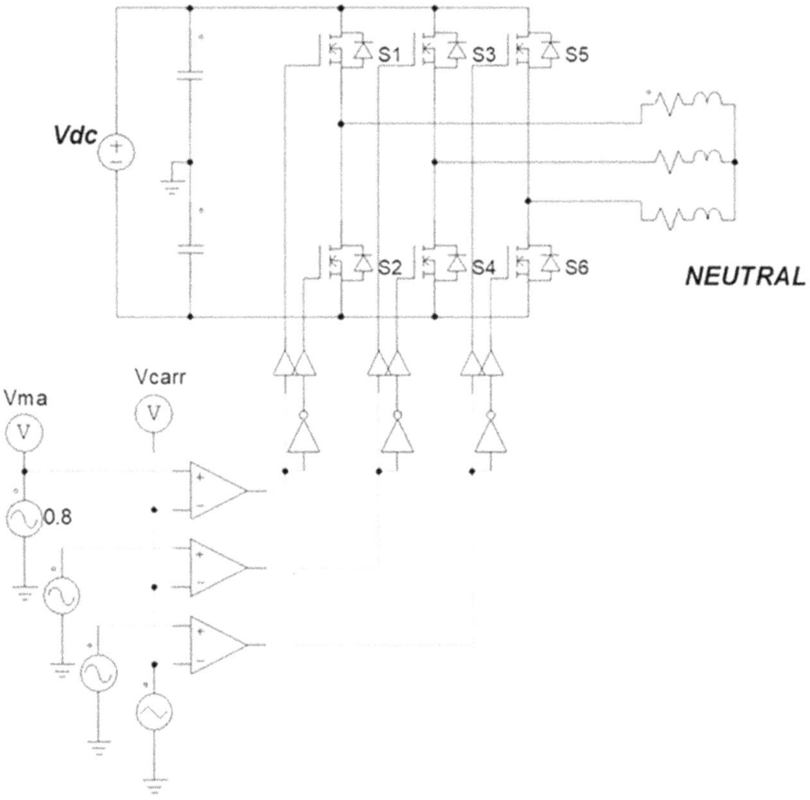

FIGURE 14.3 Three-phase inverter.

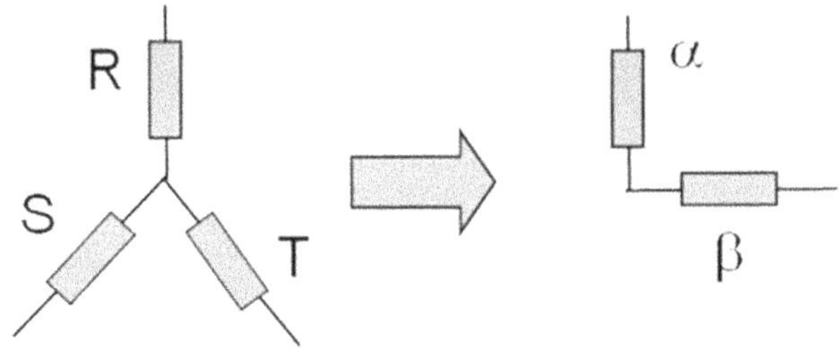

FIGURE 14.4 Symbolic representation of the Clarke transform.

illustrates the circuitry for a three-phase inverter, and more details are provided in Section 5.3.4. "Three-Phase Inverters." The *induction machine drive* uses a control based on the separation of active and reactive components of power. The mathematic apparatus for this control method is next introduced, with notations typically used in motor control applications. These may be different from the application of the same theory to grid-related applications.

Any three-phase system of measures (currents or voltages) can be transformed into an orthogonal system through a *Clarke transform*, as shown symbolically in Figure 14.4. If the three-phase system is denoted with R–S–T, and the bi-phase orthogonal system is denoted with α–β plus the homopolar component "0," the transform yields as shown in Equation (14.1) for a set of currents.

$$\begin{bmatrix} I_\alpha \\ I_\beta \\ I_0 \end{bmatrix} = \frac{2}{3} \cdot \begin{bmatrix} 1 & -\frac{1}{2} & -\frac{1}{2} \\ 0 & \frac{\sqrt{3}}{2} & \frac{-\sqrt{3}}{2} \\ \frac{1}{2} & \frac{1}{2} & \frac{1}{2} \end{bmatrix} \cdot \begin{bmatrix} i_R \\ i_S \\ i_T \end{bmatrix} \tag{14.1}$$

The coefficient 2/3 selected herein is somewhat arbitrary, in order to maintain the same length of the voltage or current vectors, also seen as the magnitude of the sinusoidal steady-state signals in both reference frames. Alternatively, a coefficient of sqrt (2/3) is selected in order to preserve the power in either reference frame.

The (α, β) system still has a sinusoidal variation of components. A *Park transform* is next used to eliminate the ac form of signals with a coordinate change able to directly provide the active and reactive components as quasi-dc measures. This transform requires knowledge of the angular coordinate (phase) of the original set of three-phase measures.

This transform is also known as a transformation to a rotating reference frame. The common frame of reference is non-rotating and can be seen as being associated with the stator, also called the *stationary frame*. The (d, q) axes can be made to rotate with the same angular velocity as the rotor circuits, and this is termed the *rotor reference frame*.

Analytically:

$$\begin{bmatrix} I_q \\ I_d \\ I_0 \end{bmatrix} = \begin{bmatrix} \cos\theta & \sin\theta & 0 \\ -\sin\theta & \cos\theta & 0 \\ 0 & 0 & 1 \end{bmatrix} \cdot \begin{bmatrix} I_\alpha \\ I_\beta \\ I_0 \end{bmatrix} \tag{14.2}$$

Sometimes, it is more convenient to combine the two transforms into a single expression, (14.3).

$$\begin{bmatrix} I_q \\ I_d \\ I_0 \end{bmatrix} = \frac{2}{3} \cdot \begin{bmatrix} \cos\theta & \cos\left[\theta - \frac{2\cdot\pi}{3}\right] & \cos\left[\theta - \frac{4\cdot\pi}{3}\right] \\ \sin\theta & \sin\left[\theta - \frac{2\cdot\pi}{3}\right] & \sin\left[\theta - \frac{4\cdot\pi}{3}\right] \\ \frac{1}{2} & \frac{1}{2} & \frac{1}{2} \end{bmatrix} \cdot \begin{bmatrix} i_R \\ i_S \\ i_T \end{bmatrix} \tag{14.3}$$

For a more visual comprehension of these transforms, the mathematical apparatus is applied to a three-phase voltage set, with 220 V rms and 50 Hz frequency. The results are shown in Figure 14.5.

It can be seen that the V_Q component is equal to the peak of the three-phase system (310 V in the example), while the V_D component is zero. This is the basis for the separation of the active and reactive power components. Any current-voltage phase shift corresponds to the existence of a reactive component and this can be seen in the *D* component. If there is no phase shift, the *D* component is zero, and all is seen

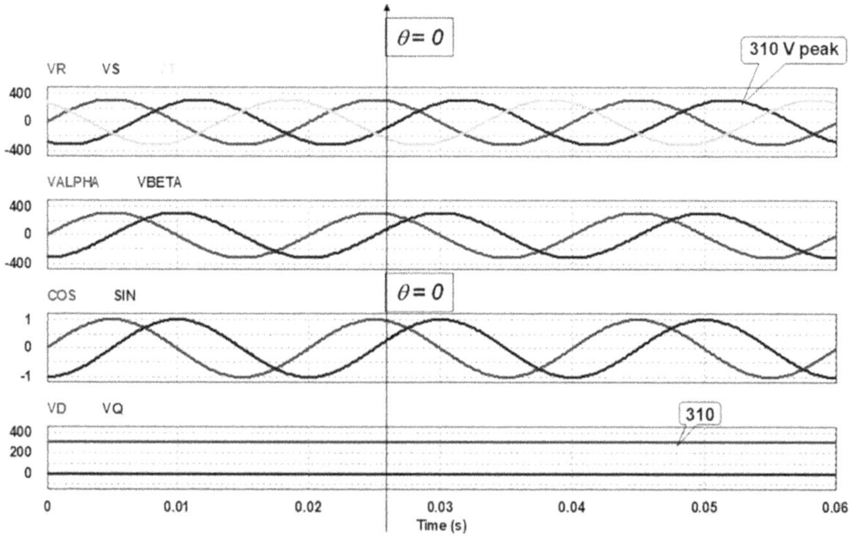

FIGURE 14.5 Application of the Clarke/Park transforms to a three-phase system.

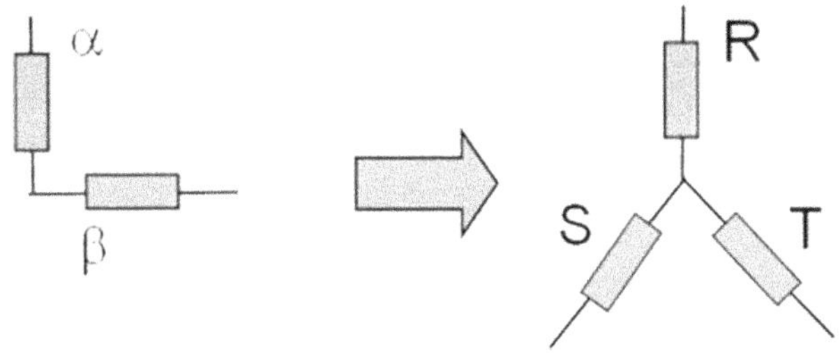

FIGURE 14.6 Reversed Clarke transform.

on the Q component. The (D, Q) components can be identified with the active and reactive components of the considered measure (current or voltage).

Analogously, a set of *reverse transforms* have been defined to go from quasi-dc components into the three-phase ac measures. The *reverse Park transform* needs the angular coordinate origin and the transform yields as follows.

$$\begin{bmatrix} I_\alpha \\ I_\beta \\ I_0 \end{bmatrix} = \begin{bmatrix} \cos\theta & -\sin\theta & 0 \\ \sin\theta & \cos\theta & 0 \\ 0 & 0 & 1 \end{bmatrix} \cdot \begin{bmatrix} I_q \\ I_d \\ I_0 \end{bmatrix} \tag{14.4}$$

The two-phase system is converted to a three-phase system with a *reversed Clarke transform*, which is defined with Equation (14.5) and symbolically shown in Figure 14.6.

$$\begin{bmatrix} i_R \\ i_S \\ i_T \end{bmatrix} = \begin{bmatrix} 1 & 0 & 0 \\ -\dfrac{1}{2} & \dfrac{\sqrt{3}}{2} & 0 \\ -\dfrac{1}{2} & \dfrac{\sqrt{3}}{2} & 0 \end{bmatrix} \cdot \begin{bmatrix} I_\alpha \\ I_\beta \\ I_0 \end{bmatrix} \tag{14.5}$$

The reversed Clarke/Park transforms can be combined into a new form as follows.

$$\begin{bmatrix} i_R \\ i_S \\ i_T \end{bmatrix} = \sqrt{\frac{2}{3}} \cdot \begin{bmatrix} \cos\theta & \sin\theta & \dfrac{\sqrt{2}}{2} \\ \cos\left(\theta - \dfrac{2\cdot\pi}{3}\right) & -\sin\left(\theta - \dfrac{2\cdot\pi}{3}\right) & \dfrac{\sqrt{2}}{2} \\ \cos\left(\theta + \dfrac{2\cdot\pi}{3}\right) & -\sin\left(\theta + \dfrac{2\cdot\pi}{3}\right) & \dfrac{\sqrt{2}}{2} \end{bmatrix} \cdot \begin{bmatrix} I_q \\ I_d \\ I_0 \end{bmatrix} \tag{14.6}$$

Using this mathematical apparatus for control of an induction motor drive is sketched in Figure 14.7.

The three-phase currents [i_R, i_S, i_T] are sensed and measured for control. Equation (14.3) is applied for a conversion [r, s, t] → [q, d, 0], into the quasi-dc quantities [q, d]

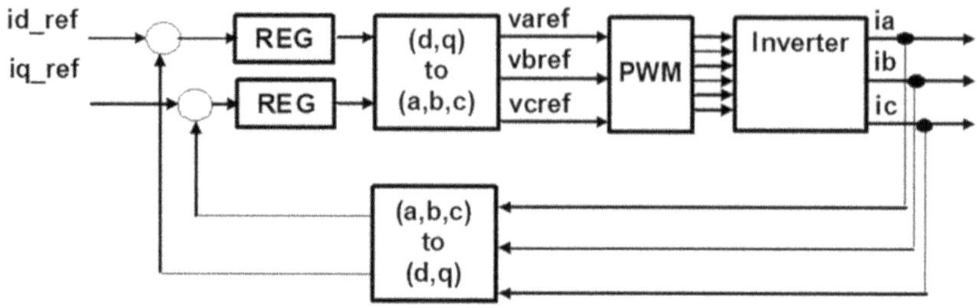

FIGURE 14.7 Principle of the current control system.

for the measured currents. The [q, d] current references are compared to the [q, d] components of the measured currents, and the difference constitutes the error to be applied to the input of regulators. In most cases, the regulators are very simple PI regulators, with a focus on the elimination of the steady-state error. The signals at the output of these regulators constitute the reference signals [q, d] for the voltage to be generated by the inverter. The reverse transform in Equation (14.6) for [q, d, 0] → [r, s, t] is employed to provide the phase voltage references to the PWM generator.

The two current references account for flux (i_d) and torque production (i_q). This is considered herein without demonstration, while more details can be found in the literature. The (i_d) component can be fixed for setting up a certain flux level within the machine, while the torque can be controlled with the (i_q) component.

The modules in Figure 14.7 are used in most motor drive systems and the software developed for each module can be highly optimized and stored as library functions within a microcontroller or DSP meant for three-phase power converter control.

The most used motor drive control method is based on field orientation or vector control, and it uses the separation of effects on the d and q axes as provided by the previous transforms. The concept has been developed by K. Hasse from Technische Universität Darmstadt and by F. Blaschke from Siemens. They pioneered vector control of ac motors starting in 1968 and the early 1970s. Their original patent drawing is copied in Figure 14.8. A simplified description is adapted herein in Figure 14.9.

The electrical angular coordinate is required for the proper operation of the transform system. It has been shown that the operation of the induction machine can produce a difference between the rotational speed of the rotor versus the magnetic field from the stator. This difference is called slip frequency. Hence, a simple measurement of the rotational speed needs to be accompanied by a calculation of the slip frequency for proper identification of the electrical frequency and angular coordinate.

It is common to express the slip as the ratio (percentage) between the shaft rotation speed and the synchronous magnetic field speed:

$$s = \frac{(n_s - n_a)}{n_s} \cdot 100\% \tag{14.7}$$

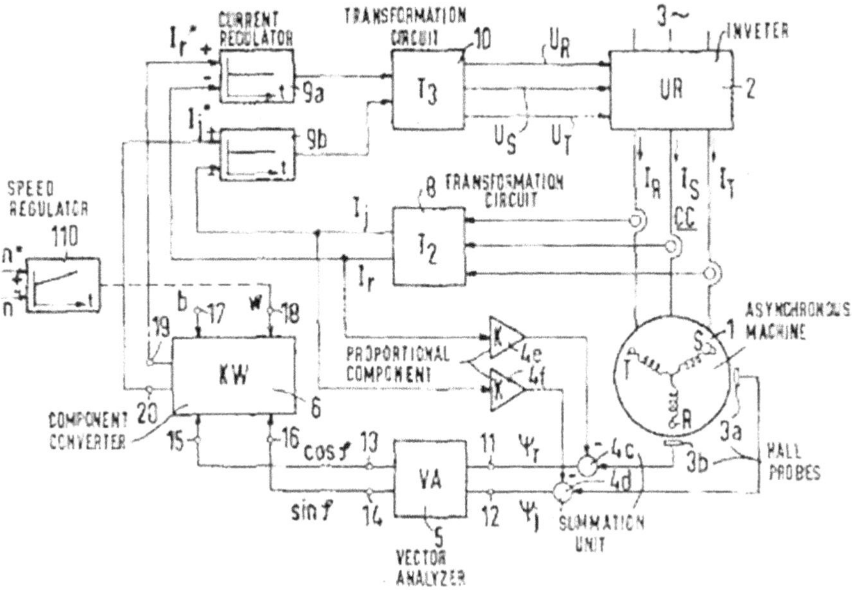

FIGURE 14.8 Copy of the original patent for the field orientation control (Blaschke's 1972 U.S. patent application, patent issued as US3,824,437 on July 16, 1974).

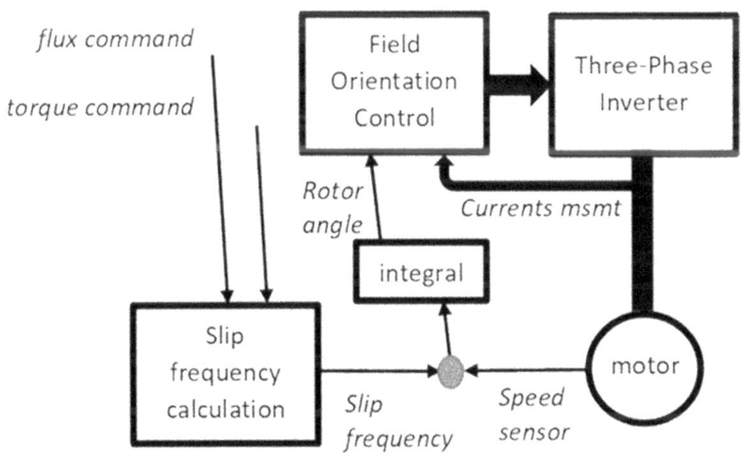

FIGURE 14.9 Simplified description of the field-oriented control.

where

n_s = synchronous speed of magnetic field (rev/min, rpm)

n_a = shaft rotating speed (rev/min, rpm).

$$s = K \cdot \frac{i_q}{i_d} \tag{14.8}$$

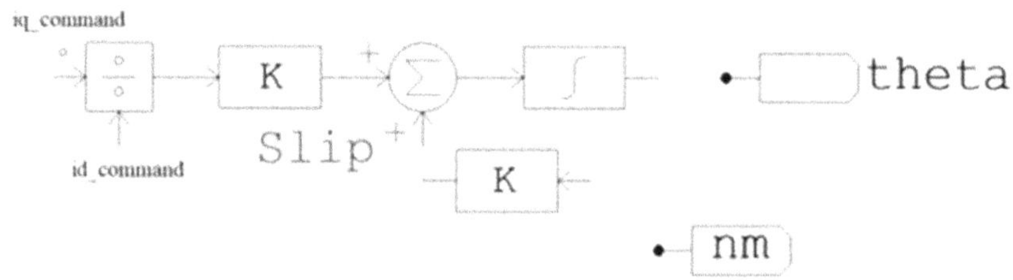

FIGURE 14.10 Calculation of the slip frequency and phase of the electrical system, for an induction motor drive.

It can be seen that the lack of torque produces zero slip frequency. As the required torque current component is increased, the slip frequency increases, which means the mechanical speed is further from the electrical frequency that produces the rotating magnetic field (Figure 14.10).

The field-oriented control method aims to establish a rotor flux in a known position (usually this position is the *d*-axis of the transformation) and then put a current on the orthogonal *q*-axis (where it will be most effective in producing torque). Figure 14.11 illustrates this. The torque equation yields as a product of the flux and applied current. If the goal of the motor drive is to control speed, it means that a speed regulator will provide the reference for the torque-producing *q*-axis, while the flux component on the *d*-axis is maintained constant.

The torque equation yields as a product of the flux ($K\,i_d$) and applied current (i_q).

If the goal of the motor drive is to control speed, it means that a speed regulator will provide the reference for the torque-producing *q*-axis current, while the flux component on the *d*-axis is maintained constant, at a predefined optimal value.

If the goal is to control torque, a torque controller will produce the current reference on the *q*-axis while the *d*-axis current component is maintained constant for a certain flux in the machine.

While advanced *sensorless control methods* are possible, circuits are needed for sensing the speed and the phase currents. The phase currents are immediately converted to (*d*,*q*) components with reversed Park/Clarke transforms using the electrical

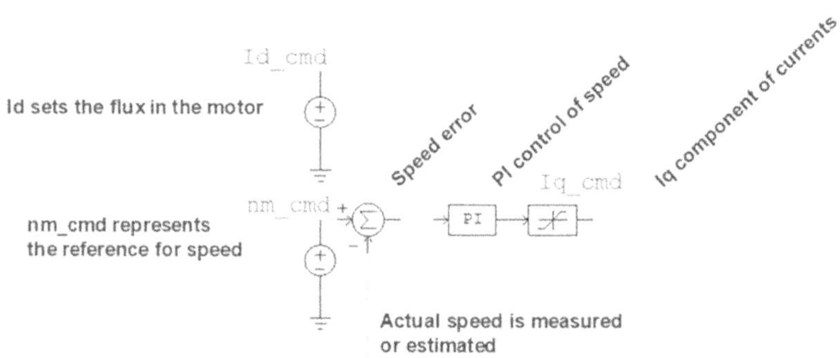

FIGURE 14.11 Principle of the field orientation control.

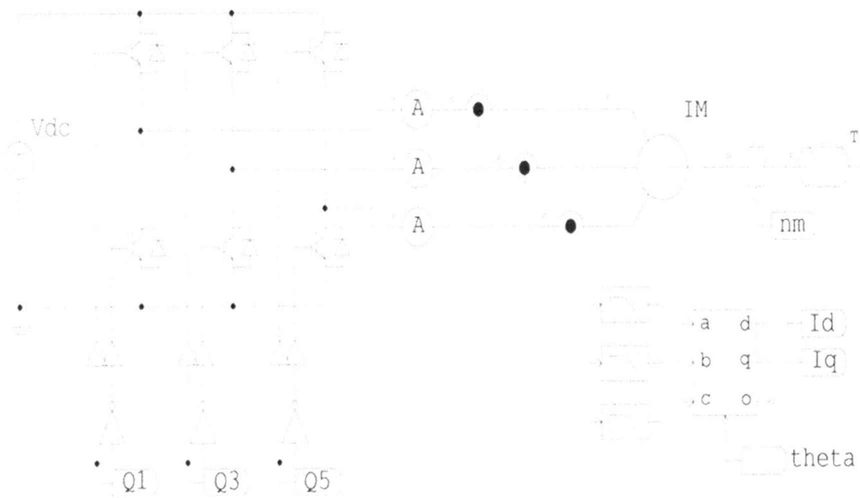

FIGURE 14.12 Sensing circuits for a three-phase inverter.

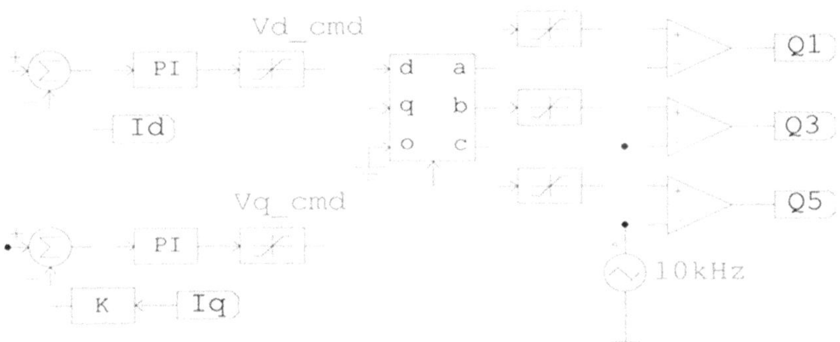

FIGURE 14.13 Current control system (can be seen as cascaded after Figure 14.12).

angular coordinate (*theta*). The position of the sensing devices and the three-phase inverter is shown in Figure 14.12.

The PI current control is performed for (*d*,*q*) components based on current references and measured values (Figure 14.13). The output of the two PI controllers are considered as (*d*,*q*) components for the voltage to be applied next to the motor. These components are converted to phase measures and applied to a three-phase PWM generator.

The overall *field orientation controller* can now be seen as a whole in Figure 14.14.

14.3 BRUSHLESS DC MOTOR DRIVE

The *brushless dc motor drive* represents the most used solution in hybrid electric automotive applications. Today, all the hybrids are powered by dc brushless drives, with no exceptions. Comparing to induction motor drives, both dc brushless and

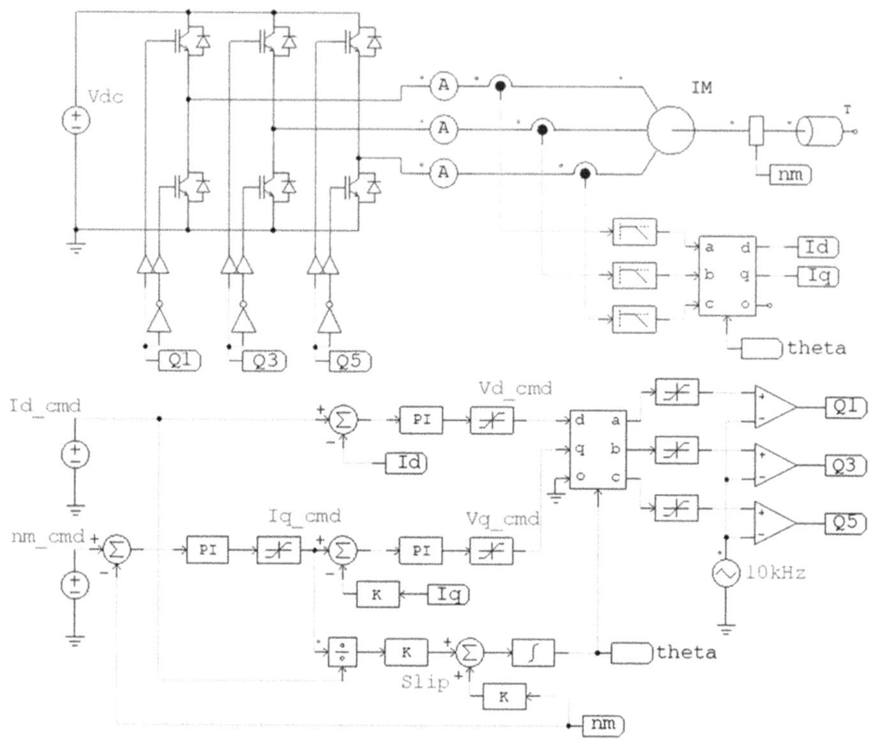

FIGURE 14.14 Overall field-oriented control system.

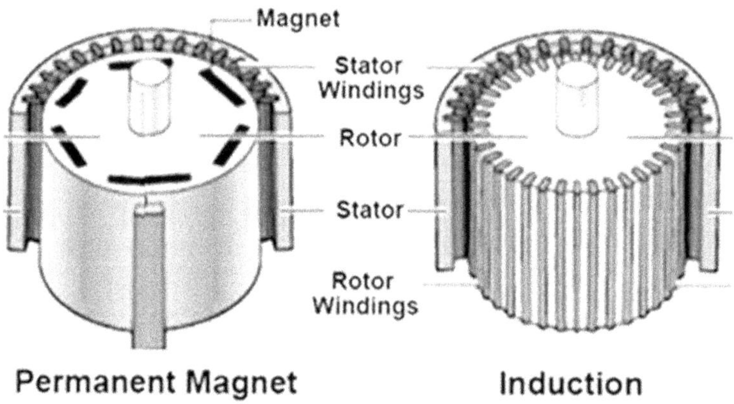

FIGURE 14.15 Comparison between induction motor and dc brushless machines.

induction drives use motors having similar stators with windings supplied with similar *three-phase modulating inverters.* While several construction options are possible, Figure 14.15 proposes a comparison between the most typical design choice.

The only differences are the rotors and the control code. In order to setup the control software, the dc brushless drives require an absolute position sensor, while

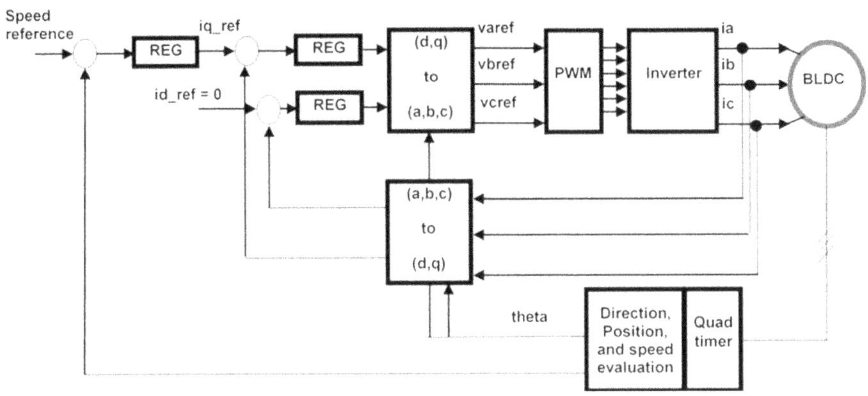

FIGURE 14.16 Complete diagram of the control system for a brushless dc drive.

induction drives require only a speed sensor. Since the magnetic field is produced by a magnet, the previous *field orientation control* principle can be applied to zero reactive current components while *theta* is provided by measurement instead of calculation of the angular frequency. The electrical control needs to be synchronous with the position of the magnets.

To rotate the brushless dc motor, the stator windings should be energized in a predefined sequence. It is therefore important to know the rotor position in order to understand which winding will be energized respecting the energizing sequence. The rotor position is sensed using *Hall effect sensors* embedded into the stator. Most BLDC motors have three Hall sensors embedded into the stator on the nondriving end of the motor. Whenever the rotor magnetic poles pass near the Hall effect sensors, these give a high or low signal, indicating the North or South magnetic pole is passing near the sensors. Based on the combination of these three Hall sensor signals, the exact sequence of commutation can be determined (Figure 14.16).

14.4 SWITCHED RELUCTANCE MOTOR DRIVE

Switched reluctance motor drives are used for propulsion due to advantages like wide speed operation area (ω_{max} = 6 to 8 times ω_{nom}), simple and robust construction, lack of the iron core, and air resistance loss seen in the rotor machines like induction or brushless dc machines.

The switched reluctance motor consists of an iron rotor and windings on the stator to create a magnetic field.

$$\varphi = \frac{i}{R_c} = \frac{i}{\dfrac{l_c}{\mu_c \cdot A_c}} \tag{14.8}$$

where R_c represents the reluctance, which depends on the winding construction data.

The stator is similar to a brushless dc motor, or an induction motor. The rotor conversely consists only of iron laminates with a different number of salient poles than

FIGURE 14.17 Rotor used within a Switched Reluctance Motor.

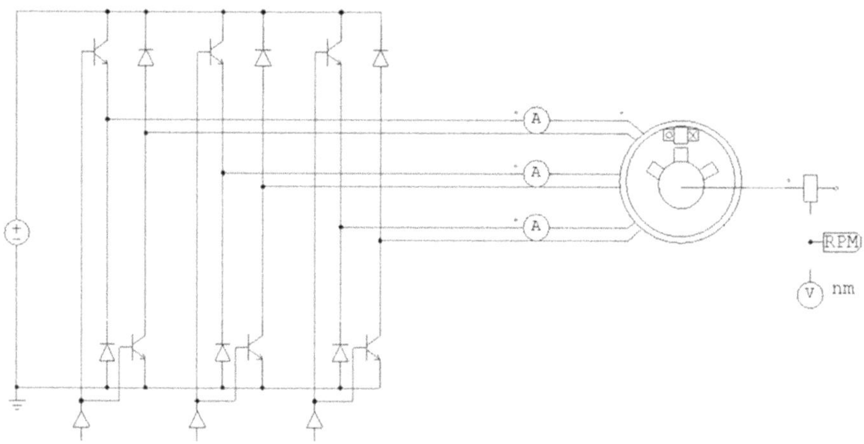

FIGURE 14.18 Power converter for the switched reluctance motor.

the stator, as shown in Figure 14.17. The iron rotor is attracted to the energized stator poles. The polarity of the stator pole does not matter therein. Torque is produced as a result of the attraction between the electromagnet on the stator and the iron rotor.

A power converter for a switched reluctance motor drive has separate circuits for each phase, able to apply a positive or negative voltage to the winding. This is illustrated in Figure 14.18.

The converter in Figure 14.18 can be controlled with a simplified circuit, as shown in Figure 14.19. The mechanical speed n_m is sensed with an encoder, and the rotor position information is calculated after scaling and integration. The rotor position is used to define the angular position of each individual phase voltage. Pulses of 60° in length are applied to the power stage based on each phase angular coordinate. The sequence of these pulses defines the direction of rotation.

While Figure 14.19 illustrates a simplified controller, the complete control system is shown in Figure 14.20. Principles of the control system are based on the

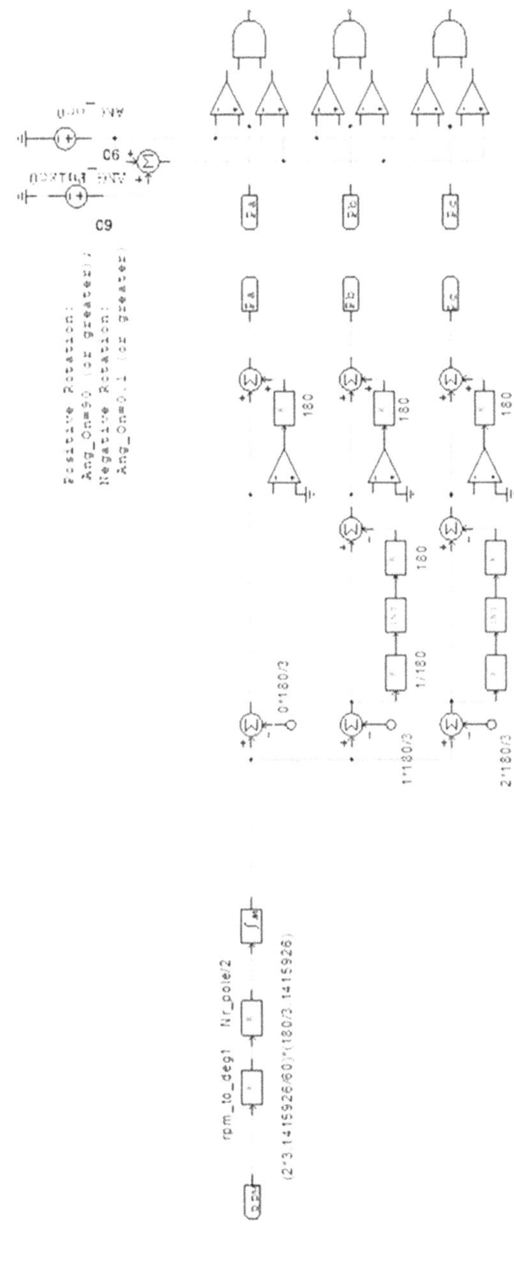

FIGURE 14.19 Example of simplified control of a switched reluctance motor.

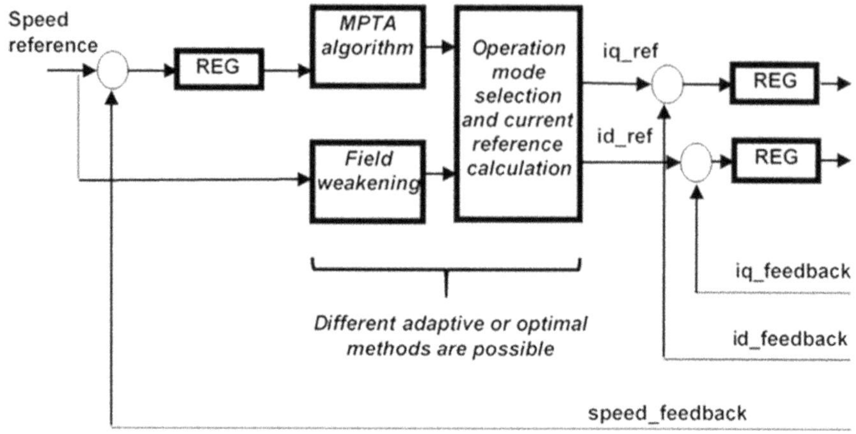

FIGURE 14.20 Current reference definition within the control system for a switched reluctance motor.

assumption that the switched reluctance motor is easily saturated due to its lack of permanent magnet material. As a result, it has nonlinear characteristics under a heavy load. To solve the problem, adaptive control algorithms are required. For this reason, this motor drive did not receive widespread acceptance.

14.5 HIGH-VOLTAGE ENERGY STORAGE

Using an electric motor drive for propulsion requires stored electrical energy. Even a hybrid arrangement—which produces electrical energy from an optimal actuation of the combustion engine—still requires temporary energy storage. The most used storage systems are based on high-voltage batteries banks, ultracapacitors or fuel cells. Each is reviewed very briefly in what follows.

The high-voltage battery banks are based on either Lithium-Ion, Nickel-Cadmium, or Nickel-Metal-Hydride technologies (Figure 14.21). While several technologies are very promising, the energy density offered by these are considerably inferior to the fossil fuel.

The ultracapacitors or supercapacitors feature a novel structure that is different to ceramic or electrolytic capacitors. The structure is based on two layers. Hence the name "double-layer capacitor." It uses electrodes and electrolytes, not unlike a battery, and is able to deliver a capacitance thousands of times larger than an electrolytic capacitor (Table 14.1).

Using batteries or supercapacitors for storage increases the battery charging problem. Through the Recovery Act of 2009, the U.S. Energy Department invested more than $115 million to help build a nationwide charging infrastructure, installing more than 18,000 residential, commercial, and public chargers across the country.

The fuel cell is an electrochemical device that combines liquid combustible (fuel) with oxygen to produce electricity, heat, and water. The fuel-cell systems are similar to a battery when concerning the chemical reaction. The hydrogen combustible (fuel) is stored in a pressurized container, while the oxygen is taken from

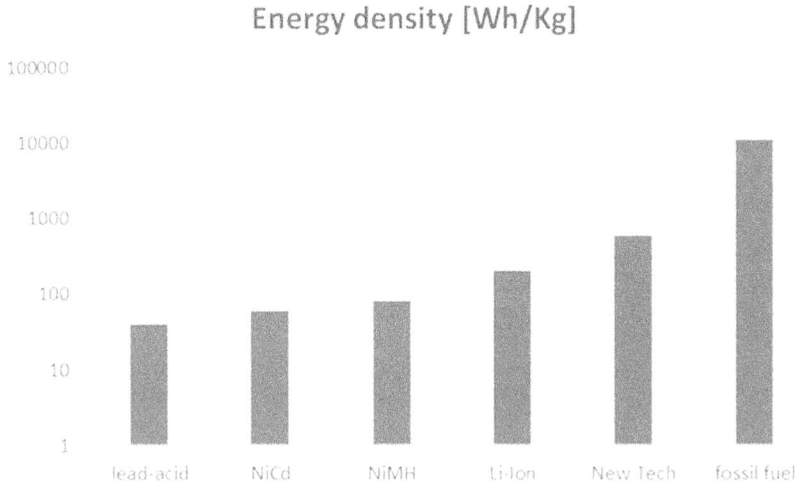

FIGURE 14.21 Comparison between battery technologies.

TABLE 14.1
Comparative Evaluation of Batteries and Ultracapacitors

Function	Supercapacitor	Lithium-Ion battery
Charging time	1 … 10 sec	10–60 minutes
Lifetime (cycles)	1mil or 30,000h	> 500
Cell voltage	2.3 ... 2.75V	3.6 ... 3.7 V
Specific energy [Wh/kg]	5 (typ)	100–200
Specific power[[W/kg]	< 10,000	1,000–3,000
Cost per Wh	$20	$0.50–$1.00

free air. Since there is no burning process, the only waste is water. The fuel-cell systems are reliable and can be 30% cheaper than batteries. They need 5 seconds to start; thus, we need another battery or ultracapacitor within the same system, for this short interval.

14.6 CONCLUSION

Conversion to an *all-electric vehicle* is based on electric propulsion. The electric propulsion is implemented with one of the possible electrical motors. Due to the power level and the performance requirements, electrical motors used for propulsion are three-phase ac motors. Digital control of electric motors was presented to fulfill this goal.

The field-oriented control is the most advanced method for the control of ac motors. A large variety of field-oriented control methods are implemented in practice.

Finally, electrical energy storage was briefly discussed.

REFERENCES

Blaschke, F. 1972. Das Prizip der Feldorientierung, Die Grundlage fur die TRNSVEKTOR-Regelung von Asynchnmachinen," Siemens Zeitschrift, / "The principle of field orientation applied to the new trans-vector closed-loop control system for rotating field machines. *Siemens-Review*, 34(3):217–220.

Denton, T. 2017. *Automobile Electrical and Electronic Systems.* 5th edition, Abingdon-on-Thames: Routledge.

Emadi, A. 2017. *Advanced Electric Drive Vehicles.* Boca Raton, FL: CRC Press.

Neacşu, D. 2004. *Power Semiconductor and Control for Automotive Applications.* Tutorial Presented at IEEE APEC, Anaheim, CA, USA.

Novotny, D.W., Lipo, T.A. 1996. *Vector Control and Dynamics of A.C. Drives.* Oxford and New York: Oxford University Press.

Kinnaird, C. 2015. *How Many Electric Motors are in Your Car?* Motor and Drive Systems. Internet reading at https://motoranddriveconference.com/2018/01/many-electric-motors-car/.

Index

A

B

C

J

L

M

N

O

P

For Product Safety Concerns and Information please contact our EU
representative GPSR@taylorandfrancis.com
Taylor & Francis Verlag GmbH, Kaufingerstraße 24, 80331 München, Germany

www.ingramcontent.com/pod-product-compliance
Ingram Content Group UK Ltd.
Pitfield, Milton Keynes, MK11 3LW, UK
UKHW021301190926
475701UK00009B/322
* 9 7 8 0 3 6 7 5 1 2 9 7 2 *